Characteristics and Applications of Boron

Edited by Chatchawal Wongchoosuk

Published in London, United Kingdom

Characteristics and Applications of Boron
http://dx.doi.org/10.5772/intechopen.102107
Edited by Chatchawal Wongchoosuk

Contributors
Gülbahar Bilgiç, Norsilawati Ngah, Nor Bahiyah Baba, Nor Azinee Said, Mohd Habir Ibrahim, Na'ain Shari, Nasim Hassani, Mehdi Neek-Amal, Mohammad Reza Hassani, Shu Han Hsu, Luqman Hakim, Linh Chi T. Cao, Shinji Kawabata, Naosuke Nnoguchi, Motomasa Furuse, Masahiko Wanibuchi, Naokado Ikeda, Naonori Hu, Kohei Tsujino, Ryo Hramatsu, Minh Tho Nguyen, Ehsan Shakerzadeh, Elham Tahmasebi, Long Van Duong, Shin Aoki, Hiroki Ueda, Tomohiro Tanaka, Taiki Itoh, Minoru Suzuki, Yoshinori Sakurai, Seiji Kojima, Thamer A. Tabbakh, Prashant Tyagi, Deepak Anandan, Michael J. Sheldon, Saeed Alshihri

First published in London, United Kingdom, 2022 by IntechOpen
IntechOpen is the global imprint of INTECHOPEN LIMITED, registered in England and Wales, registration number: 11086078, 5 Princes Gate Court, London, SW7 2QJ, United Kingdom

British Library Cataloguing-in-Publication Data
A catalogue record for this book is available from the British Library

Additional hard and PDF copies can be obtained from orders@intechopen.com

Characteristics and Applications of Boron
Edited by Chatchawal Wongchoosuk
p. cm.
Print ISBN 978-1-80356-464-7
Online ISBN 978-1-80356-465-4
eBook (PDF) ISBN 978-1-80356-466-1

For EU product safety concerns:
IN TECH d.o.o., Prolaz Marije Krucifikse Kozulić 3, 51000 Rijeka, Croatia,
info@intechopen.com or visit our website at intechopen.com.

Meet the editor

Chatchawal Wongchoosuk is an associate professor in the Department of Physics, Kasetsart University, Thailand. He is a specialist in the development of smart sensors and intelligent systems for food, agricultural, and environmental applications. He has received thirty research awards such as the TRF–OHEC–SCOPUS Young Researcher Award in physical science, the Invention Award from the National Research Council of Thailand, the Highest Citation Award for a young researcher, and others. He has served as a reviewer, guest editor, and associate editor for several scientific journals. He was listed among the top 2% of scientists in electrical and electronic engineering in 2020 and 2021 by Stanford University, USA. He has published dozens of journal articles, proceedings, book chapters, patents, and copyrights.

Contents

Preface

Boron is a group 13 (IIIA) metalloid element with three valence electrons that can be considered as sp2 hybridized for a trigonal planar geometry. Doping/integration of boron atoms into other atoms provides new wonder materials such as boron nitride, borophene, boron carbide, and others with unique physical, chemical, and electrical properties. Boron-based materials are promising materials for many real-world applications.

This book provides an overview of the research and current developments of boron-based materials such as boron nitride, boron clusters, boron doping, boron compounds, and so on. It is organized into two parts: "Principle, Properties and Characteristics of Boron-Based Materials" (Chapters 1 to 4) and "Frontiers and State-of-Art Applications of Boron-Based Materials" (Chapters 5 to 9). Chapter 1 presents synthesis techniques and properties of boron nitride. Chapter 2 focuses on the theoretical approach and physical properties of borate glass. Chapter 3 describes the modeling and simulation of boron-based clusters. Chapter 4 discusses the characterizations and application of nickel cubic boron nitride. Chapter 5 reviews the design and synthesis of new boron neutron capture therapy agents based on sugars and macrocyclic polyamines. Chapter 6 focuses on boron clusters in biomedical applications. Chapter 7 demonstrates boron-based ionic liquids for energy applications. Chapter 8 examines the doping of boron for semiconductor devices. Chapter 9 summarizes and discusses boron compounds for neutron capture therapy in the treatment of brain tumors.

I would like to express my deep appreciation to all contributors who are experts in their research fields. It is hoped that this book will be a useful tool for better understanding the basic concepts and state-of-the-art applications of boron-based materials.

Dr. Chatchawal Wongchoosuk
Associate Professor,
Faculty of Science,
Department of Physics,
Kasetsart University,
Bangkok, Thailand

Section 1

Principle, Properties and Characteristics of Boron-Based Material

Chapter 1

Boron Nitride Fabrication Techniques and Physical Properties

Thamer A. Tabbakh, Prashant Tyagi, Deepak Anandan, Michael J. Sheldon and Saeed Alshihri

Abstract

The III-nitride semiconductors are known for their excellent extrinsic properties like direct bandgap, low electron affinity, and chemical and thermal stability. Among III-nitride semiconductors, boron nitride has proven to be a favorable candidate for common dimension materials in several crystalline forms due to its sp^2- or sp^3-hybridized atomic orbitals. Among all crystalline forms, hexagonal (h-BN) and cubic (c-BN) are considered as the most stable crystalline forms. Like carbon allotropes, the BN has been obtained in different nanostructured forms, e.g., BN nanotube, BN fullerene, and BN nanosheets. The BN nanosheets are a few atomic layers of BN in which boron and nitrogen are arranged in-planer in hexagonal form. The nanostructure sheets are used for sensors, microwave optics, dielectric gates, and ultraviolet emitters. The most effective and preferred technique to fabricate BN materials is through CVD. During the growth, BN formation occurs as a bottom-up growth mechanism in which boron and nitrogen atoms form a few layers on the substrate. This technique is suitable for high quality and large-area growth. Although a few monolayers of BN are grown for most applications, these few monolayers are hard to detect by any optical means as BN is transparent to a wide range of wavelengths. This chapter will discuss the physical properties and growth of BN materials in detail.

Keywords: boron nitride, nanosheets, CVD, PLD, h-BN

1. Introduction

The III-nitride semiconductors are known for their excellent chemical and physical properties like direct bandgap, low electron affinity, and chemical and thermal stability [1–6]. Among III-nitride materials, Boron nitride (BN) is the only binary material that shows crystal polymorphism, i.e. BN can exhibit several crystal structures. This polymorphism is due to the sp^2- or sp^3-hybridized atomic orbital. The number of B and N atoms in BN structures is equal. BN exists in the form of hexagonal crystalline phase (h-BN), cubic (c-BN), wurtzite (w-BN), and rhombohedral (r-BN). Among all the crystalline phases, the hexagonal and cubic are the most stable phases [7, 8]. The c-BN exhibits a zinc blende structure consisting of boron and nitrogen atoms arranged tetrahedrally, like a diamond.

In contrast, h-BN exhibits a layered structure, with neighboring B and N atoms forming honeycomb structures for each sp^2-bonded monolayers. The layers are made up of AA' stacking configuration and bounded by weak van der Waals forces with an interlayer distance of 3.33 Å, similar to graphite structure. Because of the said reason, the h-BN is a layered material and can be easily exfoliated to even a single monolayered material. The monolayered material is also known as 2D material or low-dimensional material. The h-BN material has extraordinary properties, remarkable thermal conductivity, mechanical strength, high thermal and chemical stability, and a wide bandgap. Traditionally, the BN material is used for high-temperature applications such as furnaces insulation, furnace crucibles, metal casting molds, and high-temperature lubrication [9–13]. H-BN supplemented with extraordinary physical properties exhibits atomic smoothness and a lack of dangling bond on the surface. This material in 2D form is considered the best substrate for graphene electronics [14, 15]. Moreover, 2D h-BN sheets are being explored for their application as spacer layers for metal–insulator–metal devices and as a dielectric material for transistors and nanocapacitors [16–19].

The first growth of h-BN was reported by Paffett et al. in 1990; the group used an ultrahigh vacuum (UHV) system to deposit h-BN on Pt (111) substrates [19]. The Borazine ($B_3H_6N_3$) was used as a precursor for the growth. Various surface analysis techniques were used to characterize the grown epilayers. It was observed that h-BN monolayers were grown successfully, but thick layered h-BN growth was impossible [20]. In the year of 1995, Nagashima et al. investigated the h-BN epilayers on Ni (111), Pd (111), and Pt (111). They found that the structure of the h-BN monolayer was independent of the metal substrate [21]. Furthermore, researchers found that the BN layer formed on Ni (111) did not grow layer by layer after forming the first BN monolayer. Consequently, the non-layer-by-layer growth reduced the BN growth rate significantly due to the thermal stability of the first monolayer on the Ni (111) substrate. In addition, the bond between the BN epilayers and Ni (111) surface was weaker than graphite [22]. Later, it was discovered that h-BN forms a nanotech structure with periodic nanometer-sized holes due to the significant lattice mismatch with respect to the metal substrate [23]. In 2003, the h-BN monolayer was first observed by Auwarter et al. on Ni (111) substrates and trichloro borazine $(ClBNH)_3$. The achieved monolayer had a very low defect density and triangular domain.

Moreover, different domains with fcc and hcp boron stacking were observed. To use low-dimensional h-BN material, researchers need to find a scalable growth method. Chemical vapor deposition (CVD) is commercially used in large-area growth techniques; researchers were able to grow centimeter-scaled h-BN epilayers on various metal substrates, e.g. Ni, Cu, and Pt [24–29]. However, researchers around the globe are still working to achieve larger-sized single crystals to study the growth mechanism and ensure its feasibility in the industry. The thickest monolayer single crystal formed to date is ~500 μm [24].

The advantage of using a transition metal as a substrate is their epitaxial relationship, which enables BN films to be easily transferred to another substrate for device fabrication or material characterization [30, 31]. However, this transfer process is unreliable as it is highly dependent on the manual handling expertise of the user transferring large-area films. Furthermore, impurities induced by the solvents during the transfer process are inevitable. Therefore, to avoid considerable degradation to the h-BN film, enhancing the transfer process or introducing a direct growth method would be advantageous. At the same time, the h-BN production cost is another factor that needs to be considered before commercialization. The most common precursors for the growth of BN are ammonia and borane [32–38]. This compound is relatively stable in air, less toxic, and easy to handle. These precursors are the most popular due to their high yield and high-quality h-BN films, but the cost of these precursors is relatively high and unpredictable. Apart from ammonia and borane, researchers are working on other less toxic precursors, e.g. borazine, trichloro borazine, diborane, dimeric diborazane + trimeric triborazane, $BF_3 + N_2 + H_2$, and trimethylamine borane [32]. To be widely accepted in the industry for mass production, the precursors should be low in toxicity, provide high yield, and be economical in price. Therefore, exploring economical and low-toxic alternatives is still in high demand. In upcoming sections, we will explain structural properties, growth/fabrication technique, and BN (low dimension) application.

2. Structure

Attributed to atomic bonding, hexagonal boron nitride (h-BN) (graphite-like) and cubic polymorph (c-BN) (diamond-like) are critical materials for a wide range of material applications with small interface traps. Unlike SiO_2 and high-k materials, h-BN materials possess an excellent interface with high mobility on different 2D materials transistors. Interestingly, super-hard BN exhibits a polymorphs phase, such as zincblende BN, hexagonal BN, wurtzite BN (w-BN), BN fullerene, BN nanotubes, and amorphous BN, which can be regarded as counterpart systems of graphite, cubic-diamond (C-diamond), hexagonal-diamond (H-diamond), carbon nanotubes (CNTs), fullerene, and amorphous carbon [39]. Due to their excellent optical and mechanical properties, w-BN and c-BN have attracted massive attention for various applications. However, at different process techniques (pressure and temperature changes), BN always faces state changes, i.e. through cold-compressing h-BN exhibits metastable w-BN instead of stable c-BN. Wen et al. suggest there might be another intermediate state between hexagonal and wurtzite-phase BN and proposed a new BN-phase (bct-BN) with considerable stability and excellent mechanical properties. The resulting bonding changes their electrical and mechanical properties through the process steps, such as superconductivity and hardness [39].

2.1 Hexagonal boron nitride (h-BN)

Hexagonal BN (h-BN) structure is similar to graphite. The h-BN consists of sp^2-hybridized alternating B and N atoms, which are held together by a covalent bond in a hexagon (honeycomb) lattice structure, as shown in **Figure 1**. For a fully ordered crystal structure, the lattice constant of the boron nitride structure has lattice parameters: a = 2.504A and c = 6.661A [41]. The h-BN has six-membered rings, with boron atoms in one layer as the nearest neighbors to the N atoms in adjacent layers [40].

2.2 Cubic boron nitride (c-BN): Diamond-like

Cubic-phase BN has significant technological applications. The c-BN is the second hardest material, which is resistant to oxidation. It has a natural protective coating and is sought for its protective properties for infrared and visible spectrum applications. Cubic c-BN is also compatible with high-temperature and high-power applications; unlike diamond, c-BN can be dopped with p and n type to make a high-power photodiode. The c-BN possesses a zincblende structure with ABCABC... stacking

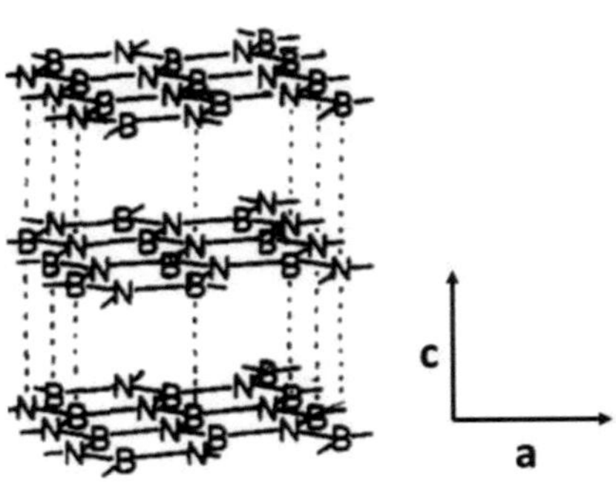

Figure 1.
Schematic diagram of the atomic configuration of layered h-BN [40].

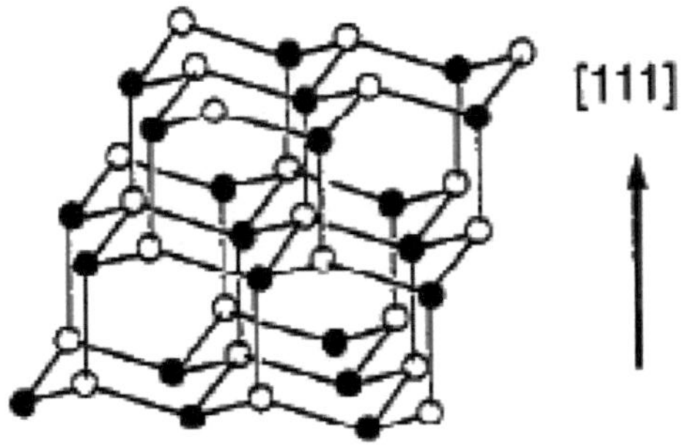

Figure 2.
Diamond structure for c-BN (lattice constant a = 3.6153A) [42].

sequence arrangements, consisting of tetrahedrally coordinated boron and nitrogen atoms with {111} plans. The atomic arrangement of B and N atoms in the c-BN lattice is represented in **Figure 2** by the ball and stick model.

2.3 Wurtzite boron nitride

Metastable-phase stacking sequence produces additional sp^3- and sp^2-bonded phase forms in polytype crystal structures. For example, the stacking relationship between wurtzite (w-BN) and sp^3-bonded cubic phases of boron nitride is analogous to that between cubic and h-BN. W-BN consists of two layers (0002 planes) structurally identical to the (111) plan c-BN. These ABC stacking arrangements yield additional metastable rhombohedral boron nitride (r-BN) crystal phase [42]. The atomic planer arrangement in w-BN and r-BN is shown in **Figure 3**.

2.4 Octahedral model: BN fullerenes

Ultrathin layer BN (typically <3 nm) forms fullerenes under in situ electron irradiation. The BN fullerenes exhibit B/N stoichiometry of ~1. Also, fullerenes revealed unusual polyhedral electron microscope images based on microscope projection, i.e. *Nested* BN fullerenes and *single* cell fullerenes square-like and rectangle-like outlines, respectively [43]. The structure of BN fullerene is shown in **Figure 4** with the help of the ball and stick model. In **Figure 4**, the black ball represents B and the white ball represents N.

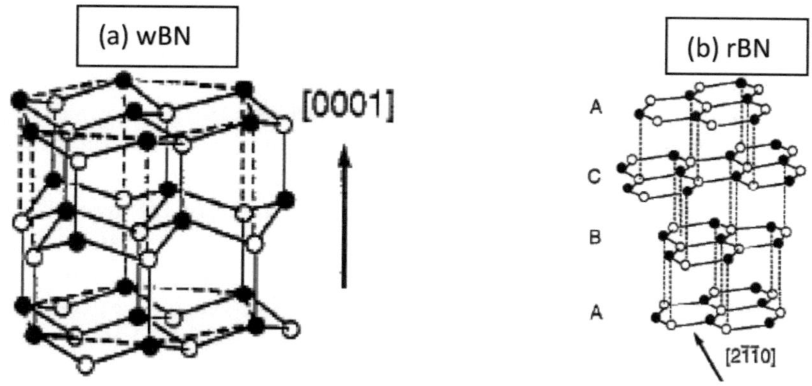

Figure 3.
(a) Sp³-bonded-phase wBN (lattice constant a = 2.5505A and c = 4.210A) (b) Sp2-bonded-phase rBN (lattice constant a = 2.5042A and c = 9.99A) [42].

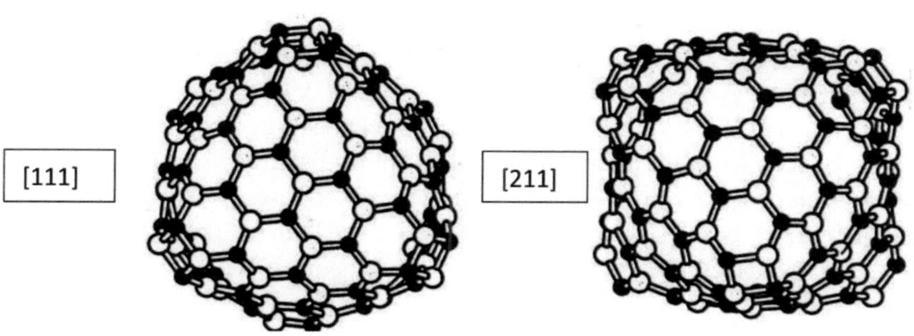

Figure 4.
3D octahedral $B_{76}N_{76}$ molecule in [111] and [211] orientation [43].

2.5 Amorphous BN

BN exhibits amorphous nature when it reacts with cesium metal. After heat treatment, the amorphous nature is transformed into turbostratic nature (tubular and corpuscles morphology) with a diameter of 3 μm for 50 to 100 μm. The researchers observed that the interlayer distance was 3.5A [44].

3. Defects

As we know, BN is a good choice for low-dimensional semiconductors (especially 2D) with remarkable thermal, mechanical, and dielectric properties. Theoretically, BN should exhibit perfect lattice structures free from defects, but as we know, the actual material is marginally different from theoretical models. Similarly, BN has some inevitable structural defects that arise due to the imperfection of the growth/preparation processes. These defects are unintentionally induced because of the dramatic influence over the material's physical properties, even in a low-dimensional state.

3.1 Point defect

The point defect, known as Stone-Wales (SW), is prevalent in semiconductor materials. It involves the connectivity change of π-bonded carbon atoms leading to their rotation by 90° [45]. This defect forms two separate vertically bonded rings instead of two rings sharing a common edge. This specific structural defect, such as graphene, is common in sp^2-bonded carbon allotropes. Similarly, SW or point defects are observed in BN material with low dimensions [46]. The formation of point defects in BN nanoribbons is due to the structural geometry. The SW defects in BN nanoribbons (zigzag and armchair) are shown in **Figure 5**. These defects decrease the bandgap regardless of BN nanoribbon orientation but maintain ultrawide bandgap behavior (insulating). At the same time, the defect site of this particular nature is far more reactive when compared to the defect-free site in BN nanoribbon [47–49].

Atomic vacancy is also a point defect observed by high-resolution transmission electron microscopy (HRTEM). The HRTEM analysis of BN monolayers reveals triangle-shaped vacancies that have been observed. It is also revealed that the monovacancy of boron (V_B) and the monovacancy of nitrogen (VN) coexist in nature. The boron atom has low knock-on energy compared to that of nitrogen. Hence, it favors the formation of boron vacancies rather than nitrogen vacancies [49]. Therefore, V_N is not observed during HRTEM observation.

On the other hand, the coexisting vacancies like V_B, V_{3B+N}, and V_{6B+3N}, etc., are evident. However, Alem et al. suggested that besides knock-on damage, there might be other mechanisms of forming coexisting vacancies, such as replacing ejected atoms with nearby atoms [50]. Furthermore, the interlayer distance with a missing boron atom was enlarged, which indicates that the dangling bonds for each N atom might be repulsive to each other. No stable divacancy (V_{BN}) was observed as V_{BN} would immediately transform into V_{3B+N} due to further removal of boron atoms [48].

3.2 Defect lines

Another defect observed in BN 2D material is *defect lines*, which form due to the difference in polarities of h-BN. This difference in polarity results in the convergence of N-terminated edges from domains with 60° rotation differences. These defects are observed when h-BN epilayers grow on lattice-matched substrates like Cu (111) or Ni (111). However, in situ observation of such defect lines is not detected yet. Therefore, determining adatom species migration during this defect needs further detailed study [51].

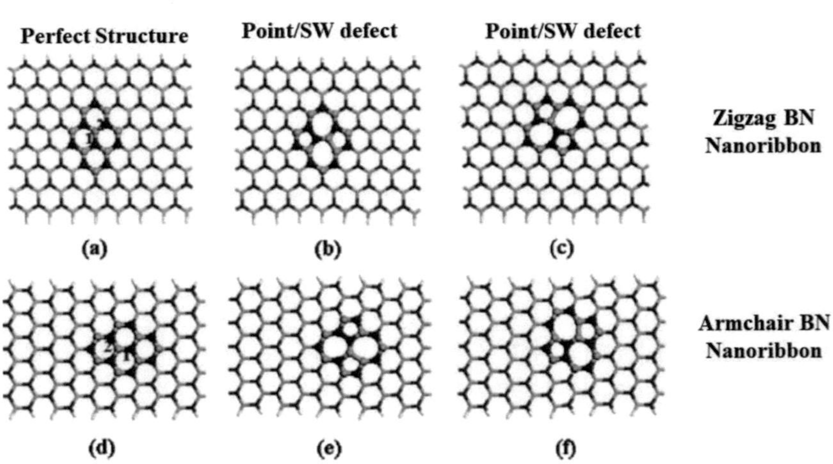

Figure 5.
Two types of B-N bonds are zigzag BN nanoribbon and armchair BN nanoribbon with their defects [46].

3.3 Adlayer defect

The *adlayer defect* is another major defect observed in low-dimensional 2D h-BN. During growth, after a few layers on the metal substrate, Cu islands form structures that are observed along the grain boundaries of h-BN films. Close observation shows that grain boundaries facilitate this island's growth. From the scanning electron microscopy observation, it was found that angular islands with truncated edges were observed. The edges resemble pyramids [51]. Some other 2D materials like MoS_2 and WS_2 were observed [52, 53]. A similar observation was found during the growth of III-V nanostructures on metal substrates. It was revealed by an X-ray photoelectron spectroscopy (XPS) study that a metal grain boundary acts as the favorable energy site for the nucleation of III-V pyramid islands. Correspondingly, during the growth of h-BN monolayers, it was observed that the orientation of adlayers is strongly affected by the crystalline substrate facets [54–56].

4. Preparation methods

4.1 Mechanical exfoliation

Mechanical exfoliation has been considered the simplest way to prepare high-quality 2D material. The mechanical exfoliation employs exterior force to overcome weak van der Waals interactions between material layers. The most frequently used method is scotch tape removal. In this method, the pulling action breaks down the weak van der Waals interactions between the layers. Pacile et al. used the same method to isolate BN nanosheets from the powdered h-BN pellet (hard-pressed). By repeating the peeling and pressing process, monolayered BN nanosheets were obtained. They have reported a strong relationship between h-BN nanosheet defects and captured excitons' recombination intensity [57]. Therefore, this technique is suitable for small-scale sample preparations, only for the laboratories, which is the main drawback of this technique. Another mechanical exfoliation method that uses shear force is ball milling. This method introduces fewer point defects compared to other mechanical methods. At the same time, this method is more efficient than the scotch tape method. According to Li et al., the controlled ball milling parameters are the key to producing relatively large in-plane BN nanosheets samples. They used h-BN powder with a benzyl benzoate ($C_{12}H_{12}O_2$) milling agent and successfully achieved 0.3–1 μm diameter flakes with thickness between 20 and 110 nm [58].

4.2 Thermal exfoliation

For the fundamental studies, mechanical exfoliation can produce small samples of a single h-BN sheet. However, the thermal exfoliation route is more convenient if some large production is needed, such as micro- or nanofillers in polymer composites. Therefore, Cui et al. attempted a large-scale thermal exfoliation of h-BN using the easy and scalable thermal oxidation approach [59]. They observed that heating h-BN in the presence of air adds oxygen to the lattice. After heating, the material was stirred in deionized water for several minutes resulting in a thick mixture that exfoliates to form hydroxylated boron nitride without the need for sonication (or mild sonication required to increase the yield). Furthermore, Yu et al. prepared the hydroxylated boron nitride by heating h-BN powder to 1000°C inside the tubular furnace. This process yields a similar material as Cui et al.; the prepared material was collected, mixed with binders, and stirred to obtain flakes of h-BN [60].

4.3 Chemical vapor deposition

Chemical vapor deposition (CVD) is widely used to grow various materials. **Figure 6** shows the schematic diagram of the basic CVD growth technique. CVD is the industrialized large-area growth process, which uses liquid precursors and process gases to grow the material on the desired surface at elevated temperatures. In earlier studies, researchers used diborane and ammonia as precursors for the deposition of h-BN nanosheets on various metal substrates [61]. CVD growth of BN nanosheets is the primary approach to achieving large-area growth. The large-area growth involves suppressing the nucleation sites and enhancing the 2D growth mode. The nucleation could further be suppressed using atomically flat surfaces and optimized CVD parameters [32, 62].

Additionally, using a metal substrate can enhance the surface migration, further enhancing the domains' size by suppressing nucleation and growth rate due to solid gas reactions involving the metal surface chemistry [63]. The type of precursors could be a separate boron or nitrogen source, e.g. for boron source BF_3 and NH_3, BCl_3 and NH_3, and B_6H_6 and NH_3). Otherwise, it could also be a single precursor like ammonium borane and borazine [64–66]. **Table 1** lists some conventional and advanced precursors in the trend to grow BN through the CVD technique. **Figure 6** illustrates the basic schematic of the horizontal CVD apparatus [76]. The apparatus consists of a horizontal quartz tube with three heater zones that provide even temperature gradient control throughout the reacting zone (inside the tube). From one side, gaseous precursors are introduced, which take part in the growth zone and precipitate (or epitaxially deposited) onto the surface of the substrate. The epitaxial growth mechanism is governed by the boundary layer adsorption phenomenon. For growth,

Precursors	Results	Reference
Ammonia borane	Single crystalline, domain size in centimeter, monolayers, and multilayers were fabricated.	[34, 36, 38, 67]
Trimethylamine borane	h-BN monolayers successfully grown on copper using organic precursor carbon doping could be achieved.	[68]
Trimethylborate, O_2, and NH_3	h-BN monolayers were successfully grown on Rh/YSZ/Si (111) multilayer substrate, an economical and less toxic process.	[69]
Borazine $(HBNH)_3$	Growth of monolayer BN nanosheets having domain size in mm.	[70, 71]
Diborane and NH_3	BN nanosheets for up to 100 layers were achieved. Crystalline quality depends on the substrate, growth time, and rate.	[72, 73]
BCl_3, NH_3, N_2, and H_2	h-BN nanosheets were vertically aligned on the substrate. The thickness of the aligned nanosheets is less than 10 nm.	[74, 75]

Table 1.
List of conventional precursors used in CVD method for BN nanosheets/nanostructure growth.

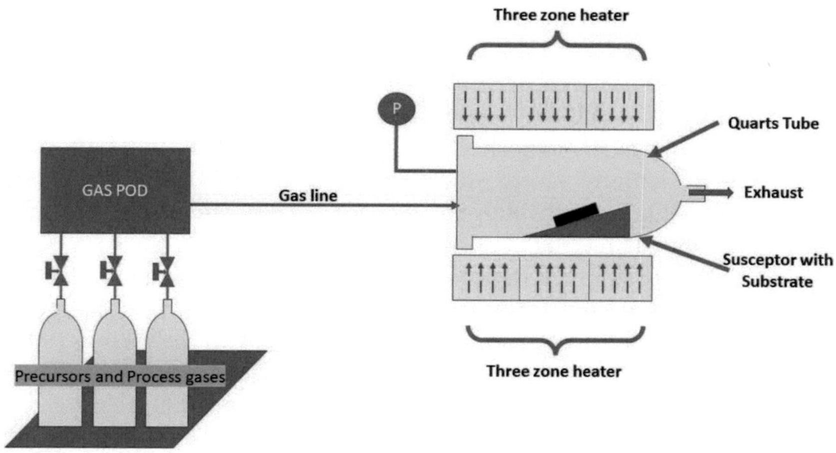

Figure 6.
Schematic diagram of basic CVD growth apparatus.

the substrate is placed over an inclined susceptor generally made up of graphite, and the inclined angle may vary from 7° to 15°. This inclination provides uniform gas flow over the susceptor and suppresses the parasitic gas-phase reactions [76, 77].

4.4 Pulsed laser deposition

Pulsed laser deposition (PLD) is a novel growth technique for the growth of III-V semiconductors. PLD is an ultrahigh vacuum technique with the base pressure ranging from ~10^{-7} to 10^{-9} torr. As the growth takes place in a high vacuum, there is a slim chance of impurity incorporation due to contamination. In this technique, the pure material is ablated with the help of a high-energy laser at the same time the process

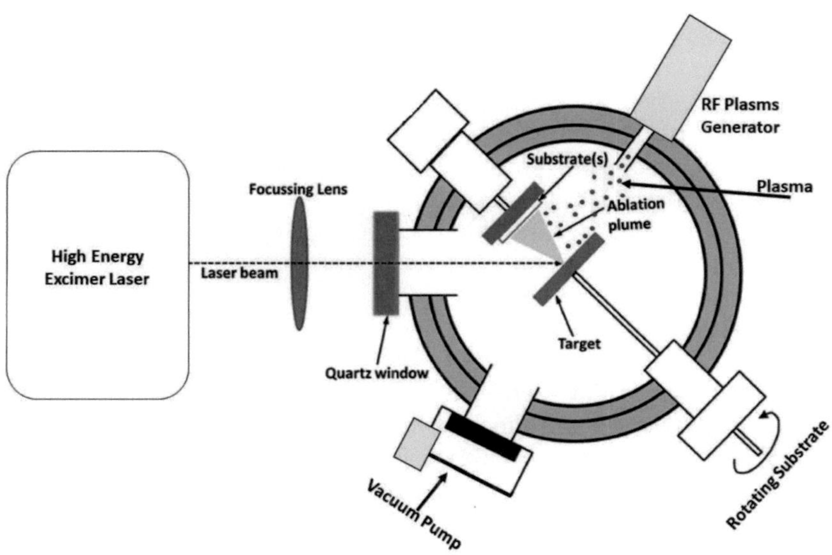

Figure 7.
Schematic diagram of PLD technique [70].

gas plasma is introduced into the chamber. The laser-ablated material reacts with plasma species and migrates on the substrate surface [55]. Recently, this technique has been used to grow h-BN nanosheets and nanostructures. **Figure** 7 shows the schematic diagram of the PLD technique for reference [78]. Glavin et al. reported this method's direct growth of BN nanosheets on the sapphire substrate. The growth was carried out using BN sintered target, which took place at 700°C growth temperature. The grown film exhibits a narrow Raman line width of ~30 cm^{-1}, providing excellent crystalline quality. Later, the prepared films were probed for their UV response, and BN thin films show deep UV detection capabilities [79]. Velázquez et al. directly grown few monolayers of h-BN on Ag/$SrTiO_3$(001). It was found that the grown h-BN monolayers were in the sub-millimeter range and scattered on the surface [80]. PLD is the immature technique for the growth of III-V semiconductors and is limited to small area fabrication. Very few reports on the growth of BN by the PLD technique are available. There is an immense need to study this technique to grow large-area films which could be of commercial use. There are several other methods to prepare 2D BN, like solvent exfoliation, solid state reactions, and substitution method, but only few potential large-scale preparation methods are described in this section.

5. Current application

The BN material shows excellent chemical, thermal, and mechanical properties, which are utilized in different applications. The industrial application of 2D BN is yet to be realized due to the lack of large-area thin film synthesis techniques. To date, large-area films with uniform thickness are hard to fabricate. The h-BN is known for its wide direct bandgap (~5.9 eV), small lattice mismatch with graphene (1.7%), and high thermal conductivity. The BN nanosheets have been employed in several semiconductors device applications, such as transparent membranes, encapsulation materials, tunneling barriers, and high dielectric materials [81]. For instance, some graphene devices fabricated on the h-BN show very high mobility in order of 60,000 cm2V-1 s-1, which is far greater than III-V devices that exploit 2DEG (two-dimensional electron gas) properties [82].

Interestingly, an ultrathin layer of BN is sandwiched between the graphene layer (C-BN-C), making a field effect tunneling transistor heterostructure. The study revealed that the h-BN nanosheet forms a good tunnel barrier [83]. Ranjan et al. studied dielectric breakdown failure of thin h-BN films. The study found that the breakdown field is 21 MV cm^{-1} for 3-nm-thick h-BN. The breakdown field suggests that h-BN is more suitable for gate dielectric than high-quality silicon dioxide [84].

6. Conclusion

The BN is expected to play a vital role in developing novel technologies in the future. Future technologies might include nanophotonics, nanoelectronics, quantum information, and microelectromechanical systems (MEMS) technology. Recently, researchers have been in pursuit of developing a way for large-area BN growth. As we have seen in an earlier section of this chapter, many researchers have achieved relatively large-area and high-quality growth. The large area of 2D BN material could be used to develop templates or substrates for epitaxy. It would be easier to remove from these substrates, so they could be used again for the epitaxy. By closely monitoring

DOI: http://dx.doi.org/10.5772/intechopen.106675

the demands of semiconductor technology, it is evident that the quest for 2D large-area growth is being pursued. Therefore, further research and studies are needed to explore 2D BN limitations to understand the potential scope for new applications.

Author details

Thamer A. Tabbakh[1], Prashant Tyagi[2], Deepak Anandan[2], Michael J. Sheldon[2] and Saeed Alshihri[1*]

1 Director of the National Center for Nanotechnology and Semiconductors, King Abdulaziz City for Science and Technology (KACST), Material Research Science Institute, Riyad, Saudi Arabia

2 Skyline Semiconductors Services, King Abdulaziz City for Science and Technology (KACST), Riyad, Saudi Arabia

*Address all correspondence to: alshihri@kacst.edu.sa

References

[1] Huang J-J, Kuo H-C, Shen S-C, editors. Nitride Semiconductor Light-Emitting Diodes (LEDs): Materials, Technologies, and Applications. Oxford: Woodhead Publishing; 2014

[2] He XG et al. GaN high electron mobility transistors with AlInN back barriers. Journal of Alloys and Compounds. 2016;**662**:16-19. DOI: 10.1016/j.jallcom.2015.12.031

[3] Jain SC, Willander M, Narayan J, Overstraeten RV. III-nitrides: Growth, characterization, and properties. Journal of Applied Physics. 2000;**87**(3):965-1006. DOI: 10.1063/1.371971

[4] Grabowski SP et al. Electron affinity of AlxGa1-xN(0001) surfaces. Applied Physics Letters. 2001;**78**(17):2503-2505. DOI: 10.1063/1.1367275

[5] Wu CI, Kahn A. Negative electron affinity and electron emission at cesiated GaN and AlN surfaces. Applied Surface Science. 2000;**162-163**:250-255. DOI: 10.1016/S0169-4332(00)00200-2

[6] Moran E, editor. Boron Nitride: Properties, Synthesis, and Applications. Hauppauge, New York: Nova Science Publishers Inc.; 2017

[7] Solozhenko VL, Turkevich VZ, Holzapfel WB. Refined phase diagram of boron nitride. The Journal of Physical Chemistry. B. 1999;**103**(15):2903-2905. DOI: 10.1021/jp984682c

[8] Sokołowska A, Wronikowski M. The phase diagram (p, T, E) of boron nitride. Journal of Crystal Growth. 1986;**76**(2):511-513. DOI: 10.1016/0022-0248(86)90401-X

[9] Liu L, Feng YP, Shen ZX. Structural and electronic properties of h -BN. Physical Review B. 2003;**68**(10):104102. DOI: 10.1103/PhysRevB.68.104102

[10] Eichler J, Lesniak C. Boron nitride (BN) and BN composites for high-temperature applications. Journal of the European Ceramic Society. 2008;**28**(5):1105-1109. DOI: 10.1016/j.jeurceramsoc.2007.09.005

[11] Jo I et al. Thermal conductivity and phonon transport in suspended few-layer hexagonal boron nitride. Nano Letters. 2013;**13**(2):550-554. DOI: 10.1021/nl304060g

[12] Li C, Bando Y, Zhi C, Huang Y, Golberg D. Thickness-dependent bending modulus of hexagonal boron nitride nanosheets. Nanotechnology. 2009;**20**(38):385707. DOI: 10.1088/0957-4484/20/38/385707

[13] Ouyang T, Chen Y, Xie Y, Yang K, Bao Z, Zhong J. Thermal transport in hexagonal boron nitride nanoribbons. Nanotechnology. 2010;**21**(24):245701. DOI: 10.1088/0957-4484/21/24/245701

[14] Järvinen P et al. Molecular self--assembly on graphene on SiO_2 and h-BN substrates. Nano Letters. 2013;**13**(7):3199-3204. DOI: 10.1021/nl401265f

[15] Garcia JM et al. Graphene growth on h-BN by molecular beam epitaxy. Solid State Communications. 2012;**152**(12):975-978. DOI: 10.1016/j.ssc.2012.04.005

[16] Barati F et al. Tuning supercurrent in Josephson field-effect transistors using h-BN dielectric. Nano Letters. 2021;**21**(5):1915-1920. DOI: 10.1021/acs.nanolett.0c03183

[17] Cui Z et al. Study of direct tunneling and dielectric breakdown in molecular

beam epitaxial hexagonal boron nitride monolayers using metal–insulator–metal devices. ACS Applied Electronic Materials. 2020;**2**(3):747-755. DOI: 10.1021/acsaelm.9b00816

[18] Monajjemi M. Metal-doped graphene layers composed with boron nitride–graphene as an insulator: A nano-capacitor. Journal of Molecular Modeling. 2014;**20**(11):2507. DOI: 10.1007/s00894-014-2507-y

[19] Shi G et al. Boron nitride–graphene Nanocapacitor and the origins of anomalous size-dependent increase of capacitance. Nano Letters. 2014;**14**(4):1739-1744. DOI: 10.1021/nl4037824

[20] Paffett MT, Simonson RJ, Papin P, Paine RT. Borazine adsorption and decomposition at Pt(111) and Ru(001) surfaces. Surface Science. 1990;**232**(3):286-296. DOI: 10.1016/0039-6028(90)90121-N

[21] Nagashima A, Tejima N, Gamou Y, Kawai T, Oshima C. Electronic structure of monolayer hexagonal boron nitride Physisorbed on metal surfaces. Physical Review Letters. 1995;**75**(21):3918-3921. DOI: 10.1103/PhysRevLett.75.3918

[22] Nagashima A, Tejima N, Gamou Y, Kawai T, Oshima C. Electronic dispersion relations of monolayer hexagonal boron nitride formed on the Ni(111) surface. Physical Review B. 1995;**51**(7):4606-4613. DOI: 10.1103/PhysRevB.51.4606

[23] Corso M, Auwärter W, Muntwiler M, Tamai A, Greber T, Osterwalder J. Boron nitride Nanomesh. Science. 2004;**303**(5655):217-220. DOI: 10.1126/science.1091979

[24] Qian Y, Kang DJ. Large-area high-quality AB-stacked bilayer graphene on h-BN/Pt foil by chemical vapor deposition. ACS Applied Materials & Interfaces. 2018;**10**(34):29069-29075. DOI: 10.1021/acsami.8b06862

[25] Zhao R, Li F, Liu Z, Liu Z, Ding F. The transition metal surface passivated edges of hexagonal boron nitride (h-BN) and the mechanism of h-BN's chemical vapor deposition (CVD) growth. Physical Chemistry Chemical Physics. 2015;**17**(43):29327-29334. DOI: 10.1039/C5CP04833H

[26] Cho H et al. Growth kinetics of white graphene (h-BN) on a planarized Ni foil surface. Scientific Reports. 2015;**5**(1):11985. DOI: 10.1038/srep11985

[27] Shi Y et al. Synthesis of few-layer hexagonal boron nitride thin film by chemical vapor deposition. Nano Letters. 2010;**10**(10):4134-4139. DOI: 10.1021/nl1023707

[28] Preobrajenski AB, Vinogradov AS, Mårtensson N. Monolayer of h-BN chemisorbed on Cu(111) and Ni(111): The role of the transition metal 3d states. Surface Science. 2005;**582**(1-3):21-30. DOI: 10.1016/j.susc.2005.02.047

[29] Li L et al. A minireview on chemical vapor deposition growth of wafer-scale monolayer h -BN single crystals. Nanoscale. 2021;**13**(41):17310-17317. DOI: 10.1039/D1NR04034K

[30] Reina A et al. Large area, few-layer graphene films on arbitrary substrates by chemical vapor deposition. Nano Letters. 2009;**9**(1):30-35. DOI: 10.1021/nl801827v

[31] Cartamil-Bueno SJ, Cavalieri M, Wang R, Houri S, Hofmann S, van der Zant HSJ. Mechanical characterization and cleaning of CVD single-layer h-BN resonators. NPJ 2D Materials and Applications. 2017;**1**(1):16. DOI: 10.1038/s41699-017-0020-8

[32] Khan MH et al. Few-atomic-layered hexagonal boron nitride: CVD growth, characterization, and applications. Materials Today. 2017;**20**(10):611-628. DOI: 10.1016/j.mattod.2017.04.027

[33] Gao Y et al. Repeated and controlled growth of monolayer, bilayer and few-layer hexagonal boron nitride on Pt foils. ACS Nano. 2013;**7**(6):5199-5206. DOI: 10.1021/nn4009356

[34] Stehle Y et al. Synthesis of hexagonal boron nitride monolayer: Control of nucleation and crystal morphology. Chemistry of Materials. 2015;**27**(23):8041-8047. DOI: 10.1021/acs.chemmater.5b03607

[35] Song L et al. Large scale growth and characterization of atomic hexagonal boron nitride layers. Nano Letters. 2010;**10**(8):3209-3215. DOI: 10.1021/nl1022139

[36] Tay RY et al. Growth of large single-crystalline two-dimensional boron nitride hexagons on Electropolished copper. Nano Letters. 2014;**14**(2):839-846. DOI: 10.1021/nl404207f

[37] Wang L, Wu B, Chen J, Liu H, Hu P, Liu Y. Monolayer hexagonal boron nitride films with large domain size and clean Interface for enhancing the mobility of graphene-based field-effect transistors. Advanced Materials. 2014;**26**(10):1559-1564. DOI: 10.1002/adma.201304937

[38] Lee KH et al. Large-scale synthesis of high-quality hexagonal boron nitride Nanosheets for large-area graphene electronics. Nano Letters. 2012;**12**(2):714-718. DOI: 10.1021/nl203635v

[39] Meng Y et al. The formation of sp3 bonding in compressed BN. Nature Materials. 2004;**3**(2):111-114. DOI: 10.1038/nmat1060

[40] Paine RT, Narula CK. Synthetic routes to boron nitride. Chemical Reviews. 1990;**90**(1):73-91. DOI: 10.1021/cr00099a004

[41] Xu B, Lv M, Fan X, Zhang W, Xu Y, Zhai T. Lattice parameters of hexagonal and cubic boron nitrides at high temperature and high pressure. Integrated Ferroelectrics. 2015;**162**(1):85-93. DOI: 10.1080/10584587.2015.1039410

[42] Mirkarimi PB, McCarty KF, Medlin DL. Review of advances in cubic boron nitride film synthesis. Materials Science & Engineering R: Reports. 1997;**21**(2):47-100. DOI: 10.1016/S0927-796X(97)00009-0

[43] Golberg D, Bando Y, Stéphan O, Kurashima K. Octahedral boron nitride fullerenes formed by electron beam irradiation. Applied Physics Letters. 1998;**73**(17):2441-2443. DOI: 10.1063/1.122475

[44] Hamilton EJM, Dolan SE, Mann CM, Colijn HO, McDonald CA, Shore SG. Preparation of amorphous boron nitride and its conversion to a Turbostratic, tubular form. Science. 1993;**260**(5108):659-661. DOI: 10.1126/science.260.5108.659

[45] Li Y, Zhou Z, Golberg D, Bando Y, von Schleyer PR, Chen Z. Stone-Wales defects in single-walled boron nitride nanotubes: Formation energies, electronic structures, and reactivity. Journal of Physical Chemistry C. 2008;**112**(5):1365-1370. DOI: 10.1021/jp077115a

[46] Chen W, Li Y, Yu G, Zhou Z, Chen Z. Electronic structure and reactivity of boron nitride nanoribbons with stone-Wales defects. Journal of Chemical Theory and Computation. 2009;**5**(11):3088-3095. DOI: 10.1021/ct900388x

[47] Pham T et al. Formation and dynamics of electron-irradiation-induced defects in hexagonal boron nitride at elevated temperatures. Nano Letters. 2016;**16**(11):7142-7147. DOI: 10.1021/acs.nanolett.6b03442

[48] Jin C, Lin F, Suenaga K, Iijima S. Fabrication of a freestanding boron nitride single layer and its defect assignments. Physical Review Letters. 2009;**102**(19):195505. DOI: 10.1103/PhysRevLett.102.195505

[49] Zobelli A, Gloter A, Ewels CP, Seifert G, Colliex C. Electron knock-on cross section of carbon and boron nitride nanotubes. Physical Review B. 2007;**75**(24):75-77. DOI: 10.1103/PhysRevB.75.245402

[50] Alem N, Erni R, Kisielowski C, Rossell MD, Gannett W, Zettl A. Atomically thin hexagonal boron nitride probed by ultrahigh-resolution transmission electron microscopy. Physical Review B. 2009;**80**(15):155425. DOI: 10.1103/PhysRevB.80.155425

[51] Tay RY. Chemical Vapor Deposition Growth and Characterization of Two-Dimensional Hexagonal Boron Nitride. Singapore: Springer Singapore; 2018. DOI: 10.1007/978-981-10-8809-4

[52] Chen L et al. Screw-dislocation-driven growth of two-dimensional few-layer and pyramid-like WSe_2 by sulfur-assisted chemical vapor deposition. ACS Nano. 2014;**8**(11):11543-11551. DOI: 10.1021/nn504775f

[53] Zhang L et al. Three-dimensional spirals of atomic layered MoS_2. Nano Letters. 2014;**14**(11):6418-6423. DOI: 10.1021/nl502961e

[54] Song X et al. Chemical vapor deposition growth of large-scale hexagonal boron nitride with controllable orientation. Nano Research. 2015;**8**(10):3164-3176. DOI: 10.1007/s12274-015-0816-9

[55] Tyagi P et al. Direct growth of self-aligned single-crystalline GaN nanorod array on flexible Ta foil for photocatalytic solar water-splitting. Journal of Alloys and Compounds. 2019;**805**:97-103. DOI: 10.1016/j.jallcom.2019.07.071

[56] Ramesh C et al. Effect of surface modification and laser repetition rate on growth, structural, electronic and optical properties of GaN nanorods on flexible Ti metal foil. RSC Advances. 2020;**10**(4):2113-2122. DOI: 10.1039/C9RA09707D

[57] Pierret A et al. Excitonic recombinations in h – BN: From bulk to exfoliated layers. Physical Review B. 2014;**89**(3):035414. DOI: 10.1103/PhysRevB.89.035414

[58] Li LH, Chen Y, Behan G, Zhang H, Petravic M, Glushenkov AM. Large-scale mechanical peeling of boron nitride nanosheets by low-energy ball milling. Journal of Materials Chemistry. 2011;**21**(32):11862. DOI: 10.1039/c1jm11192b

[59] Cui Z, Oyer AJ, Glover AJ, Schniepp HC, Adamson DH. Large scale thermal exfoliation and functionalization of boron nitride. Small. 2014;**10**(12):2352-2355. DOI: 10.1002/smll.201303236

[60] Yu B et al. Thermal exfoliation of hexagonal boron nitride for effective enhancements on thermal stability, flame retardancy and smoke suppression of epoxy resin nanocomposites via sol–gel process. Journal of Materials Chemistry A. 2016;**4**(19):7330-7340. DOI: 10.1039/C6TA01565D

[61] Rand MJ, Roberts JF. Preparation and properties of thin film boron

nitride. Journal of the Electrochemical Society. 1968;**115**(4):423. DOI: 10.1149/1.2411238

[62] Müller F, Hüfner S, Sachdev H, Laskowski R, Blaha P, Schwarz K. Epitaxial growth of hexagonal boron nitride on Ag(111). Physical Review B. 2010;**82**(11):113406. DOI: 10.1103/PhysRevB.82.113406

[63] Lu G et al. Synthesis of large single-crystal hexagonal boron nitride grains on Cu–Ni alloy. Nature Communications. 2015;**6**(1):6160. DOI: 10.1038/ncomms7160

[64] Pakdel A, Zhi C, Bando Y, Golberg D. Low-dimensional boron nitride nanomaterials. Materials Today. 2012;**15**(6):256-265. DOI: 10.1016/S1369-7021(12)70116-5

[65] Lin Y, Connell JW. Advances in 2D boron nitride nanostructures: Nanosheets, nanoribbons, nanomeshes, and hybrids with graphene. Nanoscale. 2012;**4**(22):6908. DOI: 10.1039/c2nr32201c

[66] Gautam C, Chelliah S. Methods of hexagonal boron nitride exfoliation and its functionalization: Covalent and non-covalent approaches. RSC Advances. 2021;**11**(50):31284-31327. DOI: 10.1039/D1RA05727H

[67] Bresnehan MS et al. Integration of hexagonal boron nitride with quasi-freestanding epitaxial graphene: Toward wafer-scale, high-performance devices. ACS Nano. 2012;**6**(6):5234-5241. DOI: 10.1021/nn300996t

[68] Tay RY et al. Trimethylamine borane: A new single-source precursor for monolayer h-BN single crystals and h-BCN thin films. Chemistry of Materials. 2016;**28**(7):2180-2190. DOI: 10.1021/acs.chemmater.6b00114

[69] Müller F, Hüfner S, Sachdev H, Gsell S, Schreck M. Epitaxial growth of hexagonal boron nitride monolayers by a three-step boration-oxidation-nitration process. Physical Review B. 2010;**82**(7):075405. DOI: 10.1103/PhysRevB.82.075405

[70] Wang M et al. Correction to catalytic transparency of hexagonal boron nitride on copper for chemical vapor deposition growth of large-area and high-quality graphene. ACS Nano. 2014;**8**(8):8711-8711. DOI: 10.1021/nn503982j

[71] Caneva S et al. Nucleation control for large, single crystalline domains of monolayer hexagonal boron nitride via Si-doped Fe catalysts. Nano Letters. 2015;**15**(3):1867-1875. DOI: 10.1021/nl5046632

[72] Ismach A et al. Toward the controlled synthesis of hexagonal boron nitride films. ACS Nano. 2012;**6**(7):6378-6385. DOI: 10.1021/nn301940k

[73] Bansal A, Zhang X, Redwing JM. Gas source chemical vapor deposition of hexagonal boron nitride on C-plane sapphire using B2H6 and NH3. Journal of Materials Research. 2021;**36**(23):4678-4687. DOI: 10.1557/s43578-021-00446-5

[74] Zhang C, Hao X, Wu Y, Du M. Synthesis of vertically aligned boron nitride nanosheets using CVD method. Materials Research Bulletin. 2012;**47**(9):2277-2281. DOI: 10.1016/j.materresbull.2012.05.042

[75] Oh H, Yi G-C. Synthesis of atomically thin h-BN layers using BCl3 and NH3 by sequential-pulsed chemical vapor deposition on Cu foil. Nanomaterials. 2021;**12**(1):80. DOI: 10.3390/nano12010080

[76] Pierson HO, Pierson HO. Handbook of Chemical Vapor Deposition. 2nd ed. Norwich, NY: Noyes Publications; 1999

[77] Jones AC, Hitchman ML. Chemical Vapour Deposition: Precursors, Processes and Applications. Cambridge: Royal Society of Chemistry; 2008. DOI: 10.1039/9781847558794

[78] Farhad SFU. The effect of substrate temperature and oxygen partial pressure on the properties of nanocrystalline copper oxide thin films grown by pulsed laser deposition. Data in Brief. 2021;**34**:106644. DOI: 10.1016/j.dib.2020.106644

[79] Wang G et al. Direct growth of hexagonal boron nitride films on dielectric sapphire substrates by pulsed laser deposition for optoelectronic applications. Fundamental Research. 2021;**1**(6):677-683. DOI: 10.1016/j.fmre.2021.09.014

[80] Velázquez D, Seibert R, Man H, Spentzouris L, Terry J. Pulsed laser deposition of single layer, hexagonal boron nitride (white graphene, h-BN) on fiber-oriented Ag(111)/$SrTiO_3$ (001). Journal of Applied Physics. 2016;**119**(9):095306. DOI: 10.1063/1.4943174

[81] Jang SK, Youn J, Song YJ, Lee S. Synthesis and characterization of hexagonal boron nitride as a gate dielectric. Scientific Reports. 2016;**6**(1):30449. DOI: 10.1038/srep30449

[82] Dean CR et al. Boron nitride substrates for high-quality graphene electronics. Nature Nanotechnology. 2010;**5**(10):722-726. DOI: 10.1038/nnano.2010.172

[83] Britnell L et al. Field-effect tunneling transistor based on vertical graphene Heterostructures. Science. 2012;**335**(6071):947-950. DOI: 10.1126/science.1218461

[84] Ranjan A et al. Dielectric breakdown in single-crystal hexagonal boron nitride. ACS Applied Electronic Materials. 2021;**3**(8):3547-3554. DOI: 10.1021/acsaelm.1c00469

Chapter 2

Low Energy Excitations in Borate Glass

Seiji Kojima

Abstract

The boson peak in the terahertz range is the low-energy excitations in glasses and disordered crystals. It is related to the excess part of the vibrational density of states. Borate glass is one of the typical network oxide glasses with covalent bonds and belongs to the strong type of glass formers. Alkali metal ions are well-known modifiers of the borate glass network and control various properties. The alkali metal effects are reviewed on basic physical properties such as elastic constants, density, and vibration modes in relation to the variation of structural units in a modified borate glass network. The alkali effect on a boson peak is discussed on the basis of experimental results of neutron inelastic scattering, neutron diffraction, Raman scattering, and heat capacity at low temperatures. The correlation is discussed between the boson peak frequency, the peak temperature of excess heat capacity, and shear modulus. The static and dynamical correlation lengths are also discussed.

Keywords: borate glass, boson peak, FSDP, medium range order, fragility, relaxation, excess heat capacity

1. Introduction

2022 is declared a United Nations International Year of Glass (IYOG), which celebrates the heritage and importance of this material in our lives [1]. Glass is technologically very important in industry, and it is clear that modern life would not be possible without glass. Glass is also very interesting in fundamental sciences related to random structure and non-equilibrium state. When a viscous liquid is cooled from a high temperature, it changes into a supercooled liquid state, and upon further cooling, it undergoes a liquid–glass transition into a nonequilibrium glassy state at a glass transition temperature T_g. However, a simple liquid is solidified into an equilibrium crystalline state at its melting temperature, T_m. The temperature dependence of enthalpy is shown in **Figure 1** for a liquid (AB), supercooled liquid (BD), glassy (DE), and crystalline (CG) states. The enthalpy of a liquid crosses to that of a crystal at the point, F, which is the Kauzmann temperature, T_K [1]. It is a static ideal glass transition temperature and close to the Vogel–Fulcher temperature, the dynamic ideal glass transition temperature, T_{VF} [3, 4], which is lower than the calorimetric T_g.

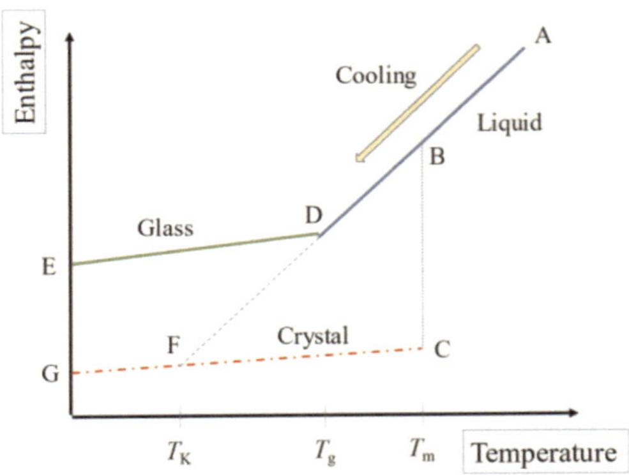

Figure 1.
Temperature dependence of the enthalpy for the liquid (AB), supercooled liquid (BD), glass (DE), and crystalline (CG) states, where T_m*,* T_g*, and* T_K *are the melting, glass transition, and Kauzmann temperatures, respectively [2].*

The temperature dependence of the relaxation time, τ_α, of α-structural relaxation process obeys the following Vogel-Fulcher law and the relaxation time diverges at T_{VF}.

$$\tau_\alpha = \tau_0 \, exp\left(\frac{B}{T - T_{VF}}\right) \text{ for } T > T_{VF} \tag{1}$$

Here, τ_0 and B are material-dependent constants, and $T_g > T_{VF}$. The fragility index m is defined by Eq. (2) using the parameters of Eq. (1). When the intermolecular interactions are weak, m is large and materials are fragile. While the interactions are strong, Eq. (1) goes to the Arrhenius law and materials are strong.

$$m = \left[\frac{d log \tau_\alpha}{d\left(\frac{T_g}{T}\right)}\right]_{T=T_g} = \frac{B}{T_g \, ln\, 10\left(1 - T_{VF}/T_g\right)^2} \tag{2}$$

In a liquid-glass transition, main three dynamical processes are the α-structural relaxation, fast β-relaxation, and boson peak. These three dynamics are interrelated with each other. According to the mode-coupling theory, the α-relaxation is related to the creation and annihiration of a molecular cage, the fast β-relaxation is related to a relaxation of molecules trapped in their cages. A boson peak is related to a damped librational motion of molecules trapped in their cages [5].

Glycerol (propane-1,2,3-triol, $C_3H_8O_3$) undergoes a liquid–glass transition at about T_g = 187 K. Glycerol is intermediate with m = 53. The melting temperature is T_m = 291 K, while it does not crystalize even in very slow cooling. It is one of the typical glass-forming materials and well-known cryoprotectants due to its strong glass-forming tendency [6]. The dominant interaction among molecules is the intermolecular hydrogen bond, and the related O-H stretching band showed a remarkable temperature dependence in the vicinity of T_g [7]. By the broadband dielectric spectroscopy, the slowing down of the α- relaxation time towards T_g, which obeys the Vogel-Fulcher law of Eq. (1), was clearly observed in the milli hertz to gigahertz range [8]. For the dynamical properties in the terahertz range, the temperature dependence of low-frequency Raman scattering spectra is shown in **Figure 2** [9]. The remarkable changes in Raman spectra were observed in the low-frequency range. In a liquid phase at 328

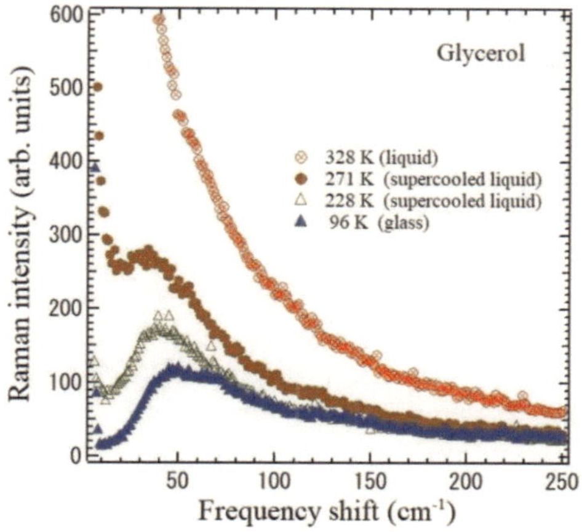

Figure 2.
Temperature dependence of low-frequency Raman scattering spectra of a liquid-glass transition of glycerol [9].

K, the broad Rayleigh wing appears, and the main contribution of this wing is the α-structural relaxation and fast β-relaxation. In supercooled liquid states at 271 and 228 K, the α-relaxation time becomes slow and the contribution to the Rayleigh wing becomes small. While the fast β-relaxation time does not depend on temperature, and its intensity gradually changes into a boson peak at about 40 cm^{-1} (=1.2 THz). In a glass state at 96 K, only a boson peak appears. The liquid-phase transition also occurs by the application of hydrostatic pressure at room temperature at P_g = 5 GPa. The boson peak was also observed in the pressure-induced glass state [10]. These dynamical properties on the boson peak, α- and fast β-relaxation processes in **Figure 1** are common in liquid-glass transitions of organic and inorganic glass-forming materials.

2. Borate glass and alkali metal modification

Borate glass is the most contemporary glass and optical material for technological and environmental applications. Pure borate glass undergoes a glass transition at T_g = 260°C. The melting temperature is T_m = 450°C [11]. The character of the temperature dependence of thermal expansibility, surface tension, and viscosity, in the range up to 1400°C, proves that, in the crystalline state and in the vitreous state below about 300°C, boron oxide consists of units held together by "weak" forces and that with increasing temperature this structure changes gradually toward a "strong" one [10]. The borate glass is strong with a fragility index of m = 30. Boron oxide glass has a random three-dimensional network of BO_3-triangles with a comparatively high fraction of six-membered rings (boroxol rings). Krogh-Moe discussed structure models for boron oxide glass and molten boron oxide with reference to spectroscopic data, diffraction data, and other physical properties for boron oxide [12]. Borate glasses for scientific and industrial applications were reviewed by Bengisu [13].

Borate glass is modified by alkali metals and physical properties remarkably change by the appearance of various structural units and structural groups. In this chapter, we discuss the dynamical properties of binary alkali metal borate glasses, $xM_2O(1-x)B_2O_3$ (M = Li, Na, K, Rb, Cs), i.e., lithium borate (LiB), sodium borate (NaB), potassium borate (KB), rubidium borate (RbB), and cesium borate (CsB) glasses. In alkali borate glasses, the physical properties are of special interest, because their alkali content dependences often show maxima and minima termed "borate anomaly" [14]. Their dependence also shows the difference among the kind of alkali metal ions. The dependences of density against the alkali content are classified to two groups. In LiB, NaB, and KB glasses, density shows a moderate increase as the alkali content increases, and they have the nature of covalent packing. While in RbB and CsB glasses, density remarkably increases against the alkali content increases, and they have the nature of ionic packing [15].

Since the physical properties are related to the structure, the variation has been discussed in terms of three kinds of structural units, $BØ_3$, $BØ_4{}^-$, and $BOØ_2{}^-$, for the alkali content below $x = 0.50$ as shown in **Figure 3** [15, 16]. Here, O and Ø denote the nonbridging and bridging oxygens, respectively. In pure borate glass, the coordination number of boron is three, and units $BØ_3$ are dominant. When borate glass is modified by alkali ions, the units $BØ_4{}^-$ and $BOØ_2{}^-$ are formed by the chemical reactions of Eqs. (3) and (4). The two units coexist by the disproportion reactions of Eq. (5). As the ionic radius of alkali ions increases, the number of nonbridging oxygen increases and the average coordination number of boron decreases. Consequently, the number of $BØ_4{}^-$ decreases, while that of $BOØ_2{}^-$ increases.

$$M_2O + 2BØ_3 \rightarrow 2M^+ + BØ_4 \tag{3}$$

$$M_2O + 2BØ_3 \rightarrow 2M^+ + BOØ_2{}^- \tag{4}$$

$$BØ_4{}^- \leftrightarrow BOØ_2{}^- \tag{5}$$

For the detailed discussion on the variation of structure, several kinds of superstructural units or structural groups such as boroxol ring, pentaborate, and diborate groups were considered [17].

For the analysis on the variation of physical properties, we introduce the quantity $V_m(B)$ and V_m (O), which denote the volume of glass containing one a mole of boron and oxygen, respectively.

$$V_m(B) = \frac{M[xM_2O(1-x)B_2O_3]}{2\rho(1-x)} \tag{6}$$

$$V_m(O) = \frac{M[xM_2O(1-x)B_2O_3]}{\rho(3-2x)} \tag{7}$$

Where ρ is density, $M[xM_2O(1-x)B_2O_3]$ is the molar mass of the entity $xM_2O(1-x)B_2O_3$. **Figure 4** shows alkali content dependences of volumes of glass containing one mole of (a) boron and (b) oxygen atoms [15].

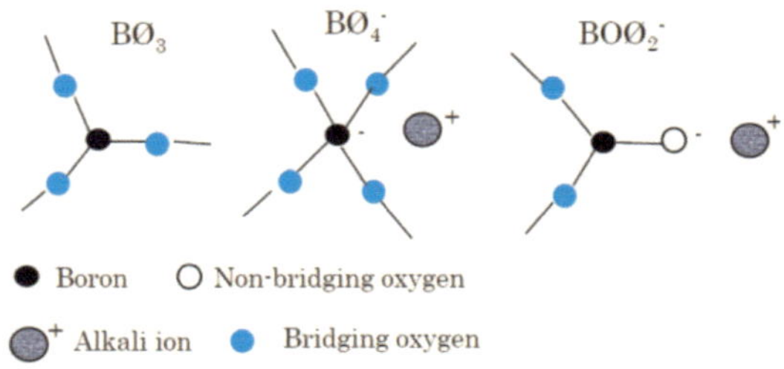

Figure 3.
Structural units of alkali borate glasses.

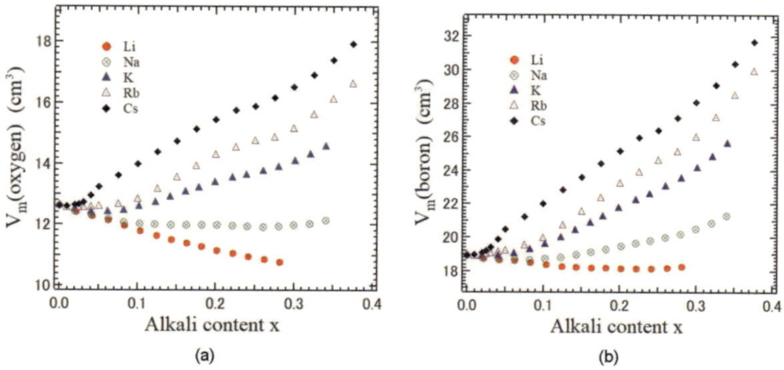

Figure 4.
Alkali content dependences of volumes of glass containing one mole of (a) boron, and (b) oxygen atoms.

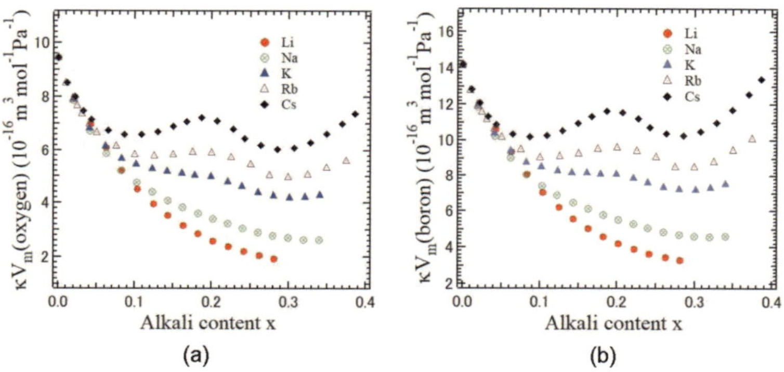

Figure 5.
Alkali content dependences of the derivatives of the molar volumes (a) $\kappa V_m(B)$, and (b) $\kappa V_m(O)$ of alkali borate glasses [15].

Both $V_{\mathrm{m}}(\mathrm{B})$ and $V_{\mathrm{m}}(\mathrm{O})$ increase with respect to alkali content in LiB, NaB, and KB glasses, while they decrease in RbB and CsB glasses. At a given alkali content, the packing of boron and oxygen in the glass structure becomes more compact as the ionic radius of alkali ion increases. The boron atoms are packed most compactly at $x = 0.20$ for LiB, $x = 0.08$ for NaB, $x = 0.02$ for KB, $x = 0.01$ for RbB, and none for CsB glasses.

The remarkable changes occur in the response to the stress such as their pressure derivatives. We discuss the derivatives of the molar volumes $V_{\mathrm{m}}(\mathrm{B})$ and V_{m} (O) with respect to pressure.

$$-\frac{d}{dp}V_m(B) = -\frac{1}{V}\frac{dV}{dP}V_m(B) = \kappa V_m(B), \tag{8}$$

$$-\frac{d}{dp}V_m(O) = -\frac{1}{V}\frac{dV}{dP}V_m(O) = \kappa V_m(O), \tag{9}$$

where $\kappa = -\frac{1}{V}\frac{dV}{dP}$, is the compressibility. The $\kappa V_{\mathrm{m}}(\mathrm{B})$ represents the effect of boron atoms on the elasticity, and the $\kappa V_{\mathrm{m}}(\mathrm{O})$ represents the effect of oxygen atoms on the elasticity. **Figure 5** shows the alkali content dependences of $\kappa V_{\mathrm{m}}(\mathrm{B})$ and $\kappa V_{\mathrm{m}}(\mathrm{O})$ [15].

Both $\kappa V_m(B)$ and κV_m (O) at a given alkali content decrease as the ionic radius of alkali ions increases. In LiB and NaB glasses with covalent packing, $\kappa V_m(B)$ and $\kappa V_m(O)$ monotonically decrease. However, in KB, RbB, and CsB glasses with ionic packing, the concave and convex shapes appear. We discuss these dependences by the division into following three alkali content ranges.

a. $0 \le x \le 0.07$: The conversion of $BØ_3 \rightarrow BØ_4^-$ in Eq. (3) is dominant and $\kappa V_m(B)$ and $\kappa V_m(O)$ decrease. The glass becomes incompressible with no alkali dependence.

b. $0.07 \le x \le 0.2$: The conversion of $BØ_3 \rightarrow BØ_4^-$ continues. However, in RbB and CsB glasses, the formation of $BØ_3 \rightarrow BOØ_2^-$ in Eq. (4) also occurs. The $\kappa V_m(B)$ and $\kappa V_m(O)$ increase by the presence of the small amount of M^+ $BOØ_2^-$ units. At a given alkali content, this conversion decreases as the ionic radius of alkali ion decreases.

c. $0.2 \le x \le 0.3$: The conversion of $BØ_3 \rightarrow BØ_4^-$ does not occur. The volume of M^+ $BOØ_2^-$ unit is larger than that of M^+ $BØ_4^-$ unit and easily deformed by stress. Thus, the increase of small amount of M^+ $BOØ_2^-$ units causes the glass incompressible, and at $x = 0.3$ RbB and CsB glasses have the most compact structure.

d. $0.3 \le x \le 0.4$: By the formation of the large amount of M^+ $BOØ_2^-$ units, the glass becomes more compressible and elastically softer.

3. Elastic properties of alkali borate glasses

Since the low-energy excitations in glass is closely related to the acoustic modes, the sound velocity was measured by the ultrasonic pulse-echo overlap method at a frequency of 10 MHz and at 298 K [15, 16]. The sound velocities of longitudinal acoustic (LA) and transverse acoustic (TA) modes of the binary alkali metal borate glasses, $xM_2O(1-x)B_2O_3$ (M = Li, Na, K, Rb, Cs) are shown in **Figure 6a** and **b**,

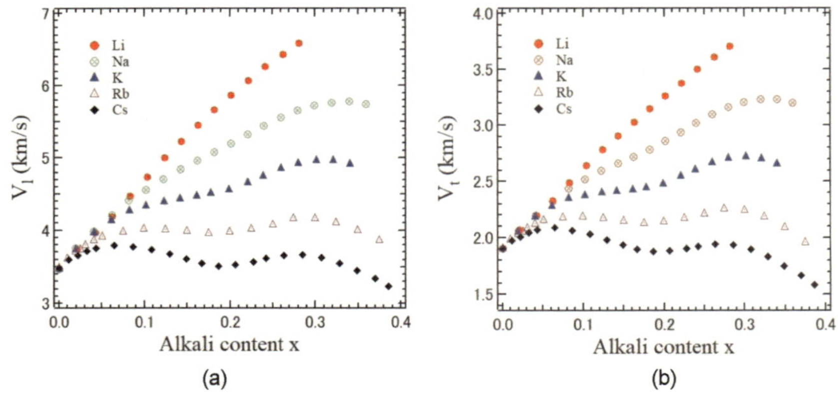

Figure 6.
Alkali content dependences of velocity of (a) longitudinal, and (b) transverse acoustic modes.

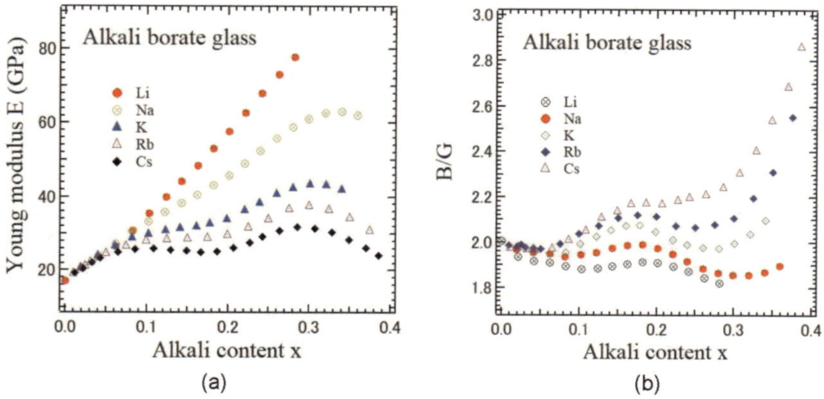

Figure 7.
Alkali content dependences of (a) Young's modulus, E, and (b) the ratio between bulk and shear moduli, B/G.

respectively [17].These dependences have a similarity with those of the reciprocal plots of $\kappa V_m(B)$ and (b) $\kappa V_m(O)$ in **Figure 5a** and **b** reflecting the variation of structural units by alkali ions.

Using the LA and TA velocities, the following elastic moduli were calculated.

$$\text{Shear modulus}: G = \rho V_T^2 \tag{10}$$

$$\text{Longitudinal modulus}: L = \rho V_L^2 \tag{11}$$

$$\text{Poisson's ratio}: \sigma = \frac{1}{2}\frac{V_L^2 - 2V_T^2}{V_L^2 - V_T^2} \tag{12}$$

$$\text{Young's modulus}: E = 2G(1+\sigma) \tag{13}$$

$$\text{Bulk modulus}: B = L - \frac{4}{3}G \tag{14}$$

$$\text{Compressibility}: \kappa = 1/B \tag{15}$$

Figure 7a shows the alkali content dependences of Young's modulus. In the Young's modulus of LiB and KB glasses, the monotonic increase was observed. However, In that of KB, RbB, and CsB glasses concave and convex shapes were observed. Such alkali dependence of Young's modulus is similar to that of $\kappa V_m(B)$ in **Figure 5a**.

The plot of bulk versus shear moduli is helpful in distinguishing ductile from brittle behavior beyond the elastic limit. When B/G>>1 (σ=0.5), materials are extremely incompressible. For ceramics, B/G≈1. 7(σ≈0.25). For polymers, B/G≈2.7(σ≈0.33). When B/G<<1 (σ = −1), materials are extremely compressible. The B/G is also related to the boson peak intensity and the fragility index [18]. The ratio between bulk and shear modulus is plotted in **Figure 7b**. The B/G is between 1.8 and 2.9. The *B* /*G* shows the concave and convex shapes were observed in all the alkali borate glasses. These dependences are explained by the division of four alkali content ranges. (a) $0 \le x \le 0.07$: The $BØ_3$ changes into $BØ_4^-$ only and does not change into $BOØ_2^-$. (b) $0.07 \le x \le 0.20$: The $BØ_3$ changes into both $BØ_4^-$ and $BOØ_2^-$. The effect of $BOØ_2^-$ is predominant over that of $BØ_4^-$. (c) $0.20 \le x \le 0.30$: The $BØ_3$ changes into both $BØ_4^-$ and $BOØ_2^-$. The effect of $BØ_4^-$ is predominant over that of $BOØ_2^-$. (d) $0.30 \le x \le 0.40$: The $BØ_3$ changes only into $BOØ_2^-$ without the further formation of $BØ_4^-$.

4. Boson peaks of alkali borate glasses

In glasses the universal features of their thermal properties at low temperatures have been observed. The heat capacity shows an excess part as the deviation from the Debye T^3 law and the thermal conductivity has a plateau at around 10 K [19]. These universal behaviors are caused by the anomalous phonon dispersion in the terahertz frequency (THz) range. In the inelastic scattering spectra, the peak of $g(E)/E^2$ has been observed, where $g(E)$ and $E=h\nu$ are the vibrational density of states (VDoS) and energy, respectively. The origin of a peak is the low-energy excess part of VDoS over the Debye model defined by $g(E)/E^2$. This THz peak is called the boson peak [20]. In the measurement of heat capacity C_p at low temperatures, a bump in C_p/T^3 at around 10 K is called a thermal boson peak. The plateau of the thermal conductivity indicates the strongly scattered of phonons above the boson peak frequency. It indicates that phonons meet the transverse Ioffe–Regel (IR) limit [21] around the boson peak.

The microscopic origin of a boson peak has been discussed by various theoretical models, such as (1) the structure and elastic constants heterogeneity [22–24]; (2) soft potential model [24–26]; (3) the resonant vibration of medium range order [27]; (4) mode-coupling theory on density fluctuations of arrested glass structures [28]; (5) broadening of the lowest van Hove singularity of the transverse phonon branch [29]; (6) the phonon-saddle transition in the energy landscape [30]; (7) the random first-order transition theory (RFOT) [31]; (8) anharmonic effects [32], and (9) recent numerical calculations reported that the boson peak originates from quasi-localized vibrations of string-like dynamical defects [33]. However, this situation has remained quite controversial because of a lack of distinct evidence.

The Stokes-component of Raman scattering intensity $I(\nu)$ is related to the imaginary part of Raman susceptibility $\chi''(\nu)$.

$$I(\nu) = I_0\chi''(\nu)\{n(\nu)+1\}, \; n(\nu) = \frac{1}{exp\left(\frac{h\nu}{k_BT}\right) - 1} \tag{16}$$

where $n(\nu)$ is the Bose-Einstein factor and I_0 is a constant which depends on the experimental condition. For the discussion of the boson peak in a Raman spectrum, the following quantity is plotted. Here, $C(\nu) = \nu^{\alpha}$ (α=0~2) is the light-vibration coupling constant and its frequency dependence shows a monotonic increase [20].

$$\frac{\chi''(\nu)}{\nu} = \frac{I(\nu)}{\nu\{n(\nu)+1\}} \propto C(\nu)\frac{g(\nu)}{\nu^2} \tag{17}$$

Figure 8a shows the boson peak spectra of lithium borate glasses observed by Raman scattering using a triple-grating spectrometer with the additive dispersion. The boson peak frequency ν_{BP} = 26 cm^{-1} at x = 0.02 increases up to 72 cm^{-1} at x = 0.26 as the lithium content increases and the increase is related to the increase of transverse sound velocity shown in **Figure 5b** [34]. In LiB glasses, the fragility index m increases from 30 at x=0.00 to 62 at x=0.28. In strong glass, the boson peak intensity is high, while the intensity of the fast β-relaxation is weak. As the fragility index m increases, the boson peak intensity becomes weak and that of the fast β-relaxation increases. As shown in **Figure 8a** the boson peak intensity decreases as the Li content increases.

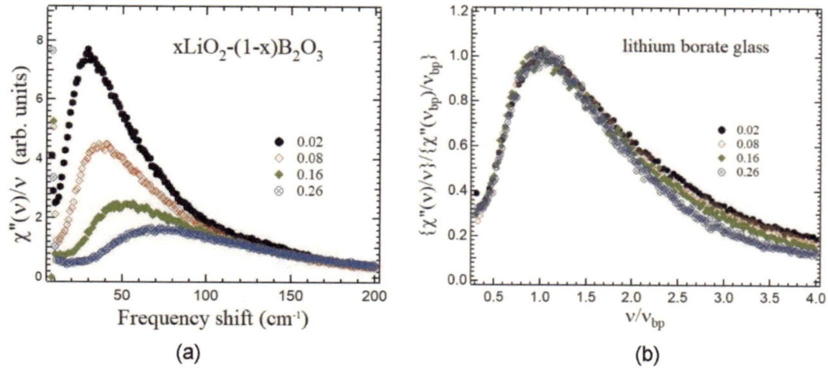

Figure 8.
(a) Reduced Raman spectra of boson peaks of lithium borate glasses observed by Raman scattering, and (b) Scaled boson peak spectra.

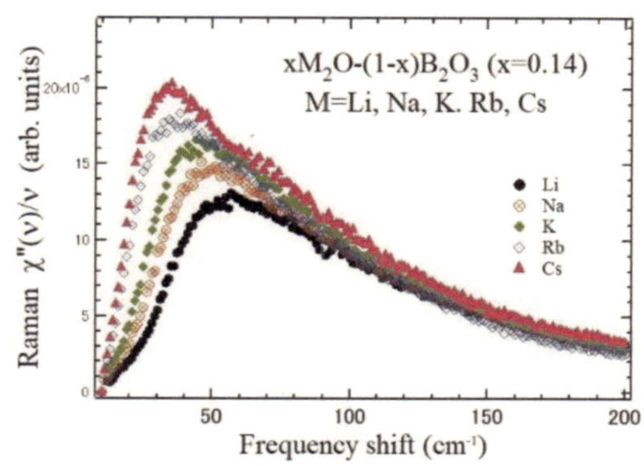

Figure 9.
Reduced Raman spectra of boson peaks of alkali borate glasses (x=0.14).

It is interesting to check that the boson peak spectra of inelastic neutron scattering, and Raman scattering are scaled by their peak positions and scattering intensity. In LiB glasses, it is found that the scaled boson peak spectra have a universal shape as shown in **Figure 8b**. The universal scaling of boson peaks indicates that the way of the distribution of VDoS basically remains the same, even though the glass structures drastically change by the alkali metal modification.

The alkali dependence of boson peak spectra at x=0.14 is shown in **Figure 9**. The origin of the boson peak in a pure borate glass was attributed to the coherent libration of several boroxol rings based on the study of the hyper-Raman scattering. As the alkali content increases, these boroxol rings change into other boron-oxygen structural units. As the ionic radius of alkali ions increases, the boson peak frequency decreases reflecting the difference in the modification of glass structure by alkali ions. Since the large Cs ions with the low charge density only slightly changes the boron-oxygen network structure. However, the small Li ions with the high charge density cause shrinking of the boron-oxygen network structure [35, 36]. By application of high pressure, it was reported that the boson peak frequency significantly increases up to 68 cm^{-1} at 4 GPa by shrinking of the boron-oxygen network structure [37]. The alkali content dependence of boson peaks of LiB glasses has the similarity with the densified borate glasses.

The boson peak frequency in a Raman spectrum includes the influence of the light-vibration coupling constant $C(\nu)$ as shown in Eq. (17). However, the boson peak in a neutron inelastic spectrum enables the direct observation of a boson peak frequency or energy even in the S/N ratio of scattering intensity is much lower than that of Raman scattering. **Figure 10** shows the alkali dependence of boson peak spectra of alkali borate glasses at x=0.22 observed by cold neutron inelastic scattering. Neutron inelastic scattering measurements of all the alkali borate glasses were carried out at 25° C (far below the T_g) using a direct geometry chopper-type ToF spectrometer AGNES belonging to the Institute for Solid State Physics, University of Tokyo [38]. The neutron boson peak energy also decreases as the ionic radius of alkali ions increases. As the charge density decreases with the increase in ionic radius, the contraction of the boron–oxygen structural units become weaker and the boson peak energy may decrease.

If we assume that the boson peak is connected to the nano-heterogeneity of the shear modulus [27]. In this approach, a dynamic length scale, L_{BP}, is given by

$$L_{BP} = V_t/2\pi\nu_{BP}, \tag{18}$$

where V_t is the transverse sound velocity, and ν_{BP} is the boson peak frequency. The L_{BP} corresponds to a medium-range scale important for characterization of structure

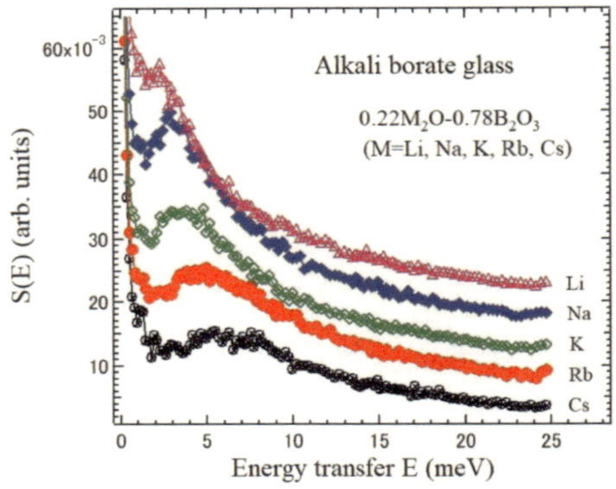

Figure 10.
Boson peaks of alkali borate glasses at x=0.22 *observed by cold neutron inelastic scattering [37].*

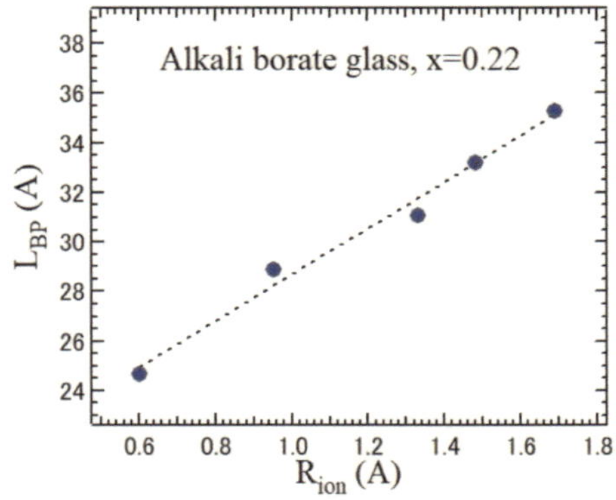

Figure 11.
The correlation between L_{BP} *and ionic radius of alkali ions in alkali borate glasses.*

correlations in glasses. The good correlation between L_{BP} and ionic radius of alkali ions is found as shown in **Figure 11**. It is found that L_{BP} is proportional to the ionic radius of alkali ions. It indicates that the dynamic length of a boson peak may relate to the size of alkali ion in the void of boron-oxygen network structure.

5. Excess heat capacity at low temperatures of alkali borate glasses

The excess heat capacity has been observed as the deviation from the Debye T^3 law at low temperatures. The broad peak in a C_p/T^3 vs. T plot is the thermal boson peak, where C_p is the heat capacity at a constant pressure. It is related to the non-Debye excess heat capacity. The alkali content dependence of C_p/T^3 in lithium borate glasses is shown in **Figure 12** [35]. In the pure borate glass, the peak of C_p/T^3 was observed at the temperature, T_m = 5.8 K [39]. As the lithium content increases, the peak value decreases and the peak temperature increases. Such a behavior of the peak value and the peak temperature of thermal boson peaks is similar to the peak intensity and peak frequency of boson peaks observed by Raman scattering and neutron inelastic scattering, respectively.

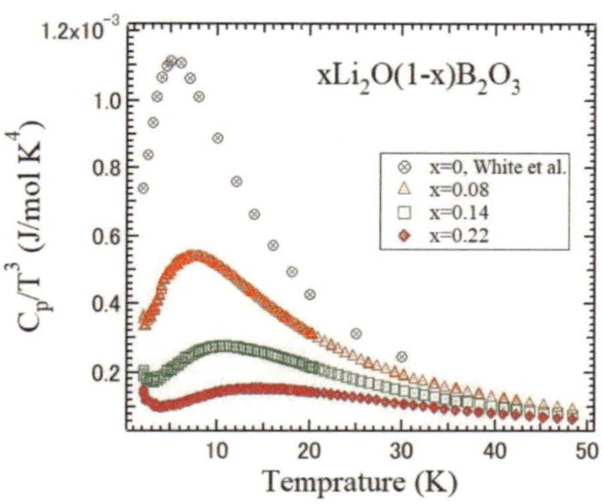

Figure 12.
Temperature dependence of C_p/T^3 *of lithium borate glasses at low temperatures.*

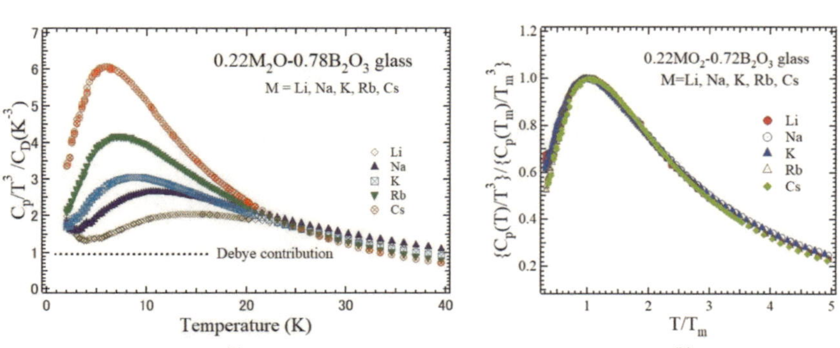

Figure 13.
(a) Temperature dependence of C_p/T^3 *of alkali borate glasses normalized by Debye contribution at low temperatures, and (b) Scaled thermal boson peaks of alkali borate glasses.*

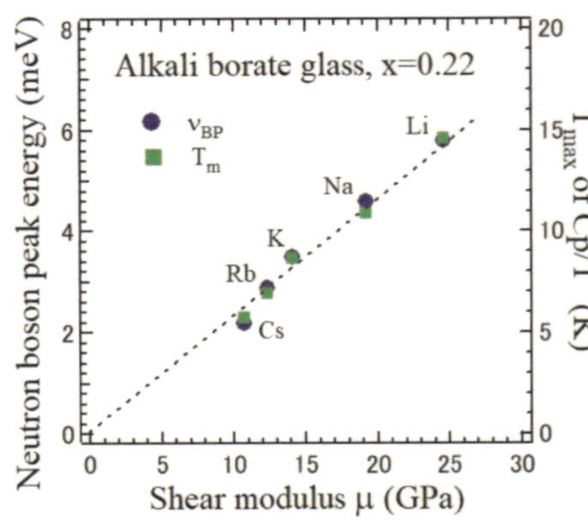

Figure 14.
Correlation between the peak temperature of C_p/T^3, boson peak frequency, and shear modulus.

The temperature dependence of C_p/T^3 of alkali borate glasses with x=0.22 below 50 K is shown in **Figure 13** [35]. The peak temperature of CsB glass is close to that of pure borate glass. However, as the ionic radius of alkali metal ions decreases, the peak temperature markedly increases. The peak temperature of LiB glass is T_m=14.6 K, which is about three times of a pure borate glass.

For all the alkali borate glasses, the universal nature of the master plot in C_p/T^3 vs. T is also observed as shown in **Figure 13b**. This universal nature is the same as that of boson peaks observed by Raman scattering and neutron inelastic scattering. This fact indicates that the distribution of the low-energy excess VDoS remains the same for all the alkali modification.

On the discussion on the origin of a boson peak, the correlation between the transverse acoustic mode and a boson peak is very interesting. According to the numerical simulation, the equality of the boson peak frequency to the Ioffe–Regel limit for "transverse" phonons was reported. The boson peak energy is proportional to the shear modulus [40]. Since the peak temperature of a thermal boson peak is proportional to the boson peak frequency, the correlation is examined between the peak temperature of C_p/T^3, boson peak energy, and shear modulus. The good correlation is found between these quantities as shown in **Figure 14**. This fact indicates that the origin of a boson peak is closely related to the Ioffe-Regal limit for transverse acoustic waves.

6. Medium range order of alkali borate glasses

In contrast to crystals with translational symmetry, glasses have disordered structure about local atomic arrangements. However, the structure of glasses has the medium range order (MRO) on a few nanometers' length scale [41, 42]. The MRO in liquid and glassy states is characterized by the first sharp diffraction peak (FSDP). The FSDP is observed by neutron and X-ray diffraction experiments in the static structure factor S(Q), where Q is the modulus of the wave vector [43]. The peak position Q_1 and the peak width ΔQ of a FSDP correspond to a periodic ordering with a periodicity of $2\pi/Q_1$ and a static structure correlation length L_{fsdp} given by

$$L_{fsdp} = 2\pi/\Delta Q \tag{19}$$

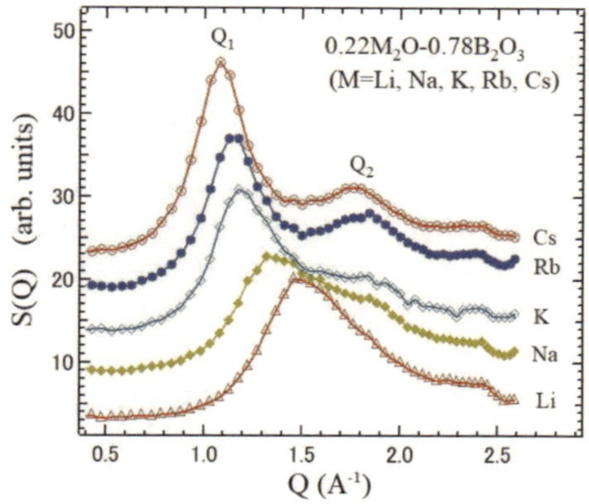

Figure 15.
Static structure factor S(Q) of alkali borate glasses observed by neutron diffraction. Q_1 is the position of a FSDP.

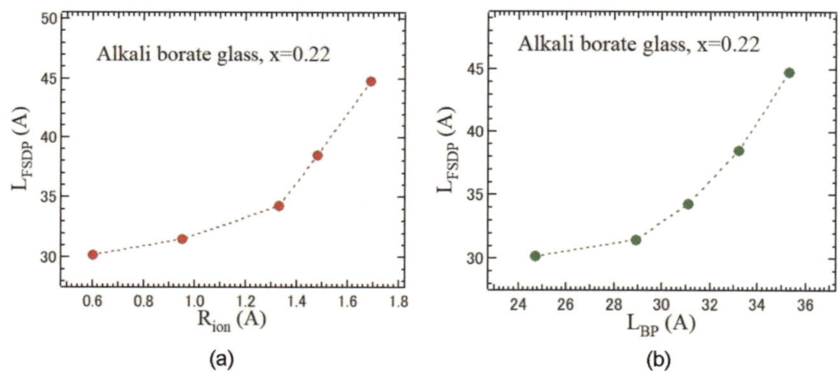

Figure 16.
(a) Correlation between the static structure correlation length of FSDP and the ionic radius of alkali ions of alkali borate glasses, and (b) Correlation between the static structure correlation lengths of a FSDP and the dynamic correlation length of a boson peak.

respectively. It was reported that the FSDP intensity and peak position can be quantified using the characteristic void distribution function, defined in terms of average void size, void distance, and void density [44].

The static structure factor S(Q) of alkali borate glasses determined by the neutron scattering is shown in **Figure 15** [38]. As the ionic radius of alkali ions decreases, the position Q_1 of a FSDP increases. Using the peak width of a FSDP, the static structure correlation lengths L_{fsdp} are determined. The correlation between the static structure correlation length and ionic radius of alkali ions is plotted in **Figure 16a**. It is found that as the ionic radius increases, the correlation length also increases. The MRO may be related to the local structure in the vicinity of an alkali ion in voids.

The correlation of various glasses between the boson peak frequency and the width of a FSDP was reported by Sokolov et al. [45]. Since the boson peak frequency is related to the dynamical correlation length L_{BP}, the relation between the static structure correlation length and the dynamical correlation length is plotted in **Figure 16b**. The good correlation between the dynamical correlation length L_{BP} and the static structure correlation length L_{fsdp} indicates that the boson peak is the vibration related to the MRO defined by the width of a FSDP.

7. Conclusions

Borate glass is most contemporary glasses and optical materials for technological and environmental applications. Borate glass is one of the typical network oxide glasses with covalent bonds and belongs to the strong type of glass formers. Alkali metal ions are well known modifiers of the borate glass network and control various properties. Basic physical properties such as elastic constants, density, and vibration modes are reviewed in relation with the variation of structural units in modified borate glass network.

The boson peak in the terahertz range is the low-energy excitations in glasses and disordered crystals. It is related to the excess part of vibrational density of states. The alkali metal effects on the boson peak are discussed on the basis of experimental results of neutron inelastic scattering, neutron diffraction, Raman scattering, low-temperature heat capacity, and ultrasonic measurements. For all the alkali borate glasses, the universal nature of the master plots in boson peak spectra and C_p/T^3 vs. T curve are observed. This fact indicates that the distribution of the low-energy excess VDoS remains the same for all the alkali modification.

The good correlation is found between the peak temperature of C_p/T^3, boson peak frequency, and shear modulus. this fact indicates that the origin of a boson peak is closely related to the Ioffe-Regal limit for transverse acoustic waves. The static and dynamical correlation lengths show also the good correlation. As the ionic radius of alkali ions increase, both correlation lengths also increase. This fact suggests that the boson peak vibration is related to the medium range order in the boron-oxygen network near the voids filled by alkali ions.

Acknowledgements

The author is thankful to Prof. M. Kodama, Prof. S. A. Feller, Prof. M. Affatigato, Prof. V. N. Novikov, Prof. M. Maczka, Prof. J.H. Ko, Prof. O. Yamamuro, Prof. H. Anwar, and Dr. Y. Matsuda for their collaboration and fruitful discussions. The author is also thankful to M. Kawashima, Y. Fukawa, K. Kaneda, S. Aramomi, and T. Sunaoshi for the experiments.

Author details

Seiji Kojima
Division of Materials Science, University of Tsukuba, Tsukuba, Japan

*Address all correspondence to: kojima@ims.tsukuba.ac.jp

References

[1] Kauzmann W. The nature of the glassy state and the behavior of liquids at low temperatures. Chemical Reviews. 1948;**43**:219-256. DOI: 10.1021/cr60135a002

[2] Kojima S. Anniversary of Brillouin scattering: Impact on materials science. Materials. 2022;**15**:3518. DOI: 10.3390/ma15103518

[3] Vogel DH. Das Temperaturabhaengigkeitsgesetz der Viskositaet von Fluessigkeiten. Physikalische Zeitschrift. 1921;**22**:645-646

[4] Fulcher GS. Analysis of recent measurements of the viscosity of glasses. Journal of the American Ceramic Society. 1925;**8**:339-355. DOI: 10.1111/j.1151-2916.1925.tb16731.x

[5] Franosch T, Götze W, Mayr MR, Singh AP. Evolution of structural relaxation spectra of glycerol within the gigahertz band. Physical Review E. 1997;**55**:3183-3190. DOI: 10.1103/PhysRevE.55.3183

[6] Polge C, Smith A, Parkes A. Revival of spermatozoa after vitrification and dehydration. Nature. 1949;**164**:666. DOI: 10.1038/164666a0

[7] Kojima S. Anomalous behaviour of the O-H stretching vibrational mode in the liquid-glass transition of glycerol. Journal of Molecular Structure. 1993;**294**:193-195. DOI: 10.1016/0022-2860(93)80347-X

[8] Lunkenheimer P, Loidl A. Dielectric spectroscopy of glass-forming materials: α-relaxation and excess wing. Chemical Physics. 2002;**284**:205-219. DOI: 10.1016/S0301-0104(02)00549-9

[9] Kojima S. Low-frequency raman investigation of the liquid-glass transition in glycerol. Physical Review B. 1993;**47**:2924-2927. DOI: 10.1103/PhysRevB.47.2924

[10] Ahart M, Aihiati D, Hemly R, Kojima S. The Boson peak of glassy glycerol under high pressure. Journal of Physical and Chemical Part B. 2017;**121**:6667-6672. DOI: 10.1021/acs.jpcb.7b01993

[11] Fajans K, Barber SW. Properties and structures of vitreous and crystalline boron oxide. Journal of the American Chemical Society. 1952;**74**:2761-2768. DOI: 10.1021/ja01131a019

[12] Krogh-Moe J. The structure of vitreous and liquid boron oxide. Journal of Non-Crystalline Solids. 1969;**1**:269-284. DOI: 10.1016/0022-3093(69)90025-8

[13] Bengisu M. Borate glasses for scientific and industrial applications: A review. Journal of Material Science. 2016;**51**:2199-2242. DOI: 10.1007/s10853-015-9537-4

[14] Shelby JE. Thermal expansion of alkali borate glasses. Journal of American Ceramic Society. 1983;**66**:225-227. DOI: 10.1111/j.1151-2916.1983.tb10023.x

[15] Kojima S, Kodama M. Velocity of sound in and elastic properties of alkali metal borate glasses. Physical and Chemical Glasses. 2014;**55**:1-12

[16] Kodama M, Kojima S. Velocity of sound and elastic properties of Li2O-B2O3 glasses. Journal of Applied Physics. 1995;**34**:2570-2574. DOI: 10.1143/JJAP.34.2570

[17] Wright AC. Borate structures: Crystalline and vitreous. Physics and Chemistry of Glasses: European Journal of Glass Science and Technology, Part B. 2010;**51**:1-39

[18] Greaves GN, Greer AL, Lakes RS, Rouxel T. Poisson's ratio and modern materials. Nature Materials. 2011;**10**: 823-837. DOI: 10.1038/NMAT3134

[19] Zeller RC, Pohl RO. Thermal conductivity and specific heat of non-crystalline solids. Physical Review B. 1971;**4**:2029-2040. DOI: 10.1103/PhysRevB.4.2029

[20] Nakayama T. Boson peak and terahertz frequency dynamics of vitreous silica. Reports on Progress in Physics. 2002;**65**:1195-1242. DOI: 10.1088/0034-4885/65/8/203

[21] Ioffe AF, Regel AR. Non-crytalline, amorphous, and liquid electronic semiconductors. Progress in Semiconductors. 1960;**4**:237-291

[22] Elliott SR. A unifed model for the low-energy vibrational behaviour of amorphous solids. Europhysics Letters. 1992;**19**:201-206. DOI: 10.1209/0295-5075/19/3/009

[23] Schirmacher W. Thermal conductivity of glassy materials and the 'boson peak'. Europhysics Letters. 2006; **73**:892-898. DOI: 10.1209/epl/i2005-10471-9

[24] Leonforte F, Tanguy A, Wittmer JP, Barrat JL. Inhomogeneous elastic response of silica glass. Physical Review Letters. 2006;**97**:055501. DOI: 10.1103/PhysRevLett.97.055501

[25] Galperin YM, Karpov VG, Kozub VI. Localized states in glasses. Advances in Physics. 1989;**38**:669-737. DOI: 10.1080/00018738900101162

[26] Buchenau U, Galperin YM, Gurevich VL, Parshin DA, Ramos MA, Schober HR. Interaction of soft modes and sound waves in glasses. Physical Review B. 1992;**46**:2798-2808. DOI: 10.1103/PhysRevB.46.2798

[27] Duval E, Garsia N, Boukenter A, Serughetti J. Correlation effects on Raman scattering from low-energy vibrational modes in fractal and disordered systems. I. Theory. Journal of Physical Chemistry. 1993;**99**: 2040-2045. DOI: 10.1063/1.465267

[28] Götze W, Mayr M. R; Evolution of vibrational excitations in glassy systems. Physical Review E. 2000;**61**:587-606. DOI: 10.1103/PhysRevE.61.587

[29] Chumakov AI et al. Equivalence of the boson peak in glasses to the transverse acoustic van Hove singularity in crystals. Physical Review Letters. 2011;**106**:225501. DOI: 10.1103/PhysRevLett.106.225501

[30] Grigera TS, Martn-Mayor V, Parisi G, Verrocchio P. Phonon interpretation of the 'boson peak' in supercooled liquids. Nature. 2003;**422**: 289-292. DOI: 10.1038/nature01475

[31] Lubchenko V, Wolynes PG. Theory of structural glasses and supercooled liquids. Annual Review of Physical Chemistry. 2007;**58**:235-266. DOI: 10.1146/annurev.physchem.58.032806.104653

[32] Baggioli M, Zaccone A. Universal origin of boson peak vibrational anomalies in ordered crystals and in amorphous materials. Physical Review Letters. 2019;**122**:145501. DOI: 10.1103/PhysRevLett.122.145501

[33] Hu YC, Tanaka H. Origin of the boson peak in amorphous solids. Nature

Physics. 2022;**8**:669-677. DOI: 10.1038/s41567-022-01628-6

[34] Kojima S, Novikov NV, Kodama M. Fast relaxation, boson peak and anharmonicity in lithium borate glass. The Journal of Chemical Physics. 2000; **113**:6344-6350. DOI: 10.1063/1.1309530

[35] Kojima S, Kawaji H. Low-temperature heat capacity of alkali metal borate glass. Journal of Thermal Analysis and Calorimetry. 2019;**135**:2759-2764. DOI: 10.1007/s10973-018-7590-3

[36] Kojima S, Novikov VN, Kofu M, Yamamuro O. Neutron scattering studies of static and dynamic correlation lengths in alkali metal borate glasses. Journal of Non-Crystalline Solids. 2019;**518**:18-23. DOI: 10.1016/j.jnoncrysol.2019.05.005

[37] Carini G, Gilioli E, Tripodo G, Vasi C. Structural changes and elastic characteristics of permanently densified vitreous B2O3. Physical Review B. 2011; **84**:024207. DOI: 10.1103/PhysRevB.84.024207

[38] Kojima S, Novikov VN, Kofu M, Yamamuro O. Nanometric fluctuations of sound velocity in alkali borate glasses and fragility of respective melts. Physica Status Solidi B. 2020;**257**:2000073. DOI: 10.1002/pssb.202000073

[39] White GK, Collecott SJ, Cook JS. Thermal expansion and heat capacity of vitreous B2O3. Physical Review B. 1984; **29**:4778-4781. DOI: 10.1103/PhysRevB.29.4778

[40] Shintani H, Tanaka H. Universal link between the boson peak and transverse phonons in glass. Nature Materials. 2008;7:870-877. DOI: 10.1038/nmat2293

[41] Lucovsky G, Galeener FL. Intermediate range order in amorphous solids. Journal of Non-Crystalline Solids. 1980;**35–36**:1209-1214. DOI: 10.1016/0022-3093(80)90362-2

[42] Greaves GN, Sen S. Inorganic glasses, glass-forming liquids and amorphizing solids. Advances in Physics. 2007;**56**:1-166. DOI: 10.1080/00018730601147426

[43] Salmon PS. Real space manifestation of the first sharp diffraction peak in the structure factor of liquid and glassy materials. Proceedings of the Royal Society of London A. 1994;**445**:351-365. DOI: 10.1098/rspa.1994.0065

[44] Zaug JM, Soper AK, Clark SM. Pressure-dependent structures of amorphous red phosphorus and the origin of the first sharp diffraction peaks. Nature Materials. 2008;7:890-899. DOI: 10.1038/nmat2290

[45] Sokolov AP, Kisliuk A, Soltwisch M, Quitmann D. Medium-range order in glasses: Comparison of Raman and diffraction measurements. Physical Review Letters. 1992;**69**:1540-1543. DOI: 10.1103/PhysRevLett.69.1540

Chapter 3

Boron-Based Cluster Modeling and Simulations: Application Point of View

Nasim Hassani, Mohammad Reza Hassani and Mehdi Neek-Amal

Abstract

Among sub-nanometer clusters, boron-based clusters and their atom-doped counterparts have attracted great attention due to their mechanical, physical, and chemical properties as well as their applications. Molecular dynamics (MDs) simulations and *ab initio* methods, including density functional theory (DFT) calculations, have been used to understand the physical and chemical properties of different materials. Much research has recently been conducted by using various methods to determine the different properties of boron clusters. In this chapter, we briefly introduce the relevant modeling and simulation methods, then review very recent theoretical researches on the application of small boron clusters, such as gas sensors, electrodes, H_2 storage, drug delivery, and catalytic applications.

Keywords: nanometric clusters, boron, application, gas sensor, electrode, H_2 storage, catalyst, drug delivery

1. Introduction

In recent years, wonderful molecular features have emerged through the study of pure and atom-doped boron clusters. Boron is a *p*-element that has three valence electrons and one of them in the *p*-orbital. Also, this element is located between the metallic and nonmetallic elements of the periodic table and, like carbon, is an exceptional element with a self-catenation character. Boron can form bonds between the same elements leading to the synthesis of the pure hydride of boron compounds. Also, it scarcely obeys the octet rule, and its valence shell typically contains only six electrons.

The main family of boron compounds is classified into two different branches: (i) the boron clusters that can form carboranes, borohydrides, metallacarboranes, and (ii) the organic compounds, in which they structures are different and their characteristics depend on the size. Although the organic compounds contain chains and rings, the boron clusters have a planar or cage-like structures. Also, the boron clusters have various shapes and symmetries that are the result of occupying vertices through different numbers of boron atoms or heteroatoms. The shape of these clusters is widespread, from unstable tetrahedral to more stable icosahedral [1, 2].

According to these intrinsic features, the chemical properties of boron clusters have gained great attention from researchers. To date, 16 polymorphs have been detected for bulk boron. Among them, the B_{12} icosahedron is a predominant motif [3]. However, it is shown that B_{12} cluster in isolated form is not stable and tends to form planar or quasi-planar structures. The anionic and cationic forms of the small boron clusters (B_n; $n \leq 36$) usually prefer to have planar and two-dimensional structures, respectively.

The planar structures in the edge consist of two-center two-electron (2c-2e) sigma bonds and between the inner atoms have multicenter two-electron (nc-2e) bonds. Multidimensional aromaticity as a result of delocalized σ and π bonds is responsible for the stability, planarity, and bioavailability of planar boron-based clusters. Furthermore, energetically favorable 1D boron nanotubes [4] and 2D boron sheets [5–7] have been produced, and their structure contains planar triangular lattices with hexagonal holes.

By discovering the cage-like structure of B_{40}/B_{40}^-, a new family is introduced to the boron cluster which is called borospherene. These hollow structures are generally interlocking boron chains formed from trigonal fragments and containing holes with hexa, hepta, and other sizes. Also, planar B_{36} is a pioneer in the borophene concept having a monolayer structure with hexagonal vacancies. The smallest borospherene that has been discovered is B_{28}^- and other motifs for borophene are B_{26}^-, B_{35}^-, B_{37}^-, and B_{38}^-.

Experimental and theoretical studies revealed that 0D boron clusters with B_n^- ($n < 38$) structure are planar and quasi-planar, and their stability is the result of delocalized multicentric bonding [8–11]. They also demonstrated that the B_n ($31 \leq n \leq 50$) clusters can have a tubular structure [10]. Series of axially chiral borospherenes structures for B_{38}^+, B_{38}^{2+}, B_{31}^+, and B_{32} clusters are also investigated [12, 13]. In a boron cluster of a certain size with a large number of boron atoms, the structures proposed for its ground state are cage-like [14], quasi-planar, and bilayer (i.e. B_{48}^-) [15–17]. Moreover, for B_n clusters with $n > 68$, the most energetically stable structure is found to be core-shell [18].

Replacing a boron atom with a specific dopant leads to the production of a new subclass that is of particular interest and diverse in structure. For example, transition metal doping of B_n ($12 \leq n \leq 22$) clusters forms metal-centered monocyclic boron rings [19–22], half-sandwich structures in metal-doped B_n clusters [23, 24], inverse-sandwich structure in La-doped B_n^- cluster [25], and endohedral boron cage in NiB_{80} cluster [26]. However, a new way to achieve intriguing features in boron clusters can be constructed by a specific combination of the number of doped atoms and boron atoms in the boron clusters.

The variety in boron clusters and their atom-doped counterparts has increased the ability of these clusters to be applied in different applications. As such, in this chapter, to generate new insights into the various applications, we review some important applications of boron clusters and their atom-doped counterparts. We will briefly introduce the most relevant computational methods to simulate these clusters and then present examples of their use in different areas, ranging from drug delivery to reaction catalysis. We hope to inspire the general community and research groups to get involved in proposing new applications for boron clusters.

2. Introducing the relevant method for modeling and simulations

The optimum structure of the materials and their corresponding applications can be predicted by modeling and simulation methods. They require analogous levels of precision and control that can also accurately describe the pertinent processes and conditions. As shown in **Figure 1**, across length and timescales, these methods can equip a wide range of opportunities to shed light on properties and phenomena that are unattainable through experimental effort. Among these methods, *ab initio* methods (such as density functional theory (DFT) calculations), standard molecular dynamics (MDs) simulations, and *ab initio* molecular dynamics simulations (AIMD) have been employed, mainly to study the properties of materials including boron clusters at the nanoscale.

Ab initio methods refer to those methods derived directly from theoretical principles, and their equations have not contained any empirical or semi-empirical parameters and the inclusion of experimental ones. The Hartree-Fock (HF) method is the simplest type of *ab initio* electronic structure calculations in which the correlated electron-electron repulsion is not explicitly included, and only its average effect is taken into account in the calculations [27, 28].

In DFT calculations, the ground-state energy is obtained as a function of a set of n one-electron Schrodinger-like equations, which are known as Kohn-Sham orbitals. This equation expresses the ground-state energy as a function of the interactions between the electrons, the nuclei, and themselves, the kinetic energy, and the exchange-correlation energy (see Eq. (1)). In these calculations, functionals (functions of another function) are employed to determine the properties of a many-electron system. There is an approximation in the hamiltonian and the expression for

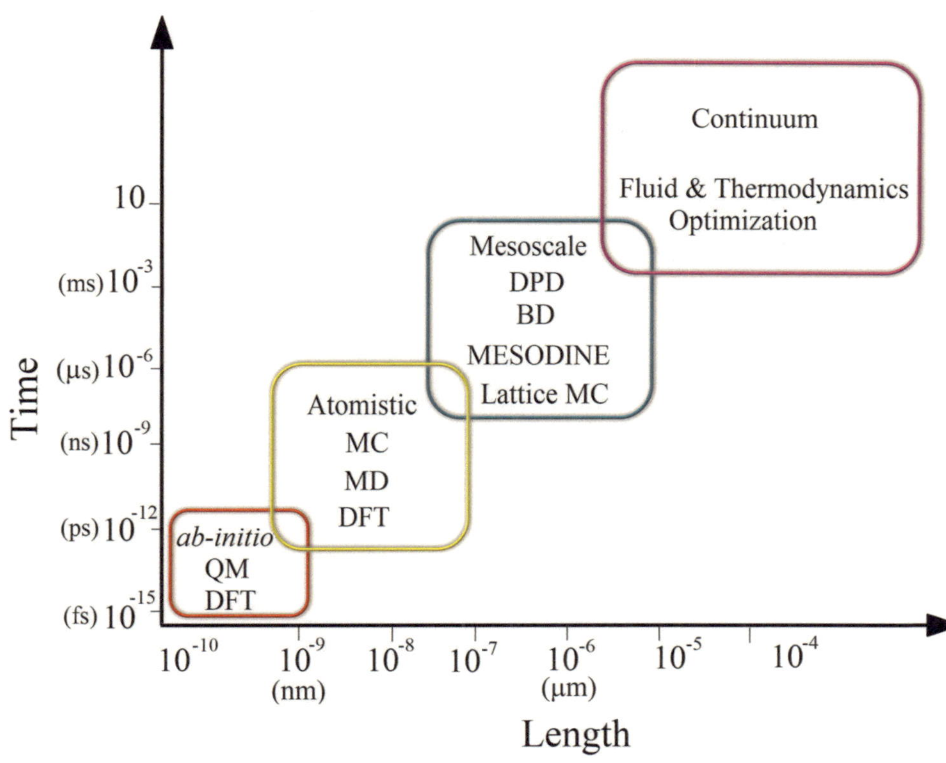

Figure 1.
Typical length and timescales in the simulation of the materials.

the total electron density in the DFT calculations. However, these type of calculations can be very accurate for little computational cost [29–32]:

$$E[n] = E_{Kin}[n] + E_{Coul}[n] + E_{xc}[n] + E_{ext}[n], \tag{1}$$

Determining the exact functionals for exchange and correlation is the main problem in DFT. Accordingly, a bunch of functionals for DFT calculations is developed which can be classified from the simplest to the most accurate functionals. The exchange-correlation energy term in the functionals is constructed based on some approximation, i.e. local density approximation (LDA, see Eq. (2)), generalized gradient approximations (GGAs, see Eq. (3)), meta-GGA (see Eq. (4)), and hybrid functionals. For example, hybrid functionals are termed based on the density functional exchange functional in combination with the Hartree-Fock exchange term:

$$E_{xc}^{LDA}[\rho] = \int \rho(r)\varepsilon_{xc}(\rho(r))dr, \tag{2}$$

$$E_{xc}^{GGA}[\rho] = \int \rho(r)\varepsilon_{xc}(\rho(r), \nabla\rho(r))dr, \tag{3}$$

$$E_{xc}^{MGGA}[\rho] = \int \rho(r)\varepsilon_{xc}(\rho(r), \nabla\rho(r), \nabla^2\rho(r))dr, \tag{4}$$

where ρ and ε_{xc} refer to the electronic density and the exchange-correlation energy per particle of a homogeneous electron gas with the charge density of ρ, respectively.

In the standard molecular dynamics (MDs) simulations, by considering classical treatment, Newton's second law is applied to the atomic coordinates. Then, force fields (FFs) which are a gradient of a prescribed interatomic potential functions are employed to calculate instantaneous force on each atom. FFs are the heart of MDs which are a function of the atomic coordinates and containing parameter sets (see Eq. (5)):

$$FF\left(\vec{R}\right) = U_{bonds} + U_{angles} + U_{dihedrals} + U_{impropers} + U_{nonbond}, \tag{5}$$

$$U_{bonds} = \sum_{bonds} k_i^{bonds}(r_i - r_0)^2,$$

$$U_{angles} = \sum_{angles} k_i^{angles}(\theta_i - \theta_0)^2,$$

$$U_{dihedrals} = \sum_{dihedrals} k_i^{dihedrals}(1 + \cos(n_i\phi_i - \delta_i)),$$

$$U_{improper} = V_{imp}$$

$$U_{nonbond} = \sum_{ij} 4k\varepsilon_{ij}\left(\frac{\sigma_{ij}^{12}}{r_{ij}^{12}} - \frac{\sigma_{ij}^{6}}{r_{ij}^{6}}\right) + \sum_{elec} \frac{q_i q_j}{r_{ij}}$$

where the local contributions to the total energy are included in the first four terms, i.e. bond stretching, angle bending, dihedral, and improper torsions. In this case, when considering a 12-6 Lennard-Jones potential, the repulsive, van der Waals, and coulombic interactions are described in the last two terms. The parameters are derived from experiments and quantum mechanics. After that, the position and velocity of the particles can be calculated by numerical integration [33–35].

In the *ab initio* MD simulations (AIMDs), at first, the interatomic forces are found at a given time instant. Then, from a quantum-mechanical perspective, the system is parameterized as a function of nuclei and electrons coordinates at a fixed time. Using the Born-Oppenheimer (BO) approximation, the nuclei are considered to be fixed at the instantaneous positions of the atoms. Consequently, the time-independent Schrodinger equation can be invoked to calculate the many-body electron wave function and the energy. In fact, the obtained energy is a function of the nuclei coordinates which were previously considered fixed. This energy can be considered as an interatomic potential to obtain the forces in Newton's equation of motion. In the other words, the gradients of the DFT energy at this fixed point can be calculated to obtain forces in which the nuclei are moved by this force to reach the next time step. This whole mentioned process is then repeated for these new atomic positions [36, 37].

The calculation way of the interatomic forces and the computational costs are the manifest and the origin of the difference between the standard MDs, AIMDs, and DFT calculations. AIMDs can be applied only for small system sizes, due to its huge computational cost. Also, AIMDs allow determining the dynamics of the systems that have no FFs. Intrinsically, AIMDs can deal with some effects such as polarization, bonding, many-body effects, and charge transfer, whereas in standard MDs these effects are artificially imposed from the data. Moreover, DFT as a quantum mechanical method for calculating energy as well as other properties of the material is a time-consuming technique. However, empirical potentials (FFs) are much faster but less accurate than the *ab initio* method like DFT. Finaly, the method selection for a specific case should be made based on these factors.

3. The application of the boron-based clusters

3.1 Gas sensor

From life safety point view, the design of sensitive materials to detect toxic gases in the environment is highly demanded. Among these hazardous gases, CO, NH_3, NO, H_2S, SO_2, SO_3, and CO_2 are mainly produced through industrial applications and automobile exhaust, which represent a harmful threat to human life and the natural environment.

Hossain et al. [38] in a theoretical study using DFT calculations investigated the quasi-planar 2D borophene B_{35} (see **Figure 2(a)**) as an efficient gas sensor toward NO, NO_2, N_2O, and NH_3 gases. Gases prefer to adsorb on the hexagonal hallow site of B_{35} where N_2O gas is chemically adsorbed, and the other gases are physically adsorbed on this nanocluster. Also, after gas adsorption, the hardness and stability of all systems increased as a result of the increased highest occupied molecular orbital (HOMO)-lowest unoccupied molecular orbital (LUMO) energy gap.

It is also demonstrated that the B_{36} (see **Figure 2(b)**) can be applied as a good detector for ammonia gas (NH_3). The minimum energy configuration of this interaction is the adsorption of NH_3 from N-head on a B atom of the B_{36}. During this interaction, the enthalpy changes −90.5 kJ/mol and 0.35 | e | charge is transferred from ammonia to the B_{36}. Also, the electrical conductance of B_{36} is found to increase after NH_3 adsorption in which the HOMO-LUMO energy gap decreased from 1.55 to 1.35 eV [39].

Ploysongsri and Ruangpornvisuti [40] studied the adsorption of gases containing sulfur on B_{36} cluster i.e. H_2S, SO_2, and SO_3. SO_2 and SO_3 gases adsorb from the oxygen

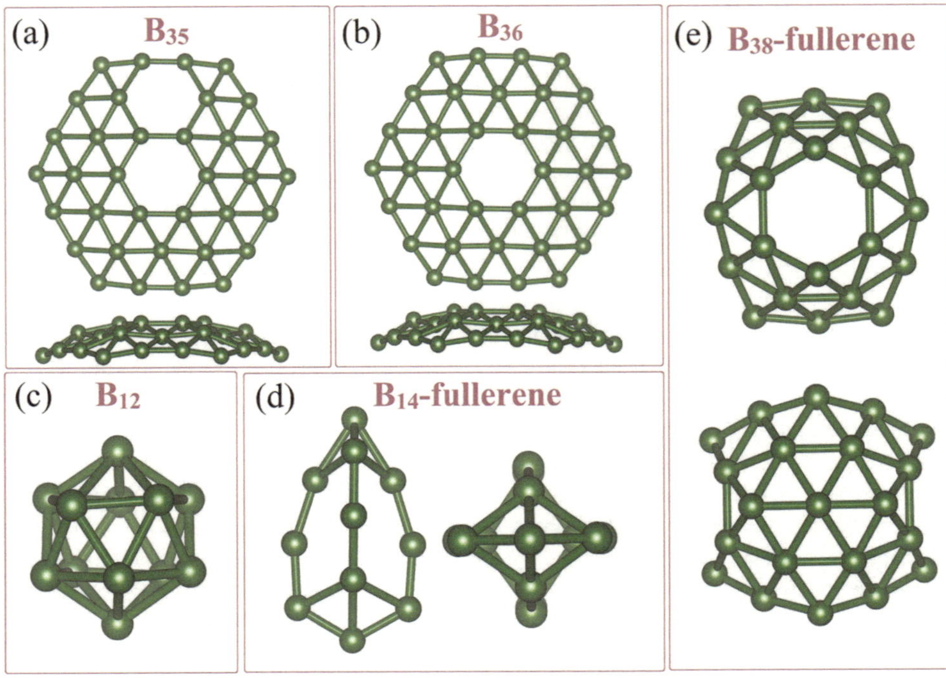

Figure 2.
The structure of some small clusters of boron. In the panels, two different views are shown and green balls represent B atoms.

side to the edge of the cluster that is thermodynamically favorable, while H_2S adsorption is not spontaneous on this cluster. The H_2S, SO_2, and SO_3 gases can be adsorbed on the edge of B_{36} with an adsorption energy of −5.29, −43.85, and − 80.57 kcal/mol, respectively.

Also, it is shown that metal-decorated B_{36} and its nitrogen-doped counterparts (M-N_x-B_{36-x} (M = Fe, Ni, and Cu; x = 0, 3)) are also sensitive to detect CO, NO, O_2, and N_2 molecules. The substitution of three nitrogen atoms in the central ring of B_{36} can increase the stability and sensitivity of the B_{36} cluster. The adsorption energy of CO, NO, O_2, and N_2 gases for the most stable configurations changes in the range of −0.32 to −3.31 eV. Among the studied gases, Fe-N_3-B_{33} and Ni-N_3-B_{33} are more sensitive to CO and NO gases, which leads to reducing the energy gap between the highest occupied molecular orbital (HOMO) and the energy of the lowest unoccupied molecular orbital (LUMO) [41].

One of the dangerous gases emitted by industrial application is nitrogen dioxide (NO_2) that puts human health at risk. Hou et al. [42] studied borophene as a highly sensitive and selective material for the NO_2 detection. The borophene-based sensor can detect NO_2 at a low concentration of 200 ppb, which has a fast response time of 30 s. The recovery time of the introduced sensor at room temperature was 200 s. The properties of this sensor were significantly better than those of other 2D materials such as phosphorene, MoS_2, and graphene. For instance, this sensor demonstrates excellent flexibility, long-time stability, and outstanding stability under different bending angles.

Wang et al. [43] using first principles density functional calculations investigated hexagonal Cr-doped borophene (CrB_6) as a potential sensor material for CO, CH_4, and

CO_2 gases. The adsorption process of these gases on the CrB_6 surface is different, in which for CO_2 and CH_4 gases it is physisorption while for CO it is chemisorption. CO adsorption remarkably affects the conduction bands of the CrB_6 monolayer, and CH_4 and CO_2 adsorption affects these bands less. Since reversibility is an important property of gas sensors, CrB_6 monolayer is recommended as a good material for CO_2 and CH_4 detection.

3.2 Electrode

One of the efficient anode materials for Li-ion batteries is 2D borophene that is not stable as free standing form. Accordingly, Khan et al. [44] using DFT calculations investigated borophene in conjugation with boron nitride (B/BN) as a good anode for Li-ion battery. Using AIMD simulation, they found that the thermal and mechanical stability of the B/BN structure was dramatically improved compared to that of pristine borophene. Also, the specific charge capacity of B/BN increased compared to the other 2D material, which was 1698 mA h g^{-1}. Moreover, Li can easily diffuses into the B/BN interlayer due to the low energy barrier (ranging from 0.06 to 0.75 eV).

Kolosov and Glukhova [45] using first-principle calculations studied how surface decoration of single-walled carbon nanotubes (CNTs) by B_{12} icosahedral clusters can affect electronic properties, capacitance, and stability. They found that the B_{12} clusters (see **Figure 2(c)**) form a chemical bond with the wall of the CNTs, and the entire system demonstrates metallic behavior. The quantum capacitance and conductivity of the CNTs increased after binding the B_{12} icosahedral clusters to the inner and outer walls of CNTs. The latter was verified by calculating the transmission function near the Fermi level. They found that increasing boron concentration decreases the heat of formation that strongyle affects the stability of the system. After increasing the boron concentration, the proposed system illustrates attributes such as an asymmetric electrode.

Xie et al. [46] studied the 3D topological porous B_4 cluster (H-boron) as a high ionic and electronic conductivity anode for lithium- and sodium-ion batteries. The electron-deficient boron atoms led to expose different adsorption sites for Li and Na ions that impose a low mass density (0.91 g/cm^3) and a high specific capacity (30 mAh/g). Li (Na) can readily migrate through this anode material with a low barrier energy of 0.15 (0.22) eV and small volume changes of 0.6% (9.8%). Suggesting that H-boron based anodes can operate with fast dynamic charge-discharge process and good cyclic life.

3.3 Hydrogen storage

Hydrogen storage as one of the clean energy sources is gaining tremendous attention from computational and experimental scientists. Hydrogen has some specific characteristics compared to gasoline, such as high energy content by weight and low energy content by volume, which offer hydrogen as a suitable fuel to obviate global energy and environmental concerns. However, there is a concern about the storage and safety of hydrogen-based technologies due to its fast burning feature. To resolve this important barrier, hydrogen can be stored on the material through chemisorption and physisorption mechanisms for future demands.

Studies indicated that metal-decorated boron clusters are potential candidates for hydrogen storage. Kumar et al. [47] studied the application of small boron clusters doped with two magnesium atoms (Mg_2B_n; n = 4–14) in hydrogen storage. The DFT results show that all the clusters are stable and H_2 molecules were adsorbed in molecular form on these clusters with an absorbtion energy in the range of 0.13–0.22 eV/H_2. Although, the Mg_2B_6 cluster indicated the maximum storage capacity of H_2, MDs analysis indicates that after 200 fs H_2 molecules are desorbed from the surface of all clusters except one H_2 molecule adsorbed on the Mg_2B_{11} cluster. Also, Liu et al. [48] predicted that the titanium-decorated B_8 cluster (Ti_2B_8) has a capacity of 6.17 wt% for hydrogen storage with the average hydrogen adsorption energy of 0.247–0.358 eV/H_2.

Kumar et al. [49] using DFT calculations investigated H_2 storage capacity of lithium-doped B_{14} clusters (Li_nB_{14}; n = 1–5, see **Figure 2(d)**). These clusters are stable at room temperature and capable of storing hydrogen in molecular form. The Li_5B_{14} cluster has a maximum H_2 storage capacity of 13.89 wt%. However, based on the AIMDs results, most of the hydrogen molecules desorb from the clusters within 400 fs.

Esrafili and Sadeghi [50] studied hydrogen storage and adsorption of yttrium-decorated B_{38} fullerene using DFT calculations (see **Figure 2(e)**). They found that the Y atoms are tightly bound to the hexagonal cavities of the cluster, which makes Y@B_{38} stable and prevents aggregation of Y atoms. This suggests that Y@B_{38} is an efficient cluster for hydrogen storage. There are six H_2 molecules per Y atom adsorbed on Y@B_{38} cluster with the gravimetric density of 4.96 wt% in which both polarization effects and Kubas mechanism play crucial role in the hydrogen adsorption process. They investigated a suitable energy range for hydrogen adsorption on Y@B_{38} cluster which is −0.180 to −0.249 eV/H_2.

Wang et al. [51] using first-principal calculations investigated the ultrahigh hydrogen storage capacity for sandwich-like beryllium-doped boron clusters B_6Be_2 and B_8Be_2. Each Be atom in these clusters can adsorb seven hydrogen molecules which convert to a hydrogen storage capacity of 25.3 and 21.1 wt% for B_6Be_2 and B_8Be_2 clusters, respectively, which far exceeds the target gravimetric density of hydrogen adsorption (5.5 wt%). Consequently, both clusters are promising for H_2 release and adsorption with adsorption energy in the range of 0.10 (0.11)–0.45 (0.50) eV/H_2 for B_6Be_2 (B_8Be_2) clusters.

3.4 Catalyst

Wang et al. [52] for the first time reported a spherical isomer of boron and phosphorus atoms that have high capability for overall water splitting (see **Figure 3 (a)**). This theoretically introduced isomer has 20 atoms and eight of them are boron, which can bare its spherical structure throughout the water-splitting process. The water molecule can adsorb on each B—P bond and strongly dissociates to OH + H. This step is the rate-limiting step with an energy barrie of 2.92 eVr.

Hamadi et al. [53] investigated the adsorption of iron atom on B_{40} fullerene (Fe@B_{40}, see **Figure 3(b)**) and its application as a catalyst for carbon monoxide oxidation by DFT calculations. The iron atom prefers to be adsorb on top of the heptagonal and hexagonal rings of B_{40} with an adsorption energy of −4.39 and − 3.45 eV, respectively. They found that when both CO and O_2 molecules are injected into the B_{40}, the surface must be covered by CO due to its higher adsorption energy. Also, the preferable mechanism of CO oxidation is predicted to be termolecular Eley-Rideal (TER) with a small energy barrier of 0.26 eV.

The most stable form of boron is α-boron which is capable of adsorbing single-metal atoms and storing hydrogen molecules. Dong et al. [54] by using DFT

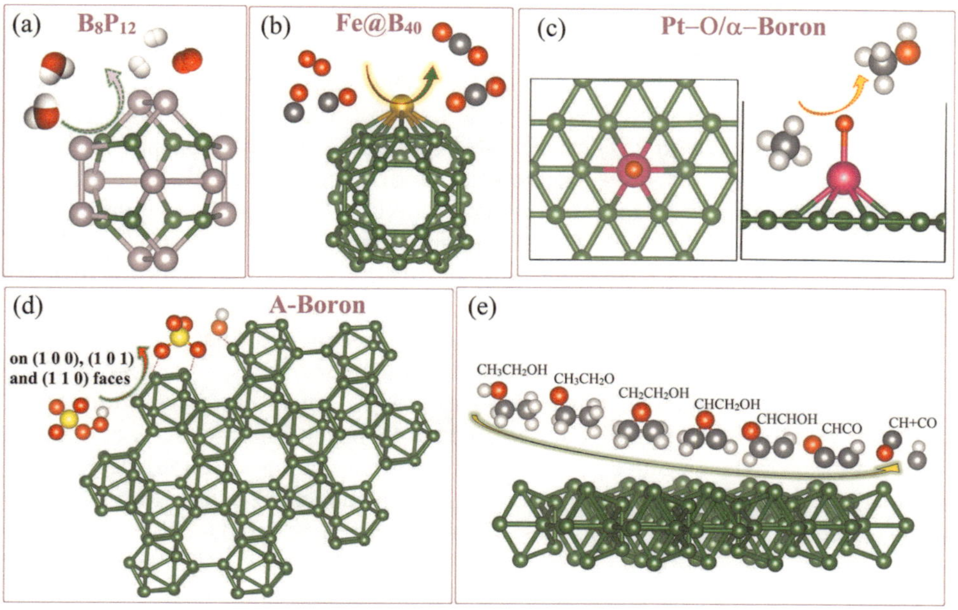

Figure 3.
The structure of some small clusters of boron that are used as a catalyst. In the figure, light gray, gray, red, green, yellow, plum, mustard, and fuchsia balls represent H, C, O, B, S, P, Fe, and Pt atoms, respectively.

calculation studied the oxidation of methane (CH_4) to form methanol on boron nanosheet/PdO (see **Figure 3(c)**). Initially, methane prefers to adsorb on the boron layer with the adsorption energy of −0.15 eV. Then, C—H bond of methane is broken through the interaction with the Pt—O moiety of the catalyst and leads to the oxidation of CH_4. This catalyst had high stability and offers excellent methanol selectivity.

Metal-free catalysts can be used instead of the toxic metal oxide catalysts. Amorphous boron (A-Boron) exhibited great catalytic merits for peroxymonosulfate (PMS) activation (see **Figure 3(d)**). The later is carried out by Duan et al. [55] to remove organic contaminants such as benzene, antibiotics, phenolics, and dyes from the water. Their results show that the performance of A-Boron is better than that of nanocarbons, transition metal oxides, and non-carbonaceous materials. They discovered through *in situ* radical capture analysis that both hydroxyl and sulfate radicals are responsible in the oxidation process of organic contaminants. In addition, a boric acid/hydroxide can form on the surface of A-Boron during heat treatment, which can further deteriorate its catalytic performance. Moreover, DFT calculations revealed that PMC decomposition and peroxide O—O bond cleavage can occur directly on boron atoms along (1 0 0), (1 0 1), and (1 1 0) faces of A-Boron crystal.

Zhao et al. [56] by using DFT calculations proposed a mechanism for the ethanol decomposition on the surface of nano-boron (0 0 1). They found that (I) the rate-limiting step is the dehydrogenation of CH to form a carbon atom (CH + CO → C + CO); (II) the oxygen dissociation can easily take place on the surface (0 0 1) of the boron crystal; and (III) the existence of O site on the surface (0 0 1) lowers the dehydrogenation energy barrier of CH_3CH_2O, CHCHO, and $CHCH_2O$ species in the ethanol decomposition pathway. The most favorable reaction pathway for the decomposition of methanol and corresponding species on this pathway is presented in **Figure 3(e)**.

3.5 Drug delivery

Boron neutron capture therapy (BNCT) is a new cancer therapy technique that allows the elimination of tumor cells without harmful side effects for other healthy tissues. Harder-Viddal et al. [57] by using MDs studied the storage of the *ortho*-carborane cluster ($C_2B_{10}H_{12}$, see **Figure 4(a)**) within the right-handed coiled-coil (RHCC) *tetrabrachion* as a nanotube carrier for BNCT. Their results of binding free energies demonstrated that $C_2B_{10}H_{12}$ can potentially enter and leave the *RHCC-tetrabrachion*, which refers to the feasibility of diffusion of $C_2B_{10}H_{12}$ cluster between solvent and carrier in the drug delivery process. They also found that there are about eight storage cavities along the central channel of the conveyor that lead to some stable configurations for this cluster within the conveyor.

The small boron-based cluster has also appeared in cancer therapies. Among the boron clusters, B_{40} (see **Figure 4(b)**) as the first all-boron fullerene has been investigated as a drug carrier in cancer therapy. For example, Zhang et al. [58] studied the adsorption of 5-fluorouracil (5-Fu) on B_{40} fullerene and M@B_{40} (M = Mg, Al, Si, Mn, Cu, Zn). The 5-Fu was adsorbed on the B atom in the corner of the B_{40} cage, forming the B—O bond. The adsorption energy of 5-Fu was −11.15 kcal mol^{-1}, which refers to the ease of release of this drug from the surface of B_{40} cage in an acidic environment of tumor tissues.

Shakerzadeh [59] studied Li- and Na-encapsulated B_{40} (Li(Na)@B_{40}) fullerenes as carrier for anticancer drug nedaplatin (NdaPt, see **Figure 4(b)**). The energy gap decreased after drug absorbtion, which refers to the formation of stable complexes. The later is a chemical signal to describe drug adsorption and its effects on the electronic properties of B_{40} cage. The results demonstrated that in both gas and water phases, the adsorption of NdaPt altered more the electronic properties of Li- and Na-encapsulated B_{40} fullerenes compared to bare B_{40} fullerene. The dipole moments of the Li(Na)@B_{40} complexes in water were high, suggesting that the solubility of these complexes in the polar medium. Moreover, the adsorption energy for NdaPt/Li(Na)@B_{40} complexes was −28 kcal/mol.

Zhang et al. [60] by using DFT calculations investigated the potential of B_{40} fullerene as a carrier for drug nitrosourea (NU, see **Figure 4(b)**). This drug was adsorbed from its N and O atoms on the fullerene surface with an adsorption energy of −25.18 kcal/mol. They showed that newly formed N—B and O—B bonds are strong polar covalent bonds. Also, it is investigated that the recovery time of NU drug under body temperature is 52 s due to the easy release of NU in the medium of cancer tissues.

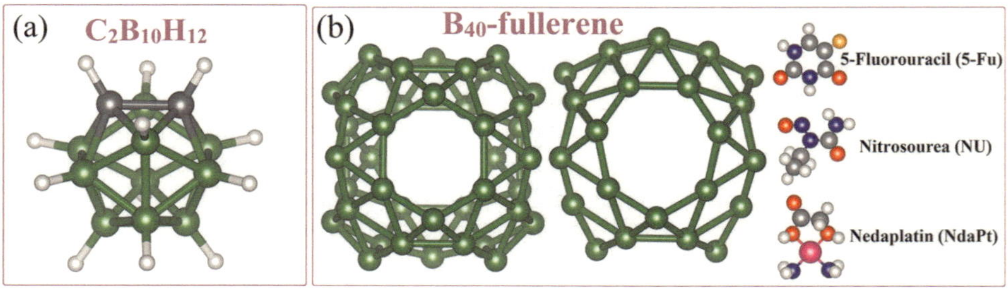

Figure 4.
The structure of some small clusters of boron that are used in drug delivery. The right panel two different views of B_{40} cluster are shown. In the figure, light gray, gray, red, green, dark blue, orange, and fuchsia balls represent H, C, O, B, N, F, and Pt atoms, respectively.

They also found that B_{40} fullerene has a high loading capacity in which it can simultaneously transport five NU drugs.

4. Conclusions

In this chapter, we briefly introduced the application of the boron clusters that can be characterized by using DFT and MD simulations. We show that both methods are useful for simulating different physical and chemical properties of small boron clusters. Based on the intrinsic characteristic of the studied systems, several groups have used MD and/or DFT techniques to model boron clusters. They were employed to model boron cluster structures for a variety of applications including H_2 storage, gas sensor, electrode, catalyst, and drug delivery. In some cases, MD and DFT were used to confirm the results of the experiment. The development of more precise nanoscale systems that can be more comparable to experimental conditions is required.

Conflict of interest

The authors declare no conflict of interest.

Author details

Nasim Hassani[1]*, Mohammad Reza Hassani[2] and Mehdi Neek-Amal[1]

1 Department of Physics, Shahid Rajaee Teacher Training University, Lavizan, Tehran, Iran

2 Faculty of Engineering, School of Civil Engineering, University of Tehran, Tehran, Iran

*Address all correspondence to: nasim.hassani.chem@gmail.com

References

[1] Barba-Bon A, Salluce G, Lostalé-Seijo I, Assaf K, Hennig A, Montenegro J, et al. Boron clusters as broadband membrane carriers. Nature. 2022;**603**: 637-642. DOI: 10.1038/s41586-022-04413-w

[2] Jian T, Chen X, Li SD, Boldyrev AI, Li J, Wang LS. Probing the structures and bonding of size-selected boron and doped-boron clusters. Chemical Society Reviews. 2019;**48**:3550-3591. DOI: 10.1039/C9CS00233B

[3] Yu X, Zhou T, Zhao Y, Lu F, Zhang X, Liu G, et al. Surface magnetism in pristine α rhombohedral boron and Intersurface exchange coupling mechanism of boron icosahedra. Journal of Physical Chemistry Letters. 2021;**16**: 6812-6817. DOI: 10.1021/acs.jpclett.1c01860

[4] Shabbir A, Azeem M. On the partition dimension of tri-hexagonal α-boron nanotube. IEEE Access. 2021;**8**: 55644-55653. DOI: 10.1109/ACCESS.2021.3071716

[5] Zhang JJ, Altalhi T, Yang JH, Yakobson BI. Semiconducting α'-boron sheet with high mobility and low all-boron contact resistance: A first-principles study. Nanoscale. 2021; **13**:8474-8480. DOI: 10.1039/D1NR00329A

[6] Mazaheri A, Javadi M, Abdi Y. Chemical vapor deposition of two-dimensional boron sheets by thermal decomposition of diborane. ACS Applied Materials & Interfaces. 2021;**13**: 8844-8850. DOI: 10.1021/acsami.0c22580

[7] Aizawa T, Suehara S, Otani S. Phonon dispersion of a two-dimensional boron sheet on Ag (111). Physical Review Materials. 2021;**5**:064004. DOI: 10.1103/PhysRevMaterials.5.064004

[8] Gribanova TN, Minyaev RM, Minkin VI, Boldyrev AI. Novel architectures of boron. Structural Chemistry. 2020;**31**:2105-2128. DOI: 10.1007/s11224-020-01606-9

[9] Chen Q, Chen TT, Li HR, Zhao XY, Chen WJ, Zhai HJ, et al. B_{31}^- and B_{32}^-: Chiral quasi-planar boron clusters. Nanoscale. 2019;**11**:9698-9704. DOI: 10.1039/C9NR01524H

[10] Wu X, Sai L, Zhou S, Zhou P, Chen M, Springborg M, et al. Competition between tubular, planar and cage geometries: A complete picture of structural evolution of B_n (n = 31-50) clusters. Physical Chemistry Chemical Physics. 2020;**22**:12959-12966. DOI: 10.1039/D0CP01256D

[11] Chkhartishvili L. Relative stability of boron planar clusters in diatomic molecular model. Molecules. 2022; **27**:1469. DOI: 10.3390/molecules27051469

[12] Liu H, Mu YW, Li SD. Axially chiral cage-like B_{38}^+ and B_{38}^{2+}: New aromatic members of the borospherene family. Journal of Cluster Science. 2020;**33**:1-7. DOI: 10.1007/s10876-020-01943-z

[13] Pei L, Yan M, Zhao XY, Mu YW, Lu HG, Wu YB, et al. Sea-shell-like B_{31}^+ and B_{32}: Two new axially chiral members of the borospherene family. RSC Advances. 2020;**10**:10129-10133. DOI: 10.1039/D0RA01087A

[14] Pei L, Yan QQ, Li SD. Predicting the structural transition in medium-sized

boron nanoclusters: From bilayer B_{64}, B_{66}, B_{68}, B_{70}, and B_{72} to core-shell B_{74}. European Journal of Inorganic Chemistry. 2021;**26**:2618-2624. DOI: 10.1002/ejic.202100328

[15] Pei L, Ma YY, Yan M, Zhang M, Yuan RN, Chen Q, et al. Bilayer B_{45}, B_{60}, and B_{62} clusters in a universal structural pattern. European Journal of Inorganic Chemistry. 2020;**2020**:3296-3301. DOI: 10.1002/ejic.202000473

[16] Yan QQ, Pei L, Li SD. Predicting bilayer B_{50}, B_{52}, B_{56}, and B_{58}: Structural evolution in bilayer B_{48}–B_{72} clusters. Journal of Molecular Modeling. 2021; **27**:1-9. DOI: 10.1007/s00894-021-04954-3

[17] Chen WJ, Ma YY, Chen TT, Ao MZ, Yuan DF, Chen Q, et al. B_{48}^-: A bilayer boron cluster. Nanoscale. 2021; **13**(6):3868-3876. DOI: 10.1039/D0NR09214B

[18] Zhang M, Lu HG, Li SD. B_{111}, B_{112}, B_{113}, and B_{114}: The most stable core-shell borospherenes with an icosahedral B_{12} core at the center exhibiting superatomic behaviors. Nano Research. 2021;**14**: 4719-4724. DOI: 10.1007/s12274-021-3411-x

[19] Chen TT, Cheung LF, Wang LS. Probing the nature of the transition-metal-boron bonds and novel aromaticity in small metal-doped boron clusters using photoelectron spectroscopy. Annual Review of Physical Chemistry. 2022;**19**:73. DOI: 10.1146/annurev-physchem-082820-113041

[20] Sun W, Kang D, Chen B, Kuang X, Ding K, Lu C. Tuning of structure evolution and electronic properties through palladium-doped boron clusters: PdB_{16} as a motif for boron-based nanotubes. The Journal of Physical Chemistry. A. 2020;**124**:9187-9193. DOI: 10.1021/acs.jpca.0c05197

[21] Celaya CA, Buendia F, Miralrio A, Paz-Borbon LO, Beltran M, Nguyen MT, et al. Structures, stabilities and aromatic properties of endohedrally transition metal doped boron clusters $M@B_{22}$, M = Sc and Ti: A theoretical study. Physical Chemistry Chemical Physics. 2020;**22**: 8077-8087. DOI: 10.1039/D0CP00307G

[22] Li SX, Zhang ZP, Long ZW, Chen DL. Structures, electronic, and spectral properties of doped boron clusters $MB_{16}^{0/-}$ (M = Li, Na, and K). ACS Omega. 2020;**5**:20525-20534. DOI: 10.1021/acsomega.0c02693

[23] Fojt L, Grüner B, Holub J, Havran L, Fojta M. Electrochemistry of icosahedral metal full and half sandwich metallacarboranes in phosphate buffers. Journal of Electroanalytical Chemistry. 2022;**910**:116165. DOI: 10.1016/j.jelechem.2022.116165

[24] Lu XQ, Wei Z, Li SD. Half-sandwich $LaB_n^{-/0}$ (n= 14–17): π dually aromatic lanthanide boride complexes with multicenter fluxional bonds. Journal of Cluster Science. 2021;**29**:1-8. DOI: 10.1007/s10876-021-02072-x

[25] Jiang ZY, Chen TT, Chen WJ, Li WL, Li J, Wang LS. Expanded inverse-sandwich complexes of lanthanum borides: $La_2B_{10}^-$ and $La_2B_{11}^-$. The Journal of Physical Chemistry. A. 2021;**125**: 2622-2630. DOI: 10.1021/acs.jpca.1c01149

[26] Xu S, Zhang Y, Huang R, Liu J, Jin W, Lefkidis G, et al. Strain manipulation of the local spin flip on $Ni@B_{80}$ endohedral fullerene. Physical Chemistry Chemical Physics. 2021;**23**:25712-25719. DOI: 10.1039/D1CP03206B

[27] Levine IN. Quantum Chemistry. Englewood Cliffs, New Jersey: Prentice Hall; 1991. pp. 455-544. ISBN 978-0-205-12770-2

[28] Parr RG, Craig DP, Ross IG. Molecular orbital calculations of the lower excited electronic levels of benzene, configuration interaction included. Chemical Physics. 1950;**18**(12): 1561-1563. DOI: 10.1063/1.1747540

[29] Bagayoko D. Understanding density functional theory (DFT) and completing it in practice. AIP Advances. 2014;**4**(12): 127104. DOI: 10.1063/1.4903408

[30] Hohenberg P, Kohn W. Inhomogeneous electron gas. Physics Review. 1964;**136**(3B):B864. DOI: 10.1103/PhysRev.136.B864

[31] Vignale G, Rasolt M. Density-functional theory in strong magnetic fields. Physical Review Letters. 1987; **59**(20):2360. DOI: 10.1103/PhysRevLett.59.2360

[32] Kohn W, Sham LJ. Self-consistent equations including exchange and correlation effects. Physics Review. 1965; **140**(4A):A1133. DOI: 10.1103/PhysRev.140.A1133

[33] Schlick T. Pursuing Laplace's vision on modern computers. In: Mathematical Approaches to Biomolecular Structure and Dynamics. New York, NY: Springer; 1996. pp. 219-247. DOI: 10.1007/978-1-4612-4066-2-13

[34] Alder BJ, Wainwright TE. Studies in molecular dynamics. I. General method. Chemical Physics. 1959;**31**(2):459-466. DOI: 10.1063/1.1730376

[35] Tuckerman ME, Berne BJ, Martyna GJ. Molecular dynamics algorithm for multiple time scales: Systems with long range forces. Chemical Physics. 1991;**94**(10): 6811-6815. DOI: 10.1063/1.460259

[36] Tuckerman ME. Ab initio molecular dynamics: Basic concepts, current trends and novel applications. Journal of Physics. Condensed Matter. 2002; **14**(50):R1297. DOI: 10.1088/0953-8984/14/50/202

[37] Paquet E, Viktor HL. Computational methods for Ab initio molecular dynamics. Advances in Chemistry. 2018; **2018**:1-4. DOI: 10.1155/2018/9839641

[38] Hossain MA, Hossain MR, Hossain MK, Khandaker JI, Ahmed F, Ferdous T, et al. An ab initio study of the B_{35} boron nanocluster for application as atmospheric gas (NO, NO_2, N_2O, NH_3) sensor. Chemical Physics Letters. 2020; **754**:137701. DOI: 10.1016/j.cplett.2020.137701

[39] Wang Z, Li Y, Sheng-Jiang G, Jing-Hui L, Mei X, Rastegar SF. Quasi-planar B_{36} boron cluster: A new potential basis for ammonia detection. Journal of Molecular Modeling. 2020;**26**(10):1-8. DOI: 10.1007/s00894-020-04486-2

[40] Ploysongsri N, Ruangpornvisuti V. Adsorption of sulfur-containing gases on B_{36} nanocluster: A DFT study. Journal of Sulfur Chemistry. 2021;**42**(4): 383-396. DOI: 10.1080/17415993.2021.1895160

[41] Hassani N, Mehdizade M. Exploring the adsorption and sensing behavior of the M-N_x-B_{36-x} (M = Fe, Ni, and Cu; x = 0, 3) bowl-shaped structures upon CO, NO, O_2, and N_2 molecules: A first-principles study. Physica E: Low-dimensional Systems and Nanostructures. 2020;**124**:114242. DOI: 10.1016/j.physe.2020.114242

[42] Hou C, Tai G, Liu Y, Liu X. Borophene gas sensor. Nano Research.

2021;**15**:1-8. DOI: 10.1007/s12274-021-3926-6

[43] Wang C, Gao C, Hou J, Duan Q. First-principle investigation of CO, CH_4, and CO_2 adsorption on Cr-doped graphene-like hexagonal borophene. J Mol Model. 2022;**28**:196. DOI: 10.1007/s00894-022-05197-6

[44] Khan MI, Majid A, Ashraf N, Ullah I. A DFT study on a borophene/boron nitride interface for its application as an electrode. Physical Chemistry Chemical Physics. 2020;**22**(6):3304-3313. DOI: 10.1039/C9CP06626H

[45] Kolosov DA, Glukhova OE. A new composite material on the base of carbon nanotubes and boron clusters B_{12} as the base for high-performance supercapacitor electrodes. C. 2021;7:26. DOI: 10.3390/c7010026

[46] Xie H, Qie Y, Muhammad I, Sun Q. B_4 cluster-based 3D porous topological metal as an anode material for both Li- and Na-ion batteries with a Superhigh capacity. Journal of Physical Chemistry Letters. 2021;**12**(5):1548-1553. DOI: 10.1021/acs.jpclett.0c03709

[47] Kumar A, Vyas N, Ojha AK. Hydrogen storage in magnesium decorated boron clusters (Mg_2B_n; n = 4-14): A density functional theory study. International Journal of Hydrogen Energy. 2020;**45**(23):12961-12971. DOI: 10.1016/j.ijhydene.2020.03.018

[48] Liu P, Zhang Y, Xu X, Liu F, Li J. Ti decorated B_8 as a potential hydrogen storage material: A DFT study with van der Waals corrections. Chemical Physics Letters. 2021;**765**:138277. DOI: 10.1016/j.cplett.2020.138277

[49] Kumar A, Ojha SK, Vyas N, Ojha AK. Light and stable Li_nB_{14} (n = 1–5) clusters for high capacity hydrogen storage at room temperature: A DFT study. International Journal of Hydrogen Energy. 2022;**47**(12):7861-7869. DOI: 10.1016/j.ijhydene.2021.12.091

[50] Esrafili MD, Sadeghi S. Y decorated all-boron B_{38} nanocluster for reversible molecular hydrogen storage: A first-principles investigation. International Journal of Hydrogen Energy. 2022; **47**(22):11611-11621. DOI: 10.1016/j.ijhydene.2022.01.160

[51] Wang YJ, Xu L, Qiao LH, Ren J, Hou XR, Miao CQ. Ultra-high capacity hydrogen storage of B_6Be_2 and B_8Be_2 clusters. International Journal of Hydrogen Energy. 2020;**45**(23): 12932-12939. DOI: 10.1016/j.ijhydene.2020.02.209

[52] Wang Y, Gong W, Zuo P, Kang L, Yin G. A novel spherical boron phosphide as a high-efficiency overall water splitting catalyst: A density functional theory study. Catalysis Letters. 2020;**150**(2):544-554. DOI: 10.1007/s10562-019-02996-0

[53] Hamadi H, Shakerzadeh E, Esrafili MD. Fe-decorated all-boron B_{40} fullerene serving as a potential promising active catalyst for CO oxidation: A DFT mechanistic approach. Polyhedron. 2020;**188**:114699. DOI: 10.1016/j.poly.2020.114699

[54] Dong A, Xu G, Dai Z, Yu A, Qiu S, Sun C. Single PdO loaded on boron nanosheet for methane oxidation: A DFT study. Progress in Natural Science. 2019; **29**(3):367-369. DOI: 10.1016/j.pnsc.2019.05.005

[55] Duan X, Li W, Ao Z, Kang J, Tian W, Zhang H, et al. Origins of boron catalysis in peroxymonosulfate activation and advanced oxidation. Journal of Materials Chemistry A. 2019;**7**(41):23904-23913. DOI: 10.1039/C9TA04885E

[56] Zhao X, Zhu B, Sun Y, Chen J, Liu J. Decomposition mechanism of ethanol molecule on the nano-boron surface: An experimental and DFT study. Fuel. 2022;**318**:123631. DOI: 10.1016/j.fuel.2022.123631

[57] Harder-Viddal C, Heide F, Roshko RM, Stetefeld J. Molecular dynamics simulations of ortho-carborane nano-diamond storage within the nonpolar channel cavities of a right-handed coiled-coil tetrabrachion nanotube. Computational and Structural Biotechnology Journal. 2021;**19**: 3531-3541. DOI: 10.1016/j.csbj.2021.06.010

[58] Zhang L, Qi ZD, Ye YL, Li XH, Chen JH, Sun WM. DFT study on the adsorption of 5-fluorouracil on B_{40}, B_{39}M, and M@B_{40} (M = Mg, Al, Si, Mn, Cu, Zn). RSC Advances. 2021;**11**(62): 39508-39517. DOI: 10.1039/D1RA08308B

[59] Shakerzadeh E. Li@B_{40} and Na@B_{40} fullerenes serving as efficient carriers for anticancer nedaplatin drug: A quantum chemical study. Computational & Theoretical Chemistry. 2021;**1202**: 113339. DOI: 10.1016/j.comptc.2021.113339

[60] Zhang L, Ye YL, Li XH, Chen JH, Sun WM. On the potential of all-boron fullerene B_{40} as a carrier for anti-cancer drug nitrosourea. Journal of Molecular Liquids. 2021;**342**:117533. DOI: 10.1016/j.molliq.2021.117533

Chapter 4

Characterisation and Application of Nickel Cubic Boron Nitride Coating via Electroless Nickel Co-Deposition

Norsilawati Ngah, Nor Bahiyah Baba, Nor Azinee Said, Mohd Habir Ibrahim and Na'ain Shari

Abstract

The chapter describes the characterisation and application of nickel cubic boron nitride (Ni-CBN) coatings using the electroless nickel co-deposition method. Two different types of substrates were used, that is, high-speed steel (HSS) and carbide. The characterisation of Ni-CBN coating was conducted using Field Emission Scanning Electron Microscope (FESEM) JSM-7800F coupled with Energy-Dispersive X-ray (EDX). As for the application, coated end mill cutting tools were inserted into DMU 50 CNC machine to conduct the machining testing. Cutting speed, feed rate, and depth of cut were chosen for the Taguchi L9 3-level factors. Taguchi analysis was employed to determine the optimal parameters for the Ni-CBN (HSS) surface finish. The ANOVA evaluation was used to identify the most significant effect on surface finish parameters. The FESEM images prove that the nano-CBN powders were embedded in the Ni-CBN coatings and are uniformly distributed. The findings show Ni-CBN-coated tool life is 195 minutes compared to the uncoated is 143 minutes. The surface roughness, Ra values using Ni-CBN-coated tools ranges between 0.251 and 0.787 μm, whereas the uncoated tools Ra values between 0.42 and 1.154 μm. It can be concluded that Ni-CBN HSS cutting tools reduce tool wear and extend tool life. The Taguchi optimum machining condition obtained is 1860 RPM spindle speed, 334 mm/min feed rate, and 2 mm depth of cut.

Keywords: coating, electroless, ceramic, Ni-CBN, high-speed steel (HSS), carbide, milling, surface roughness, high-speed machining, tool wear, tool life, optimization

1. Introduction

Boron is a very useful element that exists in compounds such as borates. Generally, boron is a non-metallic element and can be extracted into pure crystalline boron that is black in colour and conduct electricity at higher temperature and insulator at low temperature. It is as hard as carborundum but too brittle to be used as a tool. Boron is used in medicine, agriculture, decarbonisation purposes and industrial uses.

One of the outstanding compounds of boron is Cubic Boron Nitride (CBN). CBN is a synthetic abrasive material made of Cubic Boron Nitride grains bonded in ceramic material and is commonly known as Borazon™ [1]. CBN is an allotropic crystal of boron nitride (B_4N) and has a hexagonal crystal. It is the second hardest material after diamond but more chemically and thermally stable than diamond and is extensively used in cutting tools [2]. CBN has excellent thermal stability, with oxidation starting at 1000°C and finishing around 1500°C. This is aided by the presence of boron oxide layer, which allows the use of high speed of 30.5–61 ms^{-1} [3]. Polycrystalline cubic boron nitride (PCBN), an extended version of CBN, is developed for machining, superalloys, and high-temperature alloys. Besides having high-temperature resistance, it has a low coefficient of friction but low fracture toughness [4].

Cubic boron nitride (CBN) is very well known in many machining industries. CBN is man-made material that having a hardness that is second to diamond [5]. Since CBN has hardness after diamond, it has outstanding mechanical and thermal properties, for examples, having high temperatures strength and wear resistance. Multilayer CBN coatings represent a new deposition method that can improve adhesion on metal substrates. Even with high residual stress, this multilayer CBN structure showed outstanding adhesion in atmospheric conditions. A study found that the multilayer CBN films in comparison to monolayer CBN, has lower elastic moduli, but twice as high to their critical loads [6]. In recent years, the performance of CBN tools has been researched [7, 8].

Instead of pure CBN, composite coating of CBN-TiN also being used as machine cutting tools [9]. It is found that, this composite has outstanding CBN-to-TiN as well as the adhesion of composite coating-to-carbide substrate. The characterisation analysis indicates an evenly distributed CBN particles in TiN matrix [10].

1.1 Electroless nickel

Electroless nickel (EN) is an in-situ chemical reaction process where a metallic nickel is deposited onto a surface. This process is different from nickel electroplating that uses an applied current in the electrolytic bath which has effect on the current density, electrolyte composition, pH, bath agitation on the physicochemical and mechanical properties of the deposits [11, 12]. The main ingredients of EN are electroless bath, reducing agents, complexing agents, bath stabilisers and accelerators. **Table 1** describes the function and type of each EN ingredients.

Table 1 lists the three types of available EN baths, pure nickel, acid-based and alkali-based chemicals. The pure nickel bath provides pure nickel metallic deposition for semiconductor application purposes. The acid and alkali-based chemicals either produce Ni-P or Ni-B alloy deposition depending on the reducing agent used. The properties of the EN deposits strongly depend on the content of phosphorus or boron in the alloys. As seen in **Table 2**, the deposit structure changes because the phosphorus or boron content changes. EN bath concentration, temperature, pH, agitation, and bath loading effect the EN process [14].

It is known that the EN process provides exceptional standardisation and impenetrable deposition even with a coating thickness of fewer than 10 μm [15]. In manufacturing, EN deposition has been widely used for it provides excellent corrosion, lubricity, ductility, wear and abrasion resistance, high hardness, and electrical properties [16].

Ingredients	Functions	Types		
		Pure nickel	**Acid-based**	**Alkali-based**
EN bath	Provide metallic ion sources	Ni acetate	Nickel sulfate, Nickel chloride	Nickel sulfate, Nickel chloride
Reducing agents	Reduce metallic ion into metal deposit	Hydrazine	Sodium hypophosphite, sodium borohydride, dimethylamine (DMAB)	Sodium hypophosphite, sodium borohydride, dimethylamine (DMAB), hydrazine
Complexing agents	Prevent decomposition of solutions and control reaction onto the catalytic surfaces	Tetrasodium salt, glycolic acid	Citric, lactic, glycolic, propionic acids, sodium citrate, succinic acid	Citric, lactic, glycolic, propionic acids, sodium citrate, sodium acetate, sodium pyrophosphate
Bath stabilisers	Act as inhibitors, increase deposition rate and deposit brightness	—	Thiourea, lead acetate, heavy metal salts, thioorganic compound	Thiourea, lead acetate, heavy metal salts, thioorganic compound, thallium, selenium
Catalyst	Increase the deposition speed and plating rate to be economically high	—	Sodium hydroxide, sulphuric acid	Sodium hydroxide, sulphuric acid, ammonium hydroxide

Table 1.
EN process chemicals and their functions.

1.2 Electroless nickel composite

When incorporated with particles or powders of different materials, EN deposition becomes an EN composite and the process is called EN co-deposition. This incorporation of particles or powders in the EN deposit has remained extensively explored. Similar to the EN deposit, there are two EN composites upon particles incorporation, either Ni-P or Ni-B, depending on the EN reducing agent used. The particles that have been studied include ceramic, polymer and metal particles. **Table 3** summarises the particles that have been investigated for various applications. Incorporating ceramic particles into EN deposit produces a composite name cermet, which is the current issue discussed by using CBN particles for cutting tool applications.

1.3 Application of Ni-CBN

The coating technology is more demanding due to the increase in productivity rates for industry consumption, especially for cutting tool purposes. It shows the growing market of cutting tools has been developed [31]. The coated tools application is becoming more important in the machining process. These tools are produced using thermal spraying processes such as physical vapour deposition (PVD) and chemical vapour deposition (CVD). Thermal spraying processes are very reliable; however, they are costly, and the high temperature causes materials properties to degrade [32].

EN bath type	Reducing agent	Deposit alloys	Phosphorus/Boron content (%)	Structure	Properties
Acid-based	Sodium hypophosphite	Ni-P	3–5	Crystalline	Excellent wear resistance.
			6–9	Mixed Crystalline and amorphous	Good corrosion protection and abrasion resistance.
			10–14	Amorphous	Very ductile and corrosion resistant
	Dimethylamine (DMAB)	Ni-B	0.1–4	Crystalline	High melting point of approx. 1350°C for wear application.
Alkali-based	Sodium hypophosphite	Ni-P	3–6	Crystalline	Good solderability for the electronic industry. However, lower corrosion resistance and lower adhesion to steel. Suitable for plating plastics and non-metals.
	Sodium borohydride	Ni-B	4–7	Mixed Crystalline and amorphous	Low hardness and average wear resistance.
	Dimethylamine (DMAB)	Ni-B	0.2–4	Crystalline	Hardness and superior wear resistance.

Table 2.
Summary of EN baths, reducing agents and their properties [13].

Particle	Composites	Applications	References
Diamond	Ni-P-C	Cutting tool/Applied to reamers for highly abrasive applications	[17]
	Ni-B-nanodiamond	Wear & friction resistance	[18]
Silicon carbide	Ni-P-SiC	Wear resistance	[19]
Silicon oxide	Ni-P-SiO_2	Corrosion resistance	[20]
Silicon nitride	Ni-P-Si_3N_4	Water lubricated application for corrosion and wear resistance	[21]
Boron carbide	Ni-P-B_4C	Magnetic field application	[22]
Boron nitride	Ni-P-BN	Elastic–plastic behaviour	[23]
Alumina	Ni-P-Al_2O_3	Corrosion resistance	[24]
Cerium	Ni-P-CeO_2	Corrosion resistance	[25]
Titanium oxide	Ni-P-TiO2	Surgical instrument	[26]
Iron oxide	Ni–P–Fe_3O_4	High-temperature oxidation application	[27]
Yttria-stabilised zirconia	Ni-P-YSZ	Cutting tool	[28]
PTFE	Ni-P-PTFE	Dry lubrication of valve for cryogenic applications	[29]
PVP	Ni-P-PVP	Corrosion resistance	[30]

Table 3.
Investigation of various particles for EN composites and their applications.

In hard milling, the most acceptable significant representation is the cutting tool's thermal property of the material, such as thermal conductivity. The cutting tool's function ability can only be estimated via temperature tool measurements. For ferrous materials, cubic boron nitride (CBN) is one of the most demanding cutting tools. Multilayer CBN coatings provide a unique deposition method when applied to metal surfaces. Even under extreme conditions of high residual stress, the adhesion of this multilayer CBN structure was remarkable. Their heavy loads were twice as extraordinary compared to the monolayer CBN coatings, which had lower elastic moduli. It showed that stress relaxation significantly impacts the multilayer CBN structure [33]. This type of cutting tool is essential for cutting ferrous materials in a wide range of industries because of the advantages of suitable coating materials. Some of the most challenging materials to mill, such as aerospace alloys, die steels, and toughened steels, required the employment of CBN cutting tools [34, 35].

The diamond's remarkable mechanical and thermal capabilities, such as strength at elevated temperatures, abrasion resistance, and hardness, are the second property that the diamond possesses. Thus, numerous sorts of research have been undertaken in the last few years on the performance of CBN tools [36, 37]. The application of CBN as a cutting substance is a suitable method that may affect production. Nonetheless, the presentation of machining, such as progression solidity, tool wear and live performance, and surface finish quality, is significantly affected by differences in high-performance machining, which commonly requires a high material removal rate (MMR) [38, 39]. However, CBN coatings' application speeds and tool life are still lower than those of some other tools. Certain adjustments and upgrades are required, including raising the coating thickness and a rotational mechanism during the coating process. Hard coatings are typically more fragile and less lasting, whereas reinforced coatings lack strength. For real-world industrial operations, it is more critical to have coatings with a high hardness without sacrificing too much toughness.

Milling is the most common method of cutting metal. There are a variety of milling operations, but the ultimate shape and condition of the raw material dictate which ones are used. Adding features like slots or threaded holes necessitates using a milling machine. The cutting tool quality is directly proportional to the cutting process performance. In order to cut a tough workpiece materials, a harder cutting materials are needed [5]. Due to high process forces and temperatures, the first tool wear occurs in complex machining. The initial tool wear occurs in complex machining due to the high process of forces and temperatures. The machining market offers a wide variety of cutting tools, classified as coated or uncoated. Coated cutting tools typically perform better than uncoated cutting tools. Commercially available coated cutting tools include aluminium nitride (AlN), titanium nitride (TiN), titanium aluminium nitride (TiAlN), and others [28].

Due to the availability of suitable coating materials for cutting tools, this ferrous cutting material is indispensable in various industry disciplines. Certain heat-resistant CBN cutting tools are typically used on difficult-to-machine materials, such as aerospace, die steel, or hardened steel [34, 35]. CBN cutting tools have remarkable mechanical and thermal properties, including high-temperature strength, abrasion resistance, and hardness comparable to diamond. Thus, it has been demonstrated recently that CBN instruments produce excellent results in various sorts of research [37, 38]. The use of CBN as a cutting substance is a beneficial strategy that may significantly impact productivity.

CBN-based materials with bonding capabilities are frequently used to improve the machining process, which pushes researchers to continue improving coatings by utilising appropriate materials and procedures. For example, Ni-reinforced vitrified bonds are created in a high magnetic field for CBN grinding wheels. The addition of Ni does not affect the vitrified bond's refractoriness but enhances its fluidity and bending strength [40].

Additionally, CBN composites have poor machinability characteristics, such as brittleness. One way to mitigate this difficulty is to combine CBN and graphene oxide (GO) composites with the inclusion of Al-SiC at elevated temperatures and a high-pressure sintering procedure, which results in a 27.5% increase in fracture toughness compared to monolithic CBN composites. Besides this, the composites' bending strength increased from 564.2 MPa to 696.9 MPa [41]. Other studies discovered the use of ultrasonic probe sonication and spark plasma sintering (SPS) to investigate the microstructural, thermo-mechanical. Tribological properties of low-temperature sintered CBN and Ni-coated CBN reinforced bearing steel composites. It showed that these newly developed CBN and Ni-coated CBN-reinforced conducting steel composites sintered at a temperature of 100^0 C resulted in increased wear resistance with high wear and fatigue resistance [42].

This study [28] found that an electroless nickel co-deposition technique successfully coated the HSS cutting tool with Ni/YSZ composite. In another study, TiN coated surfaces with mean thickness of 59 μm shows smooth and uniform surface demonstrating consistent surface roughness measurements. For Al/SiC metal matrix composites cutting tool, the surface roughness decreased from 1.3 μm to 0.6 μm m over time when the cutting speed is increased from 300 to 450 mm/min [43].

This study was conducted to investigate the effects of a new electroless Ni-CBN composite evenly coated onto an HSS and carbide substrate. This ceramic-metal surface coating is well-known for its superior resistance to thermal wear [44]. Additionally, the layer was produced using electroless nickel co-deposition, which is more straightforward, requires less energy, and is less expensive than typical thermal spraying procedures [45].

2. Experimental methods

The methodology consists of three sections: Process of EN coating, machining process, and cutting feasibility.

2.1 EN co-deposition

In this experiment, 50 g/l CBN powder was inserted into the bath plus the substrate. Then, suspended particles near the surface were co-deposited onto the substrate surface through the agitation process. It was found that the EN solution pH range was between pH 4.9 and pH 5.4. The bath temperature was maintained at 89 ± 20°C throughout the coating process. The coating time was kept constant at 60 min. Mechanical stirring was performed with a Jenway hot plate equipped with a magnetic stirrer, and the air bubbling was supplied at 1.2 W pressure. The entire coating process is summarised in **Figure 1**.

The composite Ni-CBN deposition was carried out on a Carbide and HSS substrate with a dimension standard of ∅10 x 7.8 mm. Chemical etching and mechanical blasting were used to prepare the substrate sample's surfaces. CBN powder reinforcement ceramic particles were used. CBN powders offer superior heat conductivity and increased surface integrity when it comes to hardened alloys, nickel, cobalt-based superalloys, and tool steels. **Figure 2** displays the 7.8 mm diameter sample as a substrate for EN co-deposition.

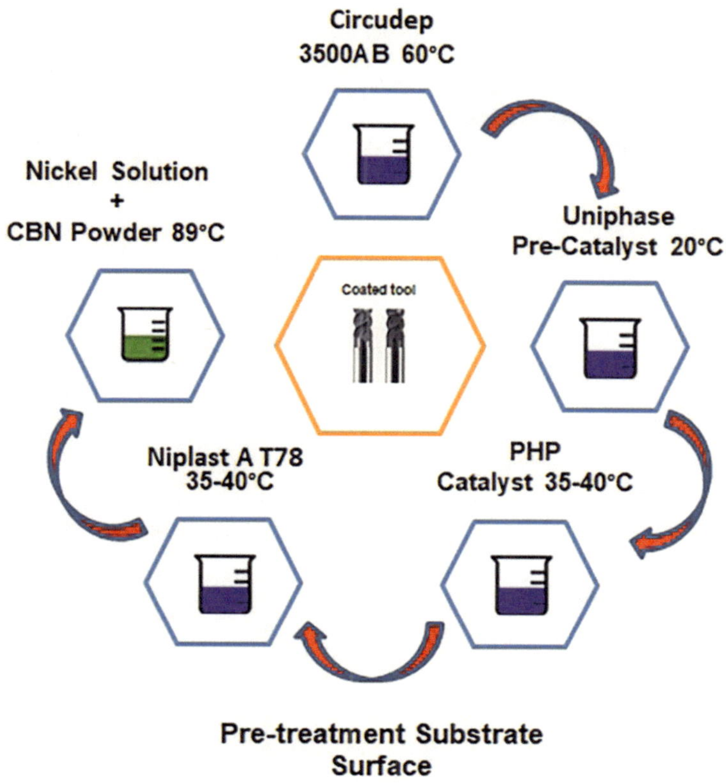

Figure 1.
Electroless nickel coating process.

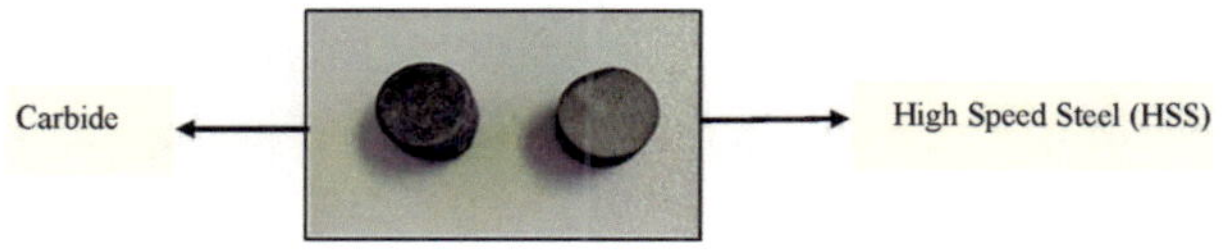

Figure 2.
Substrates for EN co-deposition: Solid carbide and HSS.

2.2 EN Co-deposition on cutting tools

The end mill cutting tool using carbide and high-speed steel (HSS) with a dimension of 6 mm was used as a substrate of Ni-CBN coating as shown in **Figure 3**. Before EN co-deposition process done on both cutting tools, chemical etching and mechanical blasting were used to modify the surface of the substrate sample to ensure better substrate-coating bonding.

Sensitising the HSS and Carbide cutting tool substrates is needed to activate the surfaces. Because of this, all non-proprietary solutions were produced using AR-grade chemicals and high purity deionised water. The EN co-deposition of Ni-CBN was conducted within 3 hours of the pre-treatment process, as shown in **Table 4**, to reduce the impacts of chemical degradation [46]. The EN chemicals produced a bright nickel deposit with a mid-phosphorous content between 6 to 9 wt.%. The optimum temperature for electroless nickel solution is at 89°C and was heated using a Jenway hotplate.

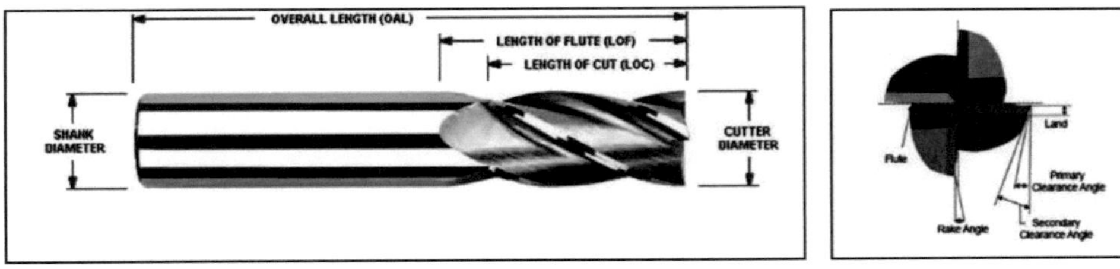

Figure 3.
End mill cutting tool.

Trade name	Soaking time (min)	Temperature (°C)
Coprolite X96DP	15	60
Uniphase PHP Pre-catalyst	15	20
Uniphase PHP Catalyst	15	40
Niplast AT78	15	40
Electroless Nickel SLOTONIP	60	89

Table 4.
EN Co-deposition materials and procedure [46].

2.3 Surface coating characterisation

The composition of the Ni-CBN composite is controlled during deposition to achieve the preferred properties. It is required to obtain a high ceramic-to-metal ratio for erosion, heat, and wear resistance. The influence of process parameters to obtain a high particle ratio was analysed. The surface characterisation and elemental composition of EN co-deposition on the substrates was performed through JSM-7800F Field Emission Scanning Electron Microscope (FESEM) in conjunction with energy dispersive X-rays (EDX) shown in **Figure 4**.

2.4 Surface roughness, tool wear and tool life

Surface roughness was measured every 0.2 mm, and each pocket had a pitch of 0.2 mm. **Figure 5** shows the Mitutoyo surface roughness tester SJ-301, a tool used to test surface roughness. The tool wear was measured using the Zeiss Stemi 20,000-C Microscope Profile optical video measuring system, as shown in **Figure 6**. Tool life is measured by the number of cuts taken by the end mill to reach average flank wear criterion 0.3 mm. All the tools failed primarily on the plank face. For all machining conditions, the machining was stopped when the flank wear land reached about 0.3 mm to ensure that the tool life data is more reliable. The flank wear was measure using Zeiss Stemi 20,000-C Microscope Profile optical video measuring system. The effect of interaction between high cutting speed and feed rate is most significant in shorten tool life. This is claimed by J.P. Urbanski et al. found that tool life decrease drastically as cutting speed is increased because at high cutting speed high temperature will be generated, which accelerates tool wear and consequently shortens tool life [47].

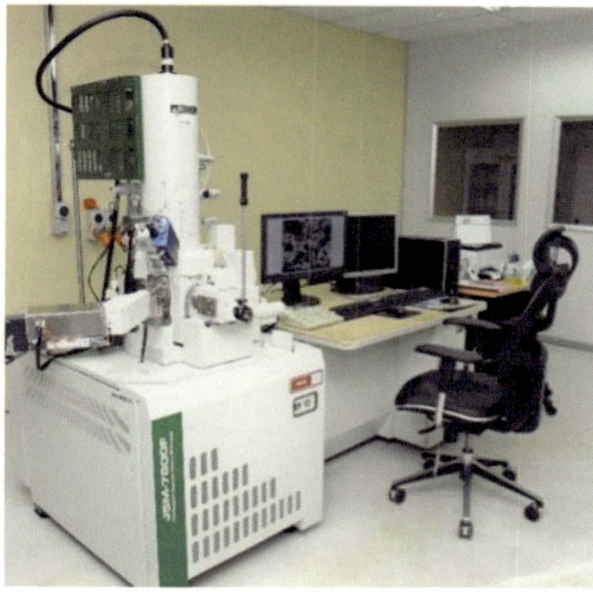

Figure 4.
Field emission scanning electron microscope (FESEM)—JSM-7800F.

Figure 5.
Mitutoyo surface roughness tester SJ-301.

2.5 Taguchi method

MINITAB 14 software was used to study the influence and range of parameters' effect on the surface roughness of 7075 Aluminium Alloy. The experiments, based on Taguchi L9, selected spindle speed, depth of cut, and feed rate as the process

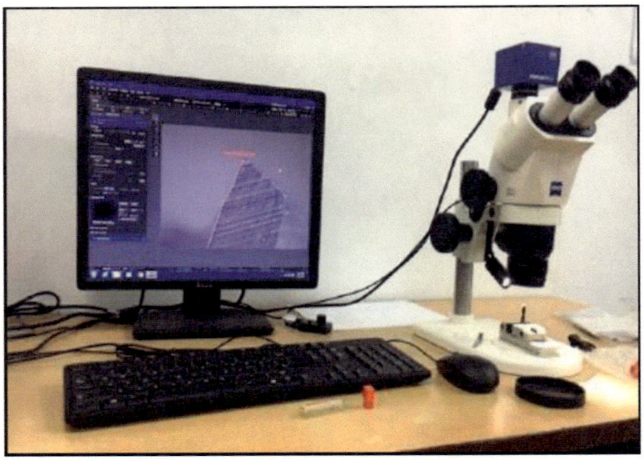

Figure 6.
Zeiss Stemi 20,000-C microscope profile optical video measuring system.

Machining parameters	Levels		
	-1	0	1
Spindle speed (RPM)	1860	2650	3450
Feed Rate (mm/min)	180	257	334
Depth of cut (mm)	1	2	3

Table 5.
Level of machining cutting parameters.

Experiment number	Spindle speed (rpm)	Feed rate (mm/min)	Depth of cut (mm)
1	1860	180	1
2	1860	257	2
3	1860	334	3
4	2650	180	2
5	2650	257	3
6	2650	334	1
7	3450	180	3
8	3450	257	1
9	3450	334	2

Table 6.
The OA arrangement of the machining process.

variables and were conducted at three different levels. The machining parameters are listed in **Table 5**.

Table 6 illustrates the Orthogonal Array (OA) L9 for each substrate was determined using the Taguchi method of experimental design (DOE) with three parameters at three levels. The Ni-CBN HSS coated end mill, and uncoated cutting tools were analysed via 18 tests in this study. The preferences of the end mill manufacturer determined the feed rate and depth of cut and had moved the experiment to the "high cutting speed" category [48, 49].

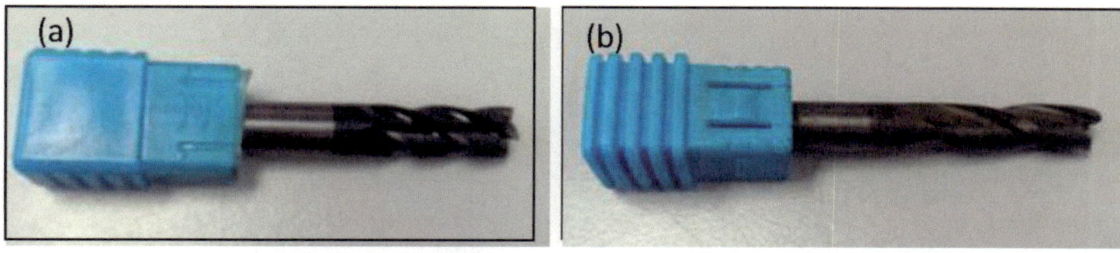

Figure 7.
Cutting tool image for HSS end mill cutting tool (a) uncoated; and (b) coated.

Figure 8.
Machining profile on aluminium alloy 7075 material.

2.6 Process of machining

The DMU 50 CNC machine was utilised in the machining process. After coating the HSS end mills with CBN composite material, the Mitotuyo digital micrometre was used to measure the thickness of the cutting tool. The average thickness was determined through the three measures taken from each tooltip.

The workpiece is an aerospace material Aluminium Alloy 7075 to determine machining performance. The cutting tools were then examined for their machining capabilities. The profile was machined with 18 pockets and two cutting tools. Both coated and uncoated HSS end mills (**Figure 7**) were used to machine nine pockets each. **Figure 8** shows the machining profile of the machine pockets with 40 mm x 35 mm dimension on the workpiece.

3. Results and discussion

3.1 Coating surface morphology

Figures 9 and **10** shows the surface morphology using Field Emission Scanning Electron Microscope (FESEM) of the Ni-CBN coating captured at different magnifications. Both figures depict microstructure with cauliflower pattern. In **Figure 9**, the coating does not display micro-cracked, coarse erection and covers the entire exterior of the substrate. HSS has a high thermal shock resistance, making it resistant to sudden and rapid temperature changes [50]. In addition, HSS can withstand large temperature fluctuations.

Figure 10 illustrates a micro-crack on the surface layer of the carbide substrate coating due to carbide low thermal resistance. High internal stress levels can cause various problems during coating use, including premature disintegration of the part due to substrate fatigue, fracture formation in the coating, and loss of deposit adhesion [50].

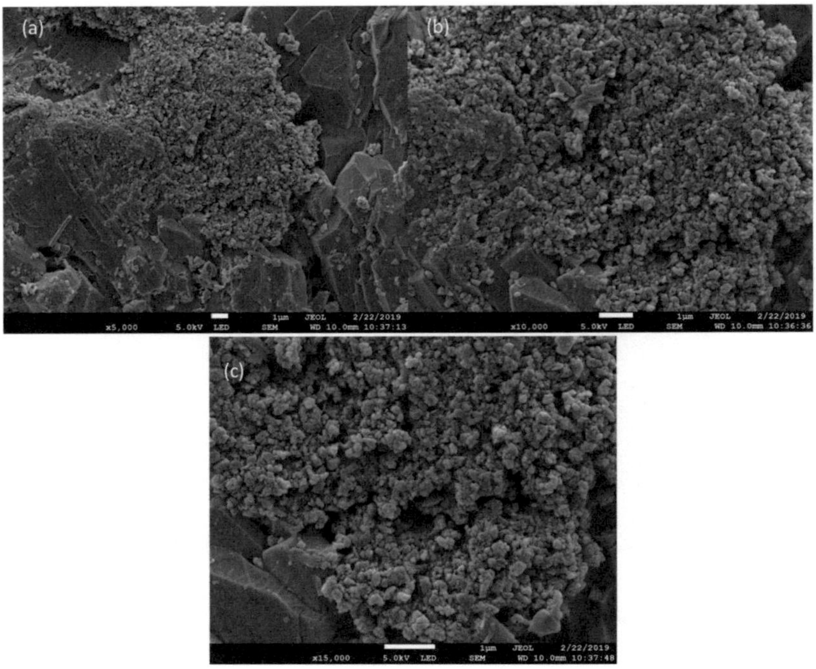

Figure 9.
Ni-CBN microstructure on HSS substrate (a) 5000X; (b) 10,000X; and (c) 15,000X.

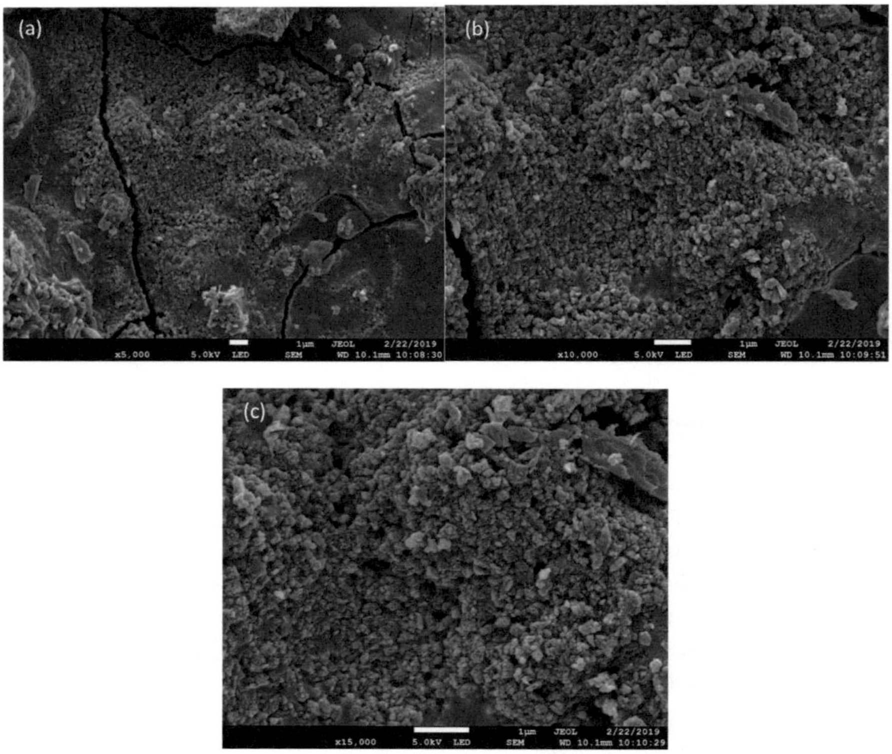

Figure 10.
Ni-CBN microstructure on carbide substrate (a) 5000X; (b) 10,000X; and (c) 15,000X.

Overall, both figures demonstrate rough surface of the coatings. The coating was mainly composed of ceramic CBN powders (white areas), metallic Ni matrix (grey areas), and pores (dark spots). The HSS coating surfaces generally showed a uniform distribution of the ceramic particles compared to the carbide substrate. The carbide substrate shows cracks on the coating surface due to thermal gradient. It is because the roughness of the EN-CBN coatings depends on the roughness of the substrate. It is also due to the growth mechanism of the coating, which forms columns locally perpendicular to the surface. The columns are parallel when the substrate is smooth, and the coating is even softer than the substrate [51].

3.2 Coating elemental composition

The as-deposited Ni-CBN coatings were subjected to energy dispersive X-ray analysis (EDX) to determine the composition of the co-deposited CBN elements in the EN matrix, as shown in **Table 7** for HSS substrate and **Table 8** for carbide substrate.

The EDX spectrum obtained for the Ni-CBN deposited on the HSS and Carbide substrate is depicted in **Figures 11** and **12**. It displays the peaks corresponding to

Element	B	C	N	O	P	Ni	Totals
Weight (%)	20.69	16.71	9.16	15.01	4.20	34.22	100.00

Table 7.
Elemental composition in weight % of electroless Ni-CBN coating on HSS substrate.

Element	B	C	N	O	P	Ni	Totals
Weight (%)	11.65	21.81	4.40	21.34	7.07	33.73	100.00

Table 8.
Elemental composition in weight % of electroless Ni-CBN coating on carbide substrate.

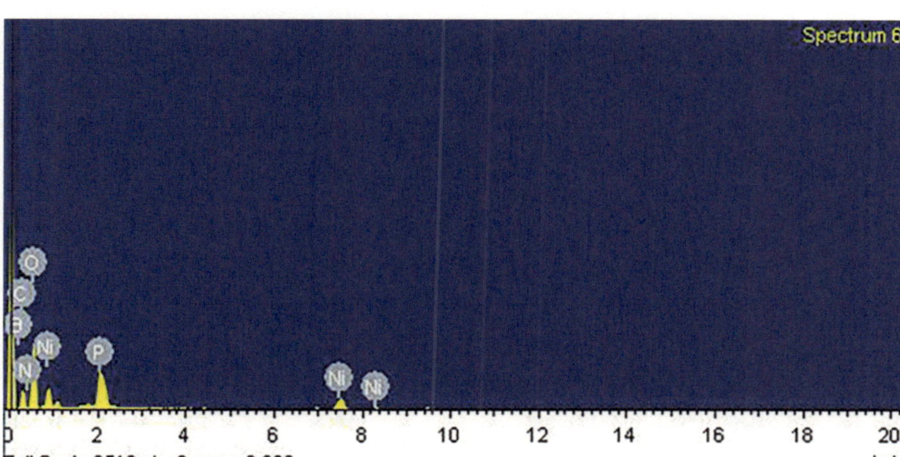

Figure 11.
EDX spectrum of as-deposited electroless Ni-CBN coating on HSS substrate.

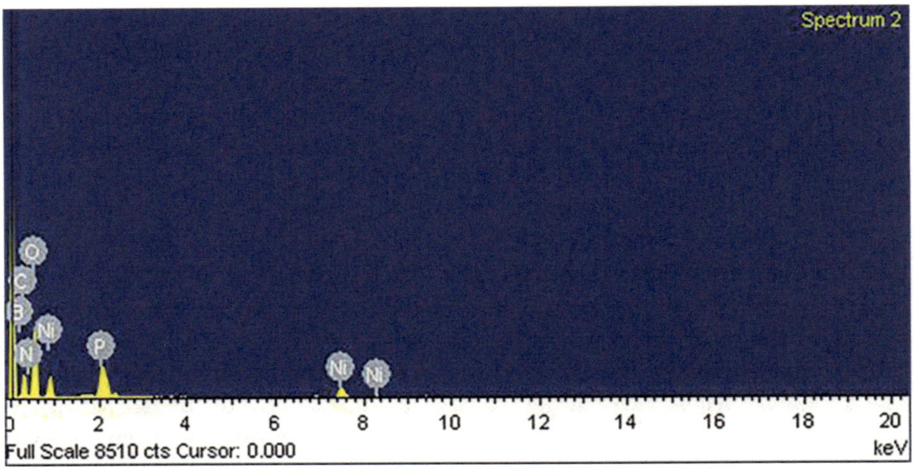

Figure 12.
EDX Spectrum of as-deposited electroless Ni-CBN coating on carbide substrate.

the CBN, approving the standard deposition of elements in the Ni matrix. There is evidence of significant peak elements of nickel (Ni), boron (B), and phosphorous (P). This proves that metallic nickel and ceramic CBN are exist. The phosphorus element in the composite indicates as one of the most critical elements in the EN hypophosphite-based bath solution [45].

3.3 Surface roughness analysis

The most critical factor in improving surface roughness analysis is the quality of the cutting tools. **Table 9** compares the Ra results of machined 7075 Aluminium Alloy for coated and uncoated cutting tools. The data indicates Test 8 of HSS coated tools; high level of cutting speed and a medium level of feed rate produced a good surface finish, Ra 0.251 μm. In comparison, the combination of feed rate at high level and cutting speed at low level in Test 3 give a high surface roughness of Ra 1.22 μm. This finding demonstrates the combination of high-value feed rate and spindle obtaining a

Surface roughness (Ra)					
Test No.	**Spindle speed (rpm)**	**Feed rate (mm/min)**	**Depth of cut (mm)**	**Coated (μm)**	**Uncoated (μm)**
1	1860	180	1	0.576	0.695
2	1860	257	2	0.787	1.154
3	1860	334	3	0.890	1.220
4	2560	180	2	0.481	0.534
5	2560	257	3	0.301	0.586
6	2560	334	1	0.412	0.619
7	3450	180	3	0.296	0.485
8	3450	257	1	0.251	0.421
9	3450	334	2	0.527	0.729

Table 9.
Data of design experiment and surface roughness.

DOI: http://dx.doi.org/10.5772/intechopen.105364

better surface finish [52, 53]. According to Mohammed [54], the interaction between cutting speed and feed rate will significantly impact the surface finish.

3.4 Tool wear and tool life analysis

Tool wear for every 0.2 mm of machining was examined using Zeiss Stemi 20,000-C Profile Optical. In accordance to ISO 8688-21:1989, the end mill cutting tool with the lowest tool wear is the best and most durable. **Figure 13** shows the tool wear on the cutting tool before and after the machining process.

Comparing the flank wear trends in **Figures 14** and **15**, the coated cutting end mill tool performed better in terms of both cutting time and tool life. Test 3 and Test 5 produced the most extended tool life, 195 min. **Figure 15** depicts an uncoated end mill's cutting time-based flank wear trend. The substrates performed better than the coated end mills in terms of scattering. Test 3 yielded the most extended tool life for the uncoated tools at 143 min. High-value of feed rate, spindle speed and depth of cut and cutting time will cause significant tool wear. The previous studies found that the cutting speed and feed rate interaction is significantly affecting the tool wear [52, 55, 56].

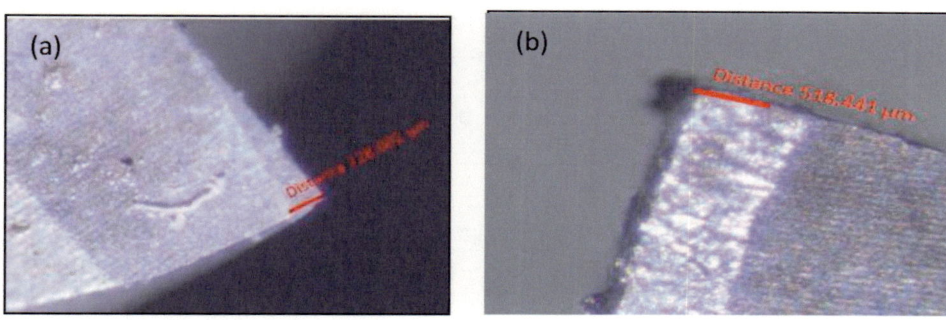

Figure 13.
Flank wear on HSS cutting tool on machining (test 9): (a) before; and (b) after.

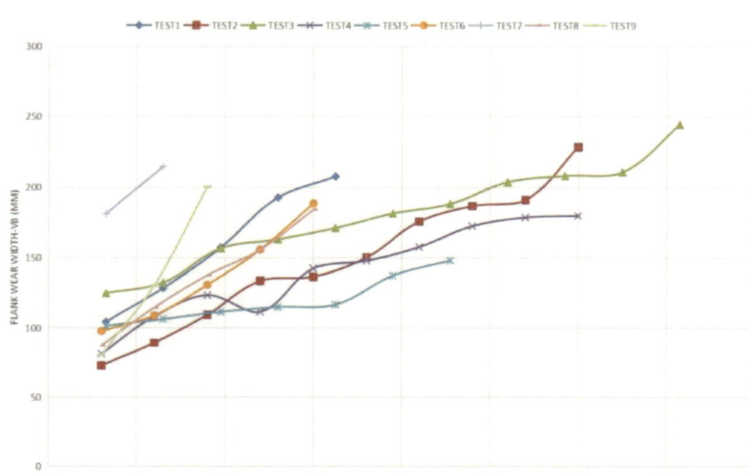

Figure 14.
Flank wear versus cutting time of uncoated HSS.

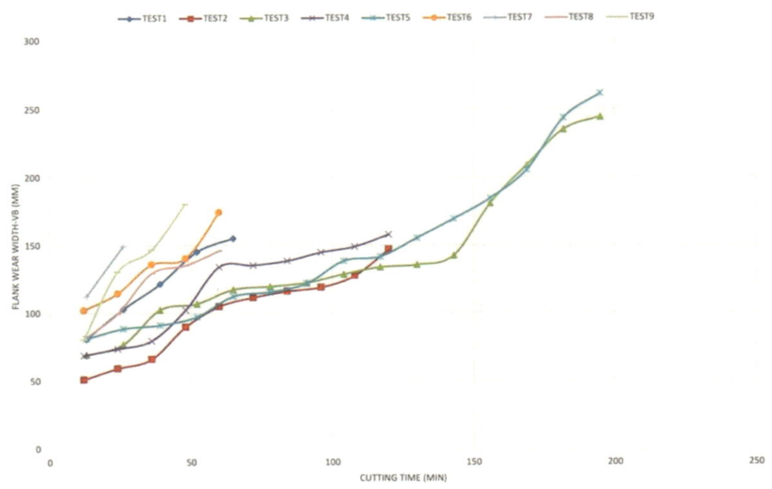

Figure 15.
Flank wear versus cutting time of coated HSS.

	Parameters			Response	
Test No.	**Spindle speed (rpm) (A)**	**Feed rate (mm/min) (B)**	**Depth of cut (mm) (C)**	**Surface roughness Ra (μm)**	**S/N ratio, d/B**
1	1860	180	1	0.576	4.7916
2	1860	257	2	0.787	2.0805
3	1860	334	3	0.890	1.0122
4	2650	180	2	0.481	6.3571
5	2650	257	3	0.301	10.4287
6	2650	334	1	0.412	7.7021
7	3450	180	3	0.296	10.5742
8	3450	257	1	0.251	12.0065
9	3450	334	2	0.527	5.5638

Table 10.
An investigation via L9 OA of Ni-CBN HSS coated end mills.

3.5 Taguchi analysis

The Taguchi L9 (3^3) Orthogonal Array (OA) was applied. The OA was generated by Minitab 14 consists of 9 runs with 3 factors at 3 levels. **Table 10** shows the Orthogonal Array (OA) of the coated HSS end mills experiment and the combinations of conditions for each control factor (A-C).

3.5.1 Regression equation

Surface roughness equations were generated using machining parameters such as spindle speed, feed rate, and depth of cut. Eq. (1) outline the main effects of surface

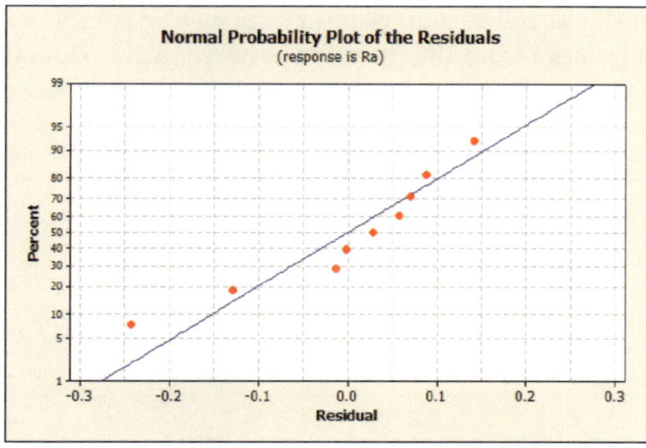

Figure 16.
Normal probability plot for Ra response.

roughness and Ra response. **Figure 16** shows the normal probability plot for Ra response based on Eq. (1).

$$Ra = 0.654 - 0.197\,\text{spindle speed} + 0.0793\,\text{Feed Rate} + 0.0413\,\text{Depth of Cut} \quad (1)$$

3.5.2 Analysis of variance (ANOVA)

The OA L9 (3^3) contains nine tests of ANOVA investigation that identify the effects of the different parameters on the response variables. A significance level of 95% was chosen in the ANOVA analysis, and the factor was considered adequate if the P-value was less than 0.05 [53]. In this study, the relation of spindle speed (A), feed rate (B), and depth of cut (C) factors on the surface roughness Ra responses are identified using ANOVA analysis. The model was formulated for a 95% confidence level. The P-value shows that the model is significant and has no influence on noise. The experiment result of surface roughness (Ra) formed the first-order model using the Minitab software.

The ANOVA results depicted in **Table 11** is the estimation for machining parameters, with a selected α-level of 0.05. The outcomes show that the spindle speed factor has the lowest p-value. This reveals that the consequence of spindle speed is significant as p-value factors that are above 0.05 are considered as insignificant [57].

3.5.3 Factor level combination and determination of optimum parameter

Based on the rank in **Table 12**, spindle speed ranks first, followed by the depth of cut and feed rate. This demonstrates spindle speed as the significant factor that affects surface roughness. Spindle speed is the most critical machining parameter affecting surface roughness because it is substantially influenced [56]. The table also represents the Taguchi response to determine the optimal factors affecting surface roughness. According to Signal to Noise (smaller is better), the optimum machining settings are 1860 RPM for spindle speed, 334 mm/min for feed rate, and 2 mm for depth of cut are. The experiment was confirmed through the S/N ratio using the optimum parameter level A1B3C2.

The surface finish was the most important influence on spindle speed and feed rate, as shown in **Figures 17** and **18**. The slope between the horizontal line and spindle speed is more pronounced than the depth of cut and feed. The changes in spindle

Parameters	DOF	Sum of squares	Mean square	F-value	P-value
Spindle speed	2	0.280658	0.140329	35.11	0.028
Feed rate	2	0.051875	0.025937	6.49	0.134
Depth of cut	2	0.051723	0.025861	6.47	0.134
Residual error	2	0.007993	0.0033996	—	—
Total	8	0.392248		—	—

Table 11.
ANOVA table for Ra response.

S. No	Level	Spindle speed (A)	Feed rate (B)	Depth of cut (C)
1	1	2.628	7.241	8.167
2	2	8.163	8.172	4.667
3	3	9.381	4.759	7.338
4	Delta	6.753	3.413	3.500
5	Rank	1	3	2

Table 12.
Response table for S/N ratio (smaller is better).

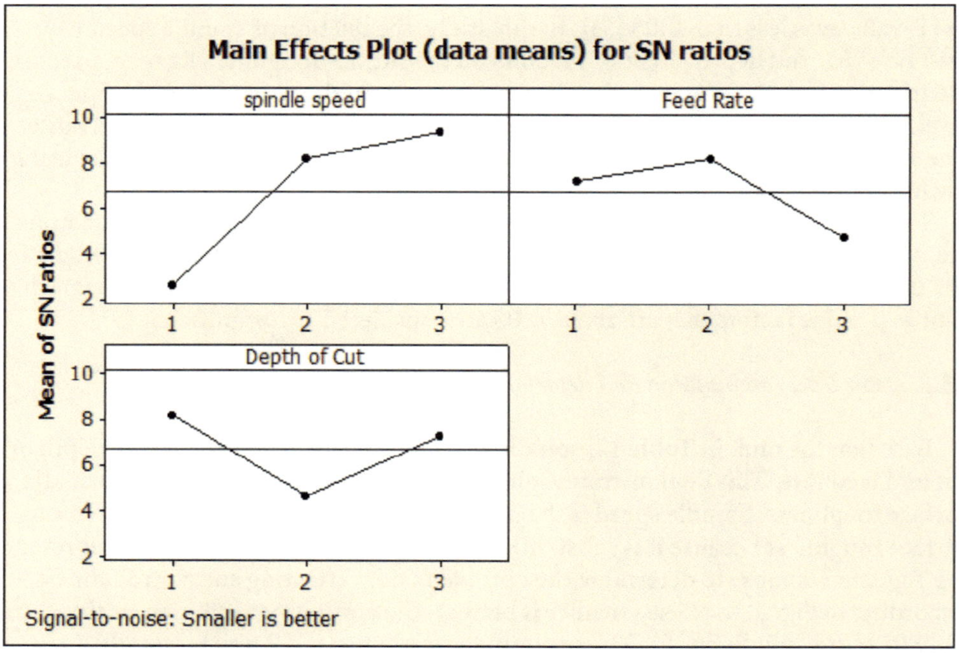

Figure 17.
Main effects plot for SN ratios.

speed significantly affect the surface roughness [56]. The optimum machining settings are determined at spindle speed value of 1860 RPM, feed rate of 334 mm/min, and depth of cut of 2 mm.

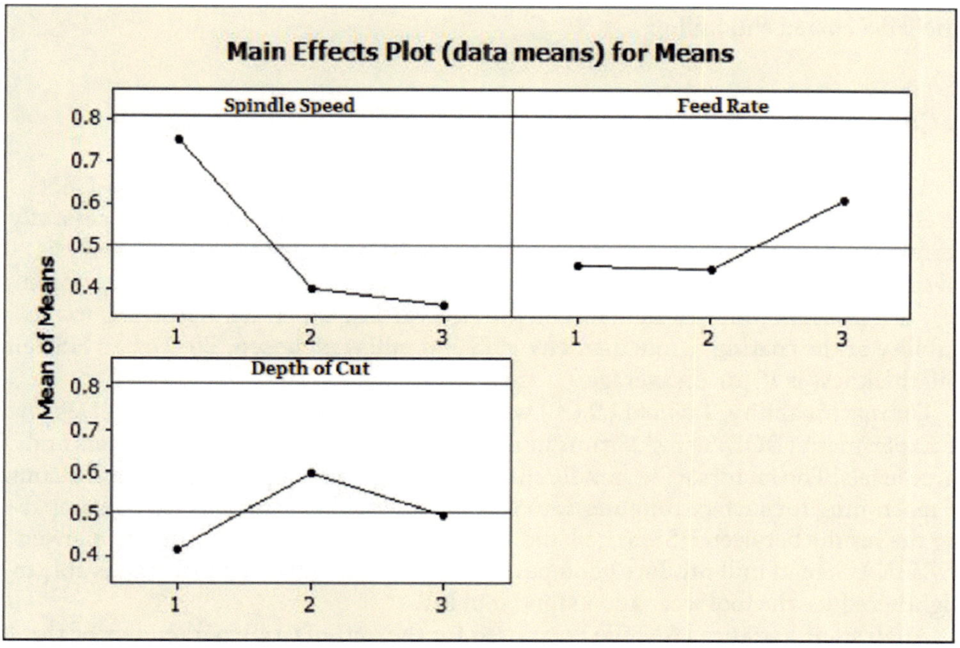

Figure 18.
Main effects plot for means.

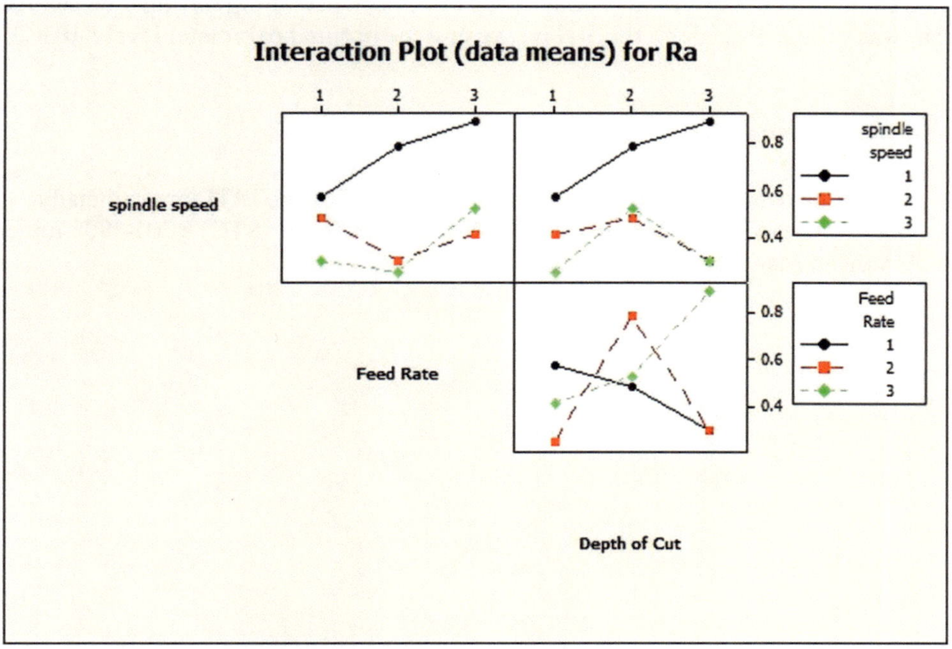

Figure 19.
Interaction plot for Ra.

Figure 19 shows the interaction plot for surface roughness, Ra in the machining process. When the lines are more non-parallel, an interaction occurs, resulting in higher strength of the interaction. The factors of spindle speed affect the surface roughness more than other factors for machining Aluminium Alloy 7075 with a Ni-CBN HSS coated end mill.

4. Conclusions

This study investigates the process of electroless and machinability of Ni-CBN on HSS and Carbide substrate. The electroless Ni-CBN coating has been successfully performed on the substrate and proven using the EDX Analysis. The EDX analysis revealed the presence of major peak for nickel (Ni), carbon (C), oxygen (O), boron (B), and phosphorous (P) elements on the HSS carbide substrate. According to the stability of the coating, 6 mm diameter HSS end mill was chosen. The coated HSS end mill thickness is 15 μm on average.

For machinability, Taguchi L9 (3^3) was used in this research to produce a Design of Experiment (DOE) using 18 runs number of experiments with three factors and three levels. The factors were spindle speed, feed rate, and depth of cut. The outcome of machining for surface roughness, tool wear and tool life was analysed by comparing the results between HSS coated and uncoated end mill. The comparison showed Ni-CBN HSS end mill produce good performance on the surface finish and is able to slightly reduce the tool wear and extend tool life.

Analysis of variance (ANOVA) was used for the optimisation parameters of the Ni-CBN HSS end mill tool. The Spindle speed is a significant factor compared to the other factors as it had the lowest P-value, that is below 0.05. For determination of optimum parameters, 1860 RPM for spindle speed, 334 mm/min for feed rate, and 2 mm for depth of cut were identified as the optimum machining settings. The experiment was validated through the S/N ratio using the optimal parameter level A1B3C2.

Acknowledgements

The authors would like to acknowledge University College TATI for financially support the research through the UC TATI Short Term Grant (STG) 9001-1808 under the Advanced Manufacturing Cluster.

Author details

Norsilawati Ngah, Nor Bahiyah Baba*, Nor Azinee Said, Mohd Habir Ibrahim and Na'ain Shari
Faculty of Engineering Technology, Department of Manufacturing, University College TATI, Kemaman, Terengganu, Malaysia

*Address all correspondence to: bahiyah@uctati.edu.my

References

[1] Doi T, Marinescu ID, Kurokawa S. Chapter 3—The Current Situation in Ultra-Precision Technology—Silicon Single Crystals as an Example. In: Advances in CMP Polishing Technologies. Norwich, NY: William Andrew Publishing; 2012. pp. 15-111. DOI: 10.1016/B978-1-4377-7859-5.00003-X

[2] Jain A, Bajpai V. In: Kapil Gupta J, Davim P, editors. Chapter 1—Introduction to High-Speed Machining (HSM). High Speed Machining. Academic Press; 2020. pp. 1-25. DOI: 10.1016/B978-0-12-815020-7.00001-1

[3] Marinescu ID, Brian Rowe W, Dimitrov B, Ohmori H. Abrasives and abrasive tools. In: Marinescu ID, Rowe WB, Dimitrov B, Ohmori H, editors. Tribology of Abrasive Machining Processes. 2nd ed. NY, Newyork: William Andrew Publishing; 2013. pp. 243-311. DOI: 10.1016/B978-1-4377-3467-6.00009-4

[4] Infante V, Vidal C. Tool and welding design. In: Givi MKB, Asadi P, editors. Woodhead Publishing Series in Welding and Other Joining Technologies, Advances in Friction-Stir Welding and Processing. Cambridge: Woodhead Publishing; 2014. pp. 199-240. DOI: 10.1533/9780857094551.199

[5] Uhlmann E, Oyanedel Fuentes JA, Keunecke M. Machining of high-performance workpiece materials with CBN coated cutting tools. Thin Solid Films. 2009;**518**(5):1451-1454. DOI: 10.1016/j.tsf.2009.09.095

[6] Park S-T, Jeon-Geon Han M, Keunecke KL. Mechanical and structural properties of multilayer c-BN coatings on cemented carbide cutting tools. International Journal of Refractory Metals and Hard Materials. 2016;**65**:52-56. DOI: 10.1016/j.ijrmhm.2016.11.009

[7] Bouacha K, Yallese MA, Mabrouki T, Rigal JF. Statistical analysis of surface roughness and cutting forces using response surface methodology in hard turning of AISI 52100 bearing steel with CBN tool. International Journal of Refractory Metals and Hard Materials. 2010;**28**:349-361. DOI: 10.1016/j.ijrmhm.2009.11.011

[8] Sahin Y, Motorcu AR. Surface roughness model in machining hardened steel with cubic boron nitride cutting tools. International Journal of Refractory Metals and Hard Materials. 2008;**26**:84-90. DOI: 10.1016/j.ijrmhm.2007.02.005

[9] Yadave SN, Malshe AP, Brown WD, Russell WC. Novel composite CBN–TiN coating: synthesis and performance analysis. Trans. NAMRI/SME. 2002;**30**:485-492. DOI: 10.1016/S1526-6125(03)70050-4

[10] Jiang W, More AS, Brown WD, Malshe AP. A cBN-TiN composite coating for carbide inserts: coating characterisation and its applications for finish hard turning. Surface and Coatings Technology. 2006;**201**:2443-2449. DOI: 10.1016/j.surfcoat.2006.04.026

[11] Mir Saman Safavi FCW. Electrodeposited Co-P alloy and composite coatings: A review of progress towards replacement of conventional hard chromium deposits. Surface and Coatings Technology. 2021;**422**(127564):127564. DOI: 10.1016/j.surfcoat.2021.127564

[12] Piotr Jenczyk HG, Milczarek M, Jarzabek DM. Mechanical and tribological properties of

co-electrodeposited particulate-reinforced metal matrix composites: A critical review with interfacial aspects. Materials. 2021;**14**(3181):1-36. DOI: 10.3390/ma14123181

[13] Sudagar J, Lian J, Sha W. Electroless nickel, alloy, composite and nano coatings—A critical review. Journal of Alloys and Compounds. 2013;**571**:183-204. DOI: 10.1016/j.jallcom.2013.03.107

[14] Nor Bahiyah Baba AD, Muneer T. YSZ-reinforced Ni-P deposit: An effective condition for high particle incorporation and porosity level. Advanced Materials Research. 2011;**214**:412-417. DOI: 10.4028/www.scientific.net/AMR.214.412

[15] Rabizadeh T, Allahkaram SR. Corrosion resistance enhancement of Ni–P electroless coatings by incorporation of nano-SiO2 particles. Materials & Design. 2011;**32**(1):133-138. DOI: 10.1016/j.matdes.2010.06.021

[16] Mallory GO, Hajdu JB. Electroless plating: fundamentals and applications. 1990

[17] Petrova M, Noncheva Z, Dobreva E. Electroless deposition of diamond powder dispersed nickel-phosphorus coatings on steel substrate. Transactions of the Institute of Metal Finishing. 2011;**89**:89-94. DOI: 10.1179/174591911X12971865404438

[18] Seyyed Saber Mirhosseini FM. Effect of plasma nitriding on tribological properties of nickel-boron-nanodiamond electroless coatings. Surface and Coatings Technology. 2022;**128216**:435. DOI: 10.1179/1743294411Y.0000000058

[19] Donya Ahmadkhaniha CZ. The effects of additives, particles load and current density on codeposition of SiC particles in NiP nanocomposite coatings. Coatings. 2019;**9**(554):1-13. DOI: 10.3390/coatings9090554

[20] Balaraju JN, Sankara Narayanan TSN, Seshadri SK. Evaluation of the corrosion resistance of electroless Ni-P and Ni-P composite coatings by electrochemical impedance spectroscopy. Journal of Solid State Electrochemistry. 2001;**5**:334-338. DOI: 10.1007/s100080000159

[21] Das CM, Limaye PK, Grover AK, Suri AK. Preparation and characterisation of silicon nitride codeposited electroless nickel composite coatings. Journal of Alloys and Compounds. 2007;**436**:1, 328-2, 334. DOI: 10.1016/j.jallcom.2006.07.036

[22] Bozzini B, Sidorov VE, Dovgopol AS, Birukov JP. Magnetic susceptibility of electroless $Ni_{84}P_{16}$-B_4C Int. Journal of Inorganic Materials. 2000;**2**:437-442. DOI: 10.1016/s1466-6049(00)00049-0

[23] Chakarova V, Georgieva M, Petrova M, Dobreva E, Stoychev D. Electroless deposition and investigation of composite coatings based on nickel or cobalt matrix, including boron nitrides as dispersoids, on polyethylene terephthalate substrate. Transactions of the IMF. 2016;**94**(5):259-264. DOI: 10.1080/00202967.2016.1208386

[24] Afroukhteh S, Dehghanian C, Emamy M. Preparation of electroless Ni–P composite coatings containing nano-scattered alumina in presence of polymeric surfactant. Progress in Natural Science: Materials International. 2012;**22**(4):318-325. DOI: 10.1016/j.pnsc.2012.06.006

[25] Chinchu KS, Riyas AH, Ameen Sha M, Geethanjali CV, Saji VS, Shibli SMA. ZrO2–CeO2 assimilated electroless Ni–P anti-corrosion coatings. Surfaces and Interfaces. 2020;**21**:100704. DOI: 10.1016/j.surfin.2020.100704

[26] Shozib IA, Ahmad A, Majdi A, Amin NT. Electroless Ni-P-TiO 2 (ENPT) Composite Coating: A Review on Microstructural Characteristics and Multifarious Properties for Surgical Instruments. Solid State Technology. 2021;**63**(6):3989-3996

[27] Zuleta AA, Galvis OA, Castaño JG, Echeverría F, Bolivar FJ, Hierro MP, et al. Preparation and characterization of electroless Ni–P–Fe3O4 composite coatings and evaluation of its high temperature oxidation behaviour. Surface and Coating Technology. 2009;**203**:3569-3578. DOI: 10.1016/j.surfcoat.2009.05.025

[28] Baba NB, Omar MF, Sharif S, Mohamed SB. Processing and properties of Ni/YSZ composite coating on high speed steel cutting tool. Journal of Fundamental and Applied Sciences. 2018;**10**(2s):688-700. DOI: 10.4314/jfas.v10i2s.51

[29] Ramalho A, Miranda JC. Friction and wear of electroless NiP and NiP+PTFE coatings. Wear. 2005;**259**:7-12. DOI: 10.1016/j.wear.2005.02.052

[30] Khalifa ORM, Sakr E. Electroless nickel-phosphorus-polymer composite coatings. The Open Corrosion Journal. 2009;**2**(1):211-215. DOI: 10.2174/1876503300902010211

[31] Bouzakis K-D, Michailidis N, Skordaris G, Bouzakis E, Biermann D, M'Saoubi R. Cutting with coated tools: Coating technologies, characterization methods and performance optimization. CIRP Annals. 2012;**61**:703-723. DOI: 10.1016/j.cirp.2012.05.006

[32] Yu D, Wang C, Cheng X, Zhang F. Optimization of hybrid PVD process of TiAlN coatings by Taguchi method. Applied Surface Science. 2008;**255**(5):1865-1869. DOI: 10.1016/j.apsusc.2008.06.204

[33] Park S-T, Han J-G, Keunecke M, Lee K. Mechanical and structural properties of multilayer c-BN coatings on cemented carbide cutting tools. International Journal of Refractory Metals and Hard Materials. 2017;**65**:12-56. DOI: 10.1016/j.ijrmhm.2016.11.009

[34] Thepsonthi T, Ozel T. Experimental and finite element simulation based investigatioms on micro-milling Ti-6Al-4V titanium alloy: Effect of cBN coating on tool wear. Journal of Materials Processing Technology. 2013;**213**:532-542. DOI: 10.1016/j.jmatprotec.2012.11.003

[35] Caggiano A, Teti R. CBN grinding performance improvement in aircraft engine components manufacture 2nd CIRP global web conference. Procedia CIRP. 2013;**9**:109-114. DOI: 10.1016/j.procir.2013.06.177

[36] Khamel S, Ouelaa N, Bouacha K. Analysis and prediction of tool wear, surface roughness and cutting forces in hard turning with CBN tool. Journal of Mechanical Science and Technology 2012;**26**:3605-3616. 10.1007/s12206-012-0853-1

[37] Sahin Y, Motorcu AR. Surface roughness model in machining hardened steel with cubic boron nitride cutting tool. International Journal of Refractory Metals and Hard Material. 2008;**26**:84-90. DOI: 10.1016/j.ijrmhm.2007.02.005

[38] Chen L, Zhou J, Bushlya V, Gutnichenko O, Stahl JE. Performnace assessment of pCBN and bCBN tools in machining of high-chromium white cast iron. Journal of Advanced Manufacturing Technology. 2015;**79**:635-644. DOI: 10.1007/s00170-015-6863-2

[39] Zhu D, Zhang X, Ding H. Tool wear characteristics in machining of nickel-based superalloys. International Journal of Machine Tools and Manufacture.

2013;**64**:60-77. DOI: 10.1016/j.ijmachtools.2012.08.001

[40] Yu T, Ma Z, Wang X, Song B, Wang W. Effects of Ni addition on properties of vitrified bond CBN composites in strong magnetic field. Ceramics International. 2018;**44**:9312-9317. DOI: 10.1016/j.ceramint.2018.02.143

[41] Sun K, Liang Y, Zhu Y, Li Z, Ji H. The effect of graphene oxide on the mechanical properties of cBN composites fabricated by HTHP method. Diamond & Related Materials. 2020;**104**:107754. DOI: 10.1016/j.diamond.2020.107754

[42] Adesina AY, Hakeem AS, Azam MU, Ahmed BA, Ibrahim AB, Ehsan MA, et al. Thermomechanical and tribological properties of spark plasma sintered bearing steel/cBN(Ni) composites for engineering applications. Journal of Materials Research and Technology. 2020;**9**(6):14645-14661. DOI: 10.1016/j.jmrt.2020.10.025

[43] Turgut Y, Şahin I, Findik T. Study of cutting force and surface roughness in milling of Al/SiC metal matrix composites. Scientific Research and Essays. 2010;**6**(10):2056-2062. DOI: 10.5897/SRE10.496

[44] Norsilawati N, Fathil CIM, Baba NB, Azinee SN, Ibrahim MH. Characterization of nickel-cubic boron nitride coating via electroless nickel deposition on high speed steel and carbide substrates. Journal of Physics: Conference Series. 2021;**1874**:012070. DOI: 10.1088/1742-6596/1874/1/012070

[45] Baba NB, Davidson A, Muneer T. Investigation of Ni-YSZ composite manufactured by electroless Ni coating. Applied Mechanics and Materials. 2011;**52**:1660-1664. DOI: 10.4028/www.scientific.net/AMM.52-54.1660

[46] Baba NB. YSZ reinforced Ni-P composite by electroless nickel co-deposition. In: Composites and Their Properties, Chapter 20. London: Intech Publisher; 2012. pp. 457-482. DOI: 10.5772/46496

[47] Urbanski JP, Koshy P, Dewes RC, Aspinwall DK. High speed machining of moulds and dies for net shape manufacture. Materials & Design. 2000;**21**(4):395-402. DOI: 10.1016/S0261-3069(99)00092-8

[48] Rama Kotaiah K, Srinivas J, Babu KJ, Srinivas K. Prediction of optimal cutting states during inward turning: an experimental approach. Materials and Manufacturing Processes. 2010;**25**(6):432-441. DOI: 10.1080/10426910903229321

[49] Gandarias A, de Luis Norberto Lopez L, Aizpitarte X, Lamikiz A. Study of the performance of the turning and drilling of austenitic stainless steels using two coolant techniques. International Journal of Machining and Machinability of Materials. 2008;**3**:1-17. DOI: 10.1504/IJMMM.2008.017621

[50] Kalpakjian S, Steven R, Musa H. Manufacturing Engineering and Technology. 6th ed. 2013

[51] Yating W, Bin S, Lei L, Wenbin H. Articial neural network modelling of plating rate and phosphorus content in the coatings of electroless nickel plating. Journal of Materials Processing Technology. 2008;**213**(205):207. DOI: 10.1016/j.jmatprotec.2007.11.095

[52] Ghani JA, Choudhury IA, Hassan HH. Application of Taguchi method in the optimization of end milling. Journal of Materials Processing Technology. 2004;**145**(1):84-92. DOI: 10.1016/S0924-0136(03)00865-3

[53] Hassan SMFBS, Shafei SB, Rashid RBA. Optimization of machining parameters in milling process for high speed machining using taguchi method for best surface roughness. IOP Conference Series: Materials Science and Engineering. 2020;**864**:012110. DOI: 10.1088/1757-899X/864/1/012110

[54] Hayajneh MT, Tahat MS, Bluhm J. A study of the effects of machining parameters on the surface roughness in the end milling process. Jordan Journal of Mechanical and Industrial Engineering. 2007;**1**(1):1-5

[55] Yıldırım ÇV, Kıvak T, Erzincanl F. Tool wear and surface roughness analysis in milling with ceramic tools of Waspaloy: A comparison of machining performance with different cooling methods. Journal of the Brazilian Society of Mechanical Sciences and Engineering. 2019;**41**:83. DOI: 10.1007/s40430-019-1582-5

[56] Azinee SN, Norsilawati N, Bahiyah Baba N, Fathil CIM, Mohd H, Ibrahim RANI. Investigation on cutting feasibility on Ni-CBN on HSS substrates for aerospace material. IOP Conference Series: Materials Science and Engineering. 2021;**1068**:012005. DOI: 10.1088/1757-899X/1068/1/012005

[57] Norsilawati N, Azinee SN, Bahiyah Baba N, Fathil CIM. Investigation and optimazation key parameters effect on cutting feasibility of electroless Ni-CBN on carbide end mill cutting tools by Taguchi method. Journal of Physics: Conference Series. 2021;**1874**:012071. DOI: 10.1088/1742-6596/1874/1/012071

Section 2

Frontiers and State-of-Art Applications of Boron-Based Materials

Chapter 5

Design, Synthesis, and Biological Applications of Boron-Containing Polyamine and Sugar Derivatives

Shin Aoki, Hiroki Ueda, Tomohiro Tanaka, Taiki Itoh, Minoru Suzuki and Yoshinori Sakurai

Abstract

Boron (B), an element that is present in ultratrace amounts in animal cells and tissues, is expected to be useful in many scientific fields. We have found the hydrolysis of C–B bond in phenylboronic acid-pendant cyclen (cyclen = 1,4,7,10-tetraazacyclododecane) and the full decomposition of *ortho*-carborane attached with cyclen and ethylenediamines in aqueous solution at neutral pH upon complexation with intracellular metals. The change in the chemical shift of the ^{11}B signals in ^{11}B-NMR spectra of these boron-containing metal chelators can be applied to the magnetic resonance imaging (MRI) of metal ions in solutions and in living cells. More important applications of B would be boron neutron capture therapy (BNCT) based on the nuclear reaction between ^{10}B atoms and thermal neutrons, yielding $^{4}He^{2+}$ (α) and $^{7}Li^{3+}$ ions, which destroy ^{10}B-containing cancer cells. The design and synthesis of new BNCT agents based on sugars and macrocyclic polyamines and their Zn^{2+} complexes are also introduced in this review.

Keywords: boron-10 (^{10}B), boron-11 (^{11}B), magnetic resonance imaging, metal probes, decomposition reactions, carborane, boron neutron capture therapy, macrocyclic polyamines, sugars

1. Introduction

Boron (B) is an element that is found in ultratrace amounts in mammalian cells and consists of two stable isotopes, boron-10 (^{10}B) and boron-11 (^{11}B), with a natural abundance ratio ($^{10}B/^{11}B$ = 19.9/80.1). The most important properties of boron compounds with respect to biological and medical sciences would be: (1) ^{11}B atoms have a higher NMR sensitivity (16.5% for ^{11}B and 2.0% for ^{10}B relative to ^{1}H NMR), thus permitting the detection of B-containing drugs themselves and analytes that react with B-containing probes in living systems [1]; and (2) the ^{10}B nucleus possesses a high reactivity with thermal neutrons resulting in the generation of two radioactive species (^{4}He and ^{7}Li particles), which induce the excitation and ionization of molecules within short path lengths [2]. For the above reasons, boron compounds can be useful in biological applications for the treatment and diagnosis of cancer and other diseases [3].

In 1936, Locher proposed the concept of boron neutron capture therapy (BNCT) based on the aforementioned nuclear reaction between ^{10}B and thermal neutrons [4]. Because the destructive effect of the two heavy particles (^{4}He and ^{7}Li particles) that are generated by the decomposition of ^{10}B lies within 5–9 μm, which is close to the size of living cells, single-cell treatment would be possible by the achievement of cancer-specific delivery of ^{10}B and irradiation with a sufficient intensity of thermal neutrons [5–7].

BNCT systems have been installed in clinical facilities as a method for the noninvasive treatment of certain types of cancers such as recurrent head and neck cancer and malignant gliomas [8]. The selective and efficient accumulation of boron into tumor tissues is one of the important clues for successful BNCT and, as described below, two boron compounds have been approved for use as BNCT drugs. In addition, monitoring the distribution of boron in patients is required for planning treatment protocols to determine the irradiation doses and positions of the patient [9].

In this review, we introduce the applications of boron compounds to ^{11}B NMR (nuclear magnetic resonance)/MRI (magnetic resonance imaging) probes for the sensing of intracellular metal ions and BNCT agents for use in the treatment of cancer. The d-block metal ion probes take advantage of changes in the chemical shift in ^{11}B NMR spectra due to the cleavage of the carbon-boron bond in phenylboronic acid-pendant cyclen (1,4,7,10-tetraazacyclododecane) and the decomposition of the *ortho*-carborane moieties of carborane-metal chelator hybrids upon complexation with metal ions in aqueous solution at neutral pH. In the second half of this review, the development of novel BNCT agents bearing sugar and macrocyclic polyamine scaffolds is described.

2. ^{11}B NMR and MRI probes for metal ions in solutions and in living cells based on carbon-boron bond cleavage and the decomposition of *ortho*-carboranes upon metal complexation of chelator units

2.1 General

Biologically essential d-block metal ions such as zinc (Zn^{2+}), copper (Cu^{2+}), manganese (Mn^{2+}), and iron (Fe^{2+}) are involved in a variety of physiological processes in living systems as cofactors for various enzymes, intracellular second messengers, and related processes [10]. It was reported that a metal imbalance in cells and tissues causes a number of disorders such as Alzheimer's disease, Parkinson's disease, Willson's disease, etc. [10]. Therefore, the development of fluorescence-based probes for the detection of these intracellular metal ions has contributed to our understanding of their functions and metabolism in living cells, while some limitations to detecting their emission from tissues remain due to their impermeability [10–12].

It is well known that MRI is one of the useful noninvasive methods for *in vivo* visualization and that it permits three-dimensional images of organisms and drug distributions to be obtained [13]. Although MRI is powerful method, there are only a few examples of MRI probes such as Gd^{3+}-based contrast agents [14, 15].

2.2 Development of d-block metal ions probes based on the cleavage of C–B bonds in B-containing probes

It is well established that macrocyclic polyamine ligands such as 1,4,7-triazacyclononane ([9]aneN_3) **1**, 1,4,7,10-tetraazacyclododecane ([12]aneN_4, cyclen) **2**, and

1,4,7,10,13-pentaazacyclopentadecane ([15]aneN_5) **3** are able to form more stable complexes **4–6** with metal ions such as Cu^{2+}, Ni^{2+}, and Zn^{2+} in aqueous solution (**Figure 1**) than metal complexes of linear polyamine types [16, 17]. In addition, metal ions in these complexes, especially the Zn^{2+} ion in Zn^{2+}-cyclen complex (**5**), possess strong Lewis acidity and the deprotonated Zn^{2+}-bound H_2O (HO^-) functions as a nucleophile and a base in aqueous solution at neutral pH [18–23].

Bendel and coworkers reported that ^{11}B NMR/MRI would be a potential technique for the imaging of boron agents in the body [24, 25]. However, a functional system for achieving this has not been established yet. In this context, we hypothesized that the sp^2 boron in **7** and **8** would be changed to the sp^3 boron due to the formation of metal complexes **9a** and **10a** and the following interaction of metal-bound H_2O (OH^-) with boron at neutral pH, resulting in change in the ^{11}B NMR signals (**Figure 2**) [26]. However, the products obtained after the addition of Zn^{2+} to **7** (L^1) (**Figure 3a**) were **11a** (ZnL^3) and boric acid ($B(OH)_3$), as confirmed by an X-ray structure analysis (**Figure 3b**). The findings strongly indicated that the Zn^{2+}-bound H_2O (**9a** and **10a**) is efficiently deprotonated

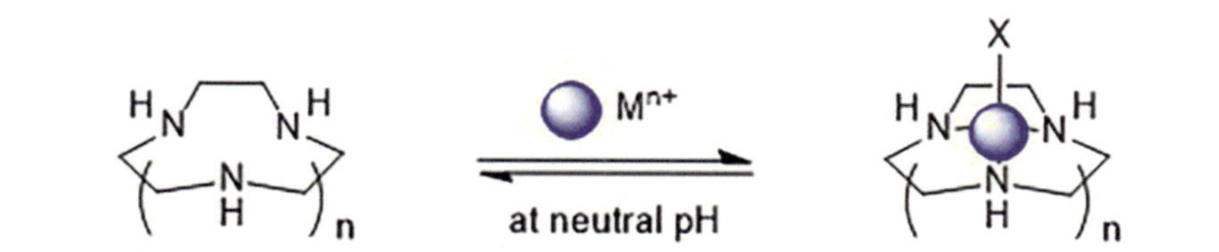

Figure 1.
The structures of 9-, 12-, and 15-membered macrocyclic polyamines **1–3** *and their metal complexes* **4–6**.

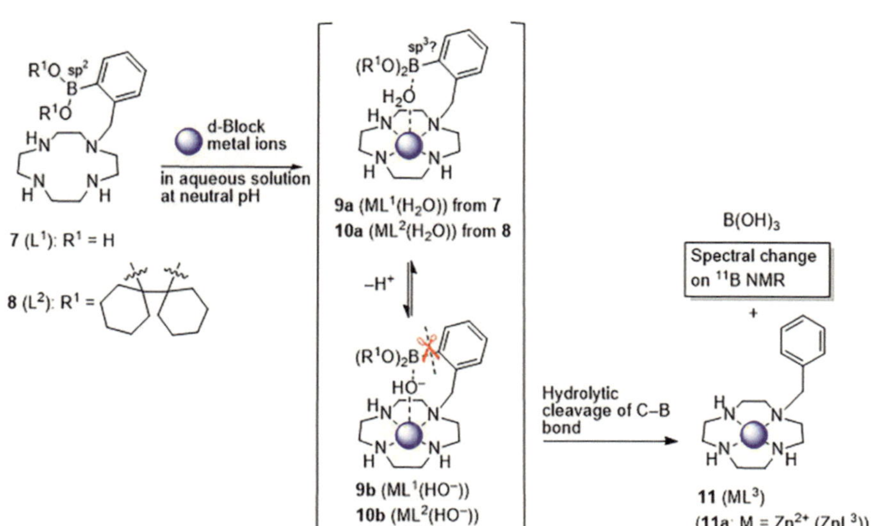

Figure 2.
The C–B bond hydrolysis of phenylboronic acid-pendant 12-membered tetraamine (cyclen) to produce inorganic boric acid.

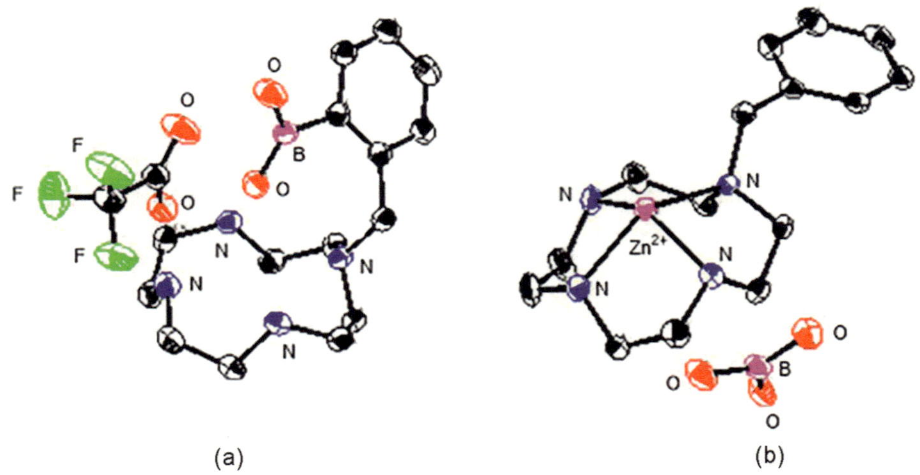

Figure 3.
X-ray crystal structures of (a) 7 *(L^1) and (b)* **11a** *(ZnL^3) with $B(OH)_3$.*

due to the double activation by Zn^{2+} and B to produce the Zn^{2+}-bound HO^- (**9b** and **10b**), which hydrolyzes the C–B bond. The hydrolytic cleavage of the C–B bond of 7 (L^1) was also observed by the measurement of ^{11}B NMR upon the addition of Zn^{2+}, in which the ^{11}B NMR signal of 7 (L^1) at 31.1 ppm was shifted to 19.4 ppm that corresponds to $B(OH)_3$.

The ^{11}B NMR spectral change of 7 (L^1) was promoted by Cu^{2+}, Fe^{2+}, Co^{2+}, and Ni^{2+} but not by Ca^{2+} and Mg^{2+} (**Table 1**). Hydrolysis of the C–B bond of 7 (L^1) with Cd^{2+} was faster than that with Zn^{2+}, possibly due to the strong nucleophilicity of the Cd^{2+}-bound HO^- [27]. Meanwhile, the C–B bond cleavage of 7 (L^1) by Mn^{2+} and Fe^{3+} was slow.

The intracellular uptake of boron in 7 and 8 into Jurkat T cells was determined by ICP-AES, and the results indicated that the uptake of 8 was higher than that of 7, possibly due to the hydrophobicity of the boronic ester group. The Zn^{2+}-induced C–B bond cleavage of 8 (L^2) by intracellular Zn^{2+} was observed in living cells. The Jurkat T cells were sequentially treated with 8 (L^2) and Zn^{2+} complex of pyrithione (Zn^{2+} ionophore to transfer Zn^{2+} into cells) for 20 min and 1 h, respectively. The cells were washed with CS-RPMI and PBS and then transferred to a quartz NMR tube, whose

	δ (ppm)	Δδ (ppm)^b	time (h)^c		δ (ppm)	Δδ (ppm)^b	time (h)^c
7 (L^1) alone	31.1	–	–	Mn^{2+}	20.6	–10.5	48
Zn^{2+}	19.4	–11.7	0.5	Ni^{2+}	19.8	–11.3	2
Cu^{2+}	19.5	–11.6	1.5	Cd^{2+}	19.2	–11.9	0.1
Fe^{2+}	19.7	–11.6	0.5	Ca^{2+}	31.7	0.6	–
Fe^{3+}	30.8	–0.3	–	Mg^{2+}	31.9	0.8	–
Co^{2+}	19.6	–11.5	1				

^{a}*All data are referenced to external BF_3·Et_2O in $CDCl_3$ (δ = 0 ppm).*
^{b}*Δδ = δ (7 (L^1) with metal ions) – δ (7 (L^1)).*
^{c}*Approximate reaction time for the completion of C–B bond cleavage.*

Table 1.
^{11}B NMR spectral change of 7 *(L^1) (20 mM) upon the addition of d-block metal ions (20 mM) in 1 M HEPES buffer at pD 7.4 and 25 °C [26].^{a}*

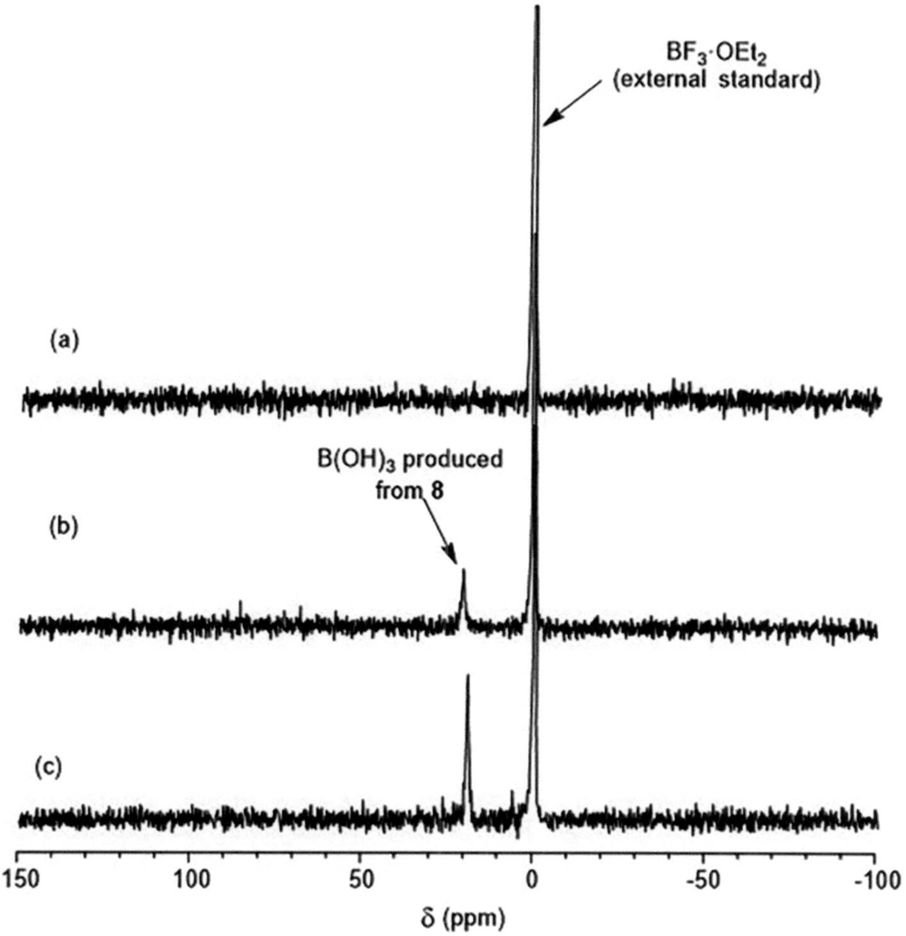

Figure 4.
*In-cell ^{11}B NMR spectra of **8** (L^2) in the absence of Zn^{2+}–pyrithione (ionophore) and in the presence of Zn^{2+}–pyrithione ($BF_3 \cdot Et_2O$ was used as an external references). The Jurkat T cells (4 × 10^8 cells) were incubated with 33 μM **8** (L^2) in culture medium at 37 °C for 1 h, and then (a) DMSO (as negative control), (b) 2.5 μM Zn^{2+}–pyrithione, and (c) 10 μM Zn^{2+}–pyrithione at 37 °C for 20 min.*

^{11}B NMR spectra were measured in D_2O containing PBS. As shown in **Figure 4**, the ^{11}B signal for $B(OH)_3$ (ca. 19 ppm) in Jurkat T cells was observed with a positive correlation to the concentrations of Zn^{2+}-pyrithione complex, indicating the successful detection of the intracellular Zn^{2+} ions. It should be noted that the ^{11}B signal for **8** (ca. 31 ppm) in the absence of Zn^{2+} was observed as a broad signal.

2.3 Development of Cu^{2+} ion probes based on decomposition reaction of *ortho*-carborane–metal chelator hybrids

It is known that the reaction of the *o*-carborane **12** with Brønsted or Lewis bases affords the corresponding *nido*-form **13** and $B(OH)_3$ and that the further degradation of **13** proceeds slowly under harsh conditions such as in acidic solutions and/or at high temperatures (**Figure 5**) [28]. On the other hand, we found that *o*-carborane derivatives such as **12, 14,** and **15a–c** generate 4–9 equiv. of $B(OH)_3$ upon the reaction with

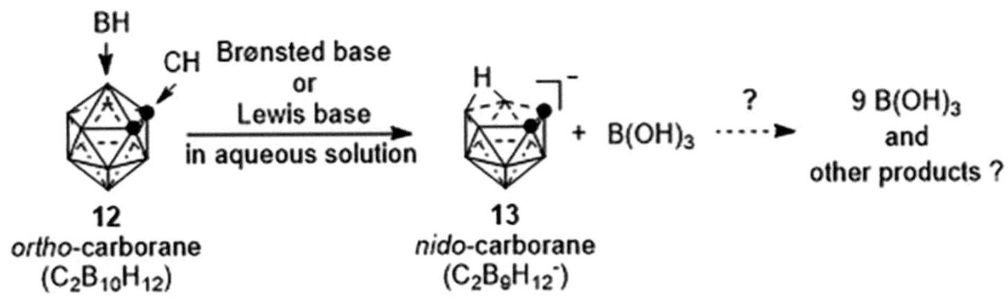

Figure 5.
Decomposition of o-carborane **12** *in the presence of a Brønsted or Lewis base.*

Cu^{2+} and Mn^{2+} via the corresponding *nido*-forms **12'**, **14'**, and **15a'–c'** under physiological conditions (**Figure 6a**) [29]. Our studies also indicated that the modification of *nido-o*-carborane (**16** (L^5)) with *N,N,N'*-trimethylethylenediamine (TriMEDA) as

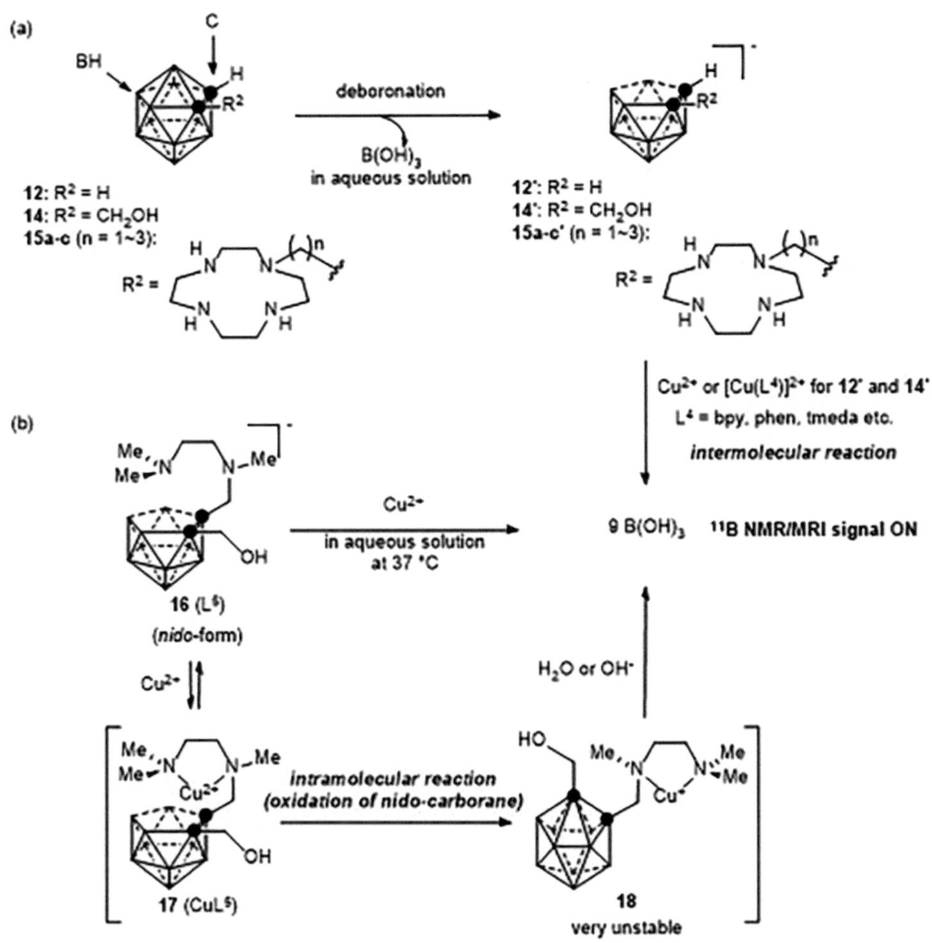

Figure 6.
Decomposition of o-carborane-pendant chelators (a) the ^{11}B *NMR/MRI detection of* Cu^{2+} *ion based on decomposition reaction of o-carborane derivatives and (b).*

a chelator unit facilitates the Cu-promoted decomposition of the molecule (**Figure 6b**) via the Cu^{2+}-complex **17** (CuL^5) to produce 9 $B(OH)_3$ in aqueous solution [30].

Changes in the ^{11}B NMR spectra of **16** (L^5) in the presence of various d-block metal ions are shown in **Figure 7**. A strong ^{11}B signal at ca. 20 ppm corresponding to $B(OH)_3$ was observed in the presence of Cu^{2+}, while, in the presence of other metal ions, the change was negligible. These results showed good agreement with the results of an azomethine-H assay, which also indicate the Cu^{2+} selectivity.

As shown in **Figure 8**, the oxidation potentials of **12′**, **14′**, and **16** are +0.57, +0.51, and +0.38 V (vs Ag/AgCl), respectively (determined by cyclic voltammetry), which are less positive than +0.7 V (vs Ag/AgCl) for $[Cu(TMEDA)]^+/[Cu(TMEDA)]^{2+}$. These data may explain the reasons why **12′**, **14′**, and **16** are oxidized by $Cu(TMEDA)]^{2+}$ complex. More efficient oxidation of **16** by Cu^{2+} than that of **12′** and **14′** is possibly due to the order of oxidation potentials (+0.38 V for **16** vs +0.57 and +0.51 V for **12′** and **14′**) and the close contact between the *o*-carborane unit and stable Cu^{2+}-TMEDA complex part in **17** and **18** (**Figure 6**).

In addition, the chemical yields of $B(OH)_3$ from **16** (L^5) with Cu^+ were decreased when antioxidants (sodium ascorbate, NaAsc) were added to the reaction mixture. According to these results and DFT calculations, a proposed mechanism for the decomposition of *o*-carborane moieties by Cu^{2+} is shown in **Figure 9**. Initially, the *nido*-form **20** is generated from the *closo*-form **19** by reaction with a nucleophile such

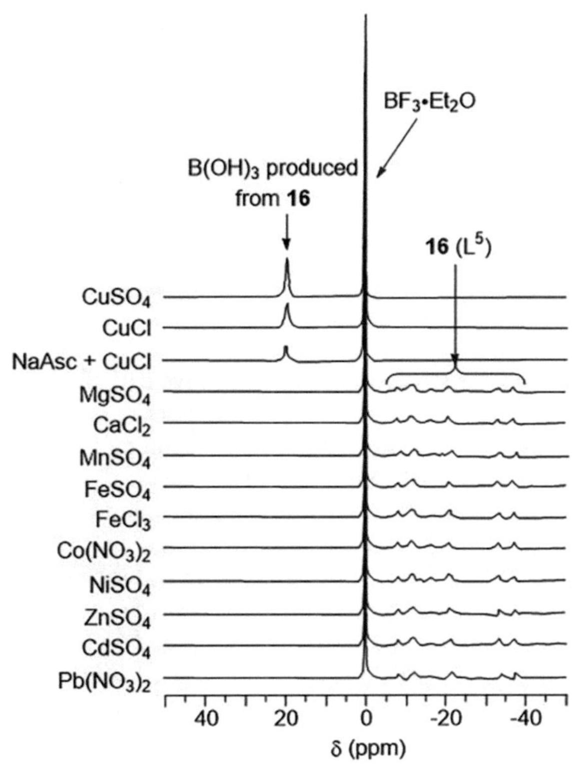

Figure 7.
*Decomposition of **16** (L^5) (1.4 mM) in the presence of Cu^{2+}, Cu^+, Cu^++NaAsc, Mg^{2+}, Ca^{2+}, Mn^{2+}, Fe^{2+}, Fe^{3+}, Co^{2+}, Ni^{2+}, Zn^{2+}, Cd^{2+} and Pb^{2+} (2 mM) in DMSO/0.5 M HEPES buffer (pH 7)/D_2O (5:4:1, 0.5 mL in total) at 37 °C after incubation for 4 h measured by $^{11}B\{^1H\}$ NMR. For $^{11}B\{^1H\}$ NMR experiments, 2.5% $BF_3 \cdot Et_2O$ in $CDCl_3$ was used for an external reference.*

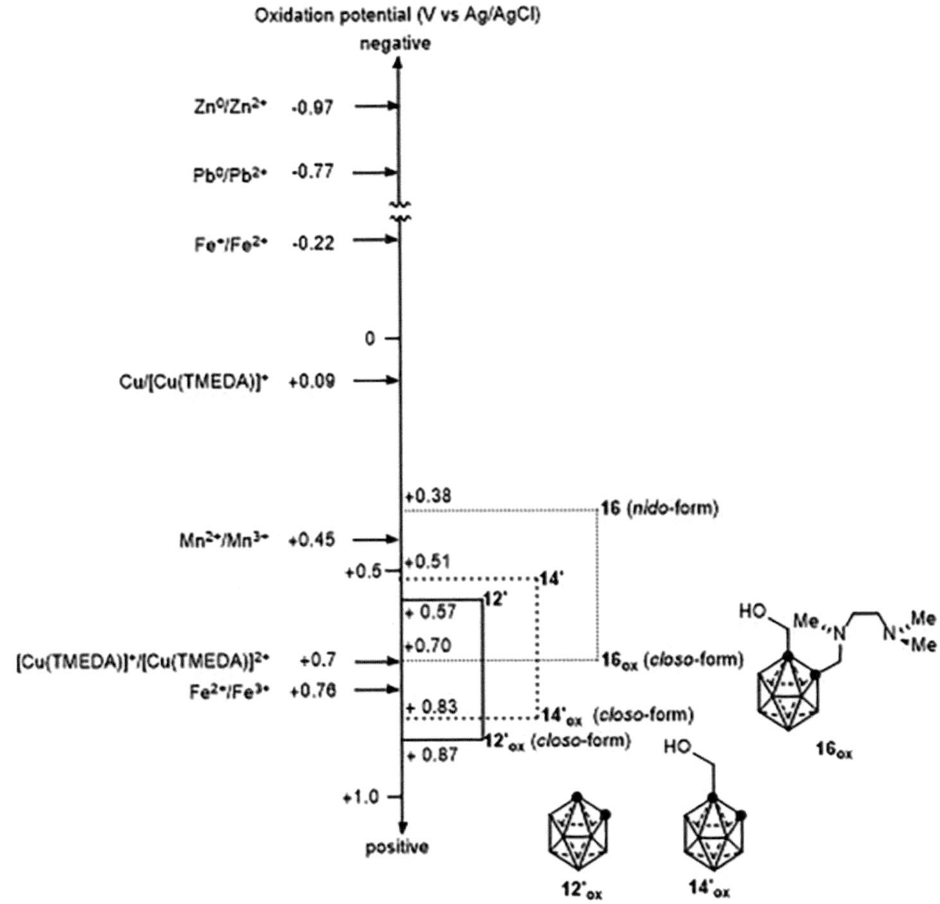

Figure 8.
Summary of the oxidation potentials of **12'**, **14'**, *and* **16** *(nido-form) with redox potentials of Cu, Fe, Pb, and Zn.*

as HO^-. Following the oxidation of the electronegative B10 (B at the 10 position) of **20** by Cu^{2+}, the *closo*-form **21** is produced by a ring-closure reaction. The unstable intermediate **21** would react with H_2O at the B9 position and is then completely decomposed to 9 equiv. of $B(OH)_3$ and other products via the transition state **22**.

^{11}B MRI experiments were conducted by using an aqueous solution of $B(OH)_3$ (10 mM) and Cu(bpy) (1 mM) in a larger vial (S_{out}) and a *o*-carborane analogue **14** (**Figure 6**) (1 mM) in a smaller vial (S_{in}) that was nested in the larger vial (**Figure 10**). To detect these boron compounds separately, BF1 (the basic transmitter frequency) values for $B(OH)_3$ and **14** are set ca. 128.392 and 128.387 MHz, respectively, because they have different chemical shifts (a-i and b-i in **Figure 10**). Besides, ^{11}B NMR images are obtained by using a two-dimensional ultra-short echo time sequence (UTE2D) with TE (echo time) of 199 μsec and TR (repetition time) of 30 msec. The ^{11}B signals for both $B(OH)_3$ and the *o*-carborane derivatives **14** were clearly observed, as shown in **Figure 10** (a-ii and b-ii).

The detection of Cu^{2+} by a ^{11}B NMR probe **16** (L^5) (2 mM) was carried out by the measurement of ^{11}B MRI and NMR at the increasing concentrations of Cu^{2+} (0, 0.02, 0.1, 0.2, 1.0, and 2.0 mM) in aqueous solution at neutral pH. The ^{11}B MRI/

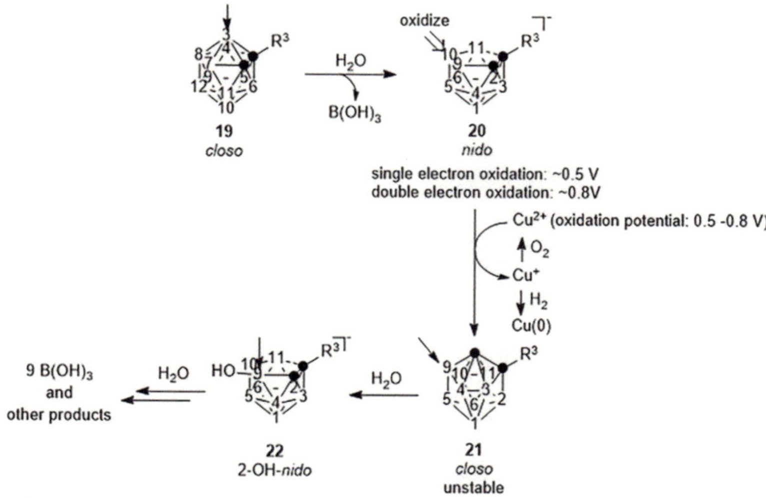

Figure 9.
Proposed mechanism for the decomposition reaction (arrows indicate positively charged boron atoms, which are susceptible to attack by H_2O or HO^-).

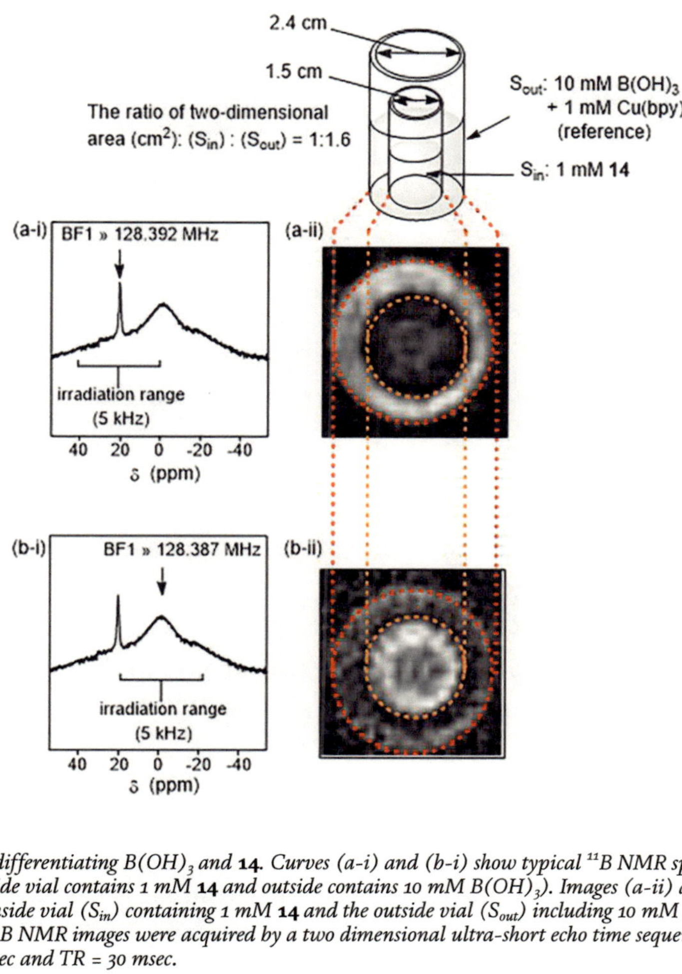

Figure 10.
*^{11}B MRI images differentiating $B(OH)_3$ and **14**. Curves (a-i) and (b-i) show typical ^{11}B NMR spectra of solutions in two vials (inside vial contains 1 mM **14** and outside contains 10 mM $B(OH)_3$). Images (a-ii) and (b-ii) show ^{11}B MRI of the inside vial (S_{in}) containing 1 mM **14** and the outside vial (S_{out}) including 10 mM $B(OH)_3$ + 1 mM Cu(bpy). Both ^{11}B NMR images were acquired by a two dimensional ultra-short echo time sequence (UTE2D) with TE = 199 μsec and TR = 30 msec.*

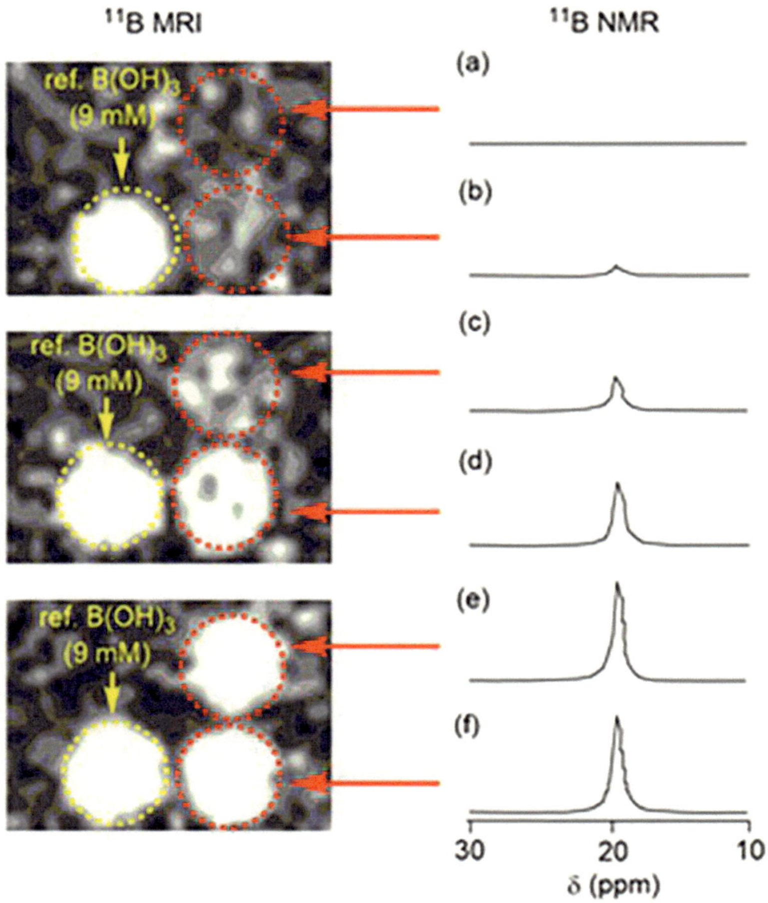

Figure 11.
^{11}B MRI and ^{11}B{^{1}H} NMR (128 MHz) spectra of **16** *(L^5) (2 mM) in DMSO/0.5 M HEPES buffer (pH 7)/D_2O (5:4:1, 0.5 mL in total) after incubation with various concentrations (0 (a), 0.02 (b), 0.1 (c), 0.2 (d), 1 (e), 2 mM (f)) of Cu^{2+} at 37 °C for 8 h (A 2.5% solution of $BF_3{\cdot}Et_2O$ in $CDCl_3$ was used as the external reference). ^{11}B NMR images were acquired by a two dimensional ultra-short echo time sequence (UTE2D) with BF1 values ≈ 128.392 MHz, TE = 199 μsec and TR = 30 msec.*

NMR signals of $B(OH)_3$ were successfully observed, and the signal intensities were increased in a dose-dependent manner due to the Cu^{2+}-promoted decomposition of **16** (L^5), as shown in **Figure 11**.

3. Design and synthesis of boron-containing agents for boron neutron capture therapy (BNCT)

3.1 General

As described in the Introduction, BNCT is one of the powerful cancer treatment methods utilizing two heavy particles, ^{4}He and ^{7}Li, which are produced from ^{10}B by a neutron capture reaction [^{10}B (n, α)^{7}Li] and induce the damage of biomolecules such as DNA, RNA, and so on within a short range of 5–9 μm [4–8]. For this BNCT to be achieved, the development of cancer-specific ^{10}B carriers is urgently needed. To date,

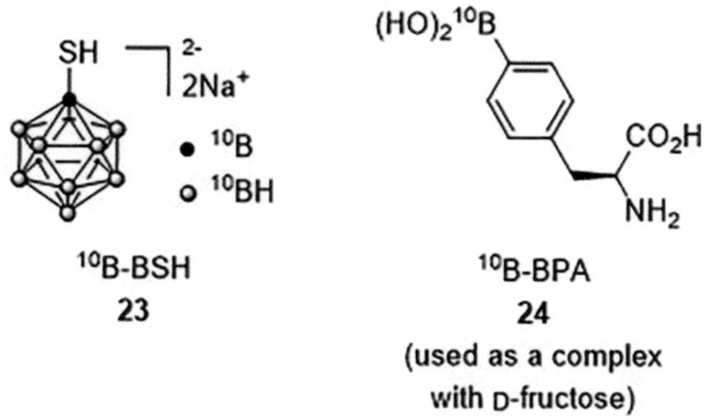

Figure 12.
Structures of representative BNCT agents.

only two boron compounds, namely disodium mercaptoundecahydrododecaborate (BSH) **23** and L-4-boronophenylalanine (BPA) **24** (used as a complex with D-fructose), have been approved for use as BNCT agents in clinical settings (**Figure 12**) [31, 32], but they are not sufficiently effective for the treatment of various tumor types. Because more selective and more efficient BNCT agents are required, the design and synthesis of new boron carriers based on sugar and macrocyclic polyamine scaffolds were conducted.

3.2 Design and synthesis of boron-containing sugars for BNCT

Sulfoquinovosyl acylglycerol (SQAG) **25** was isolated from sea algae and characterized by Sakaguchi et al., and **25** and its derivative sulfoquinovosyl acylpropanediol (SQAP) **26** were reported to be accumulated in cancer cells and exhibit weak toxicity against normal cells (**Figure 13a**) [33]. Because the modification of the long alkyl

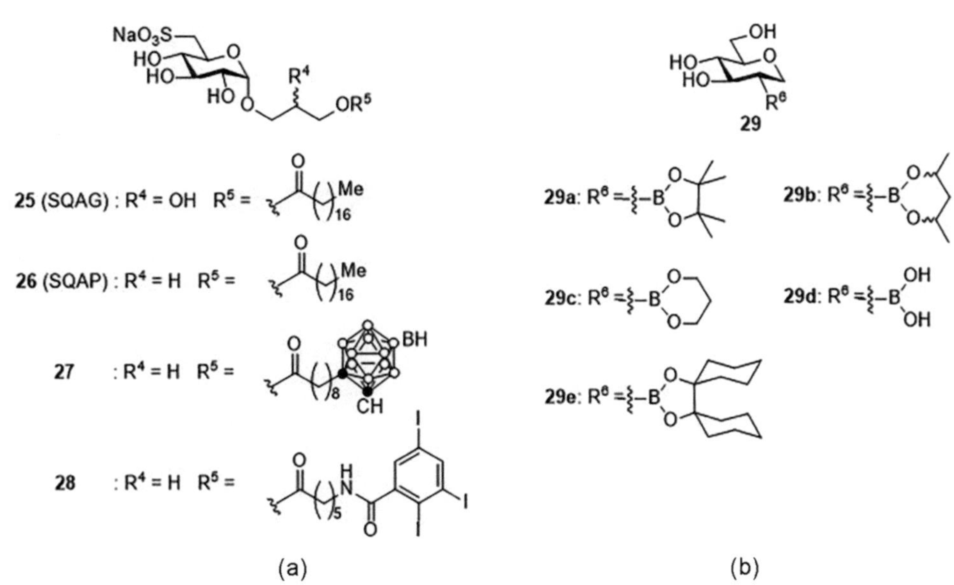

Figure 13.
Structures of (a) SQAG and SQAP derivatives and (b) 2-boryl-1,2-dideoxy-D-glucose derivatives.

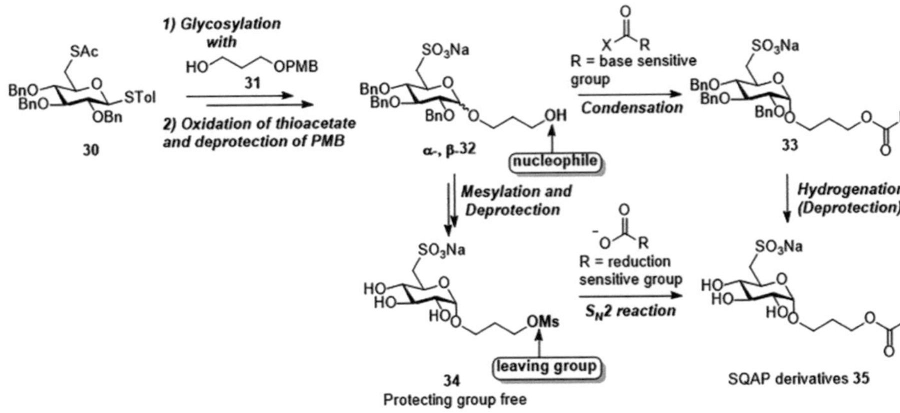

Figure 14.
The synthetic route of SQAP derivatives developed by Aoki et al.

chain of SQAG has negligible effect on its biological activity, the design and synthesis of SQAP derivatives **27** and **28** containing a boron cluster unit and iodine atoms as BNCT agents and imaging agents for X-ray computed tomography (CT) were conducted [34, 35].

The synthesis route for preparing SQAG analogues **27** and **28** is presented in **Figure 14**. The intermediate **32** was obtained by the α selective glycosylation of **30** with **31** in CH_2Cl_2/*tert*-butyl methyl ether (1/3), followed by the oxidation of thioacetate and the deprotection of *p*-methoxybenzyl (PMB) group. The condensation of **32** with a long chain fatty acid unit and subsequent deprotection of the benzyl groups could give the desired product **35**, which would be ideal for the synthesis of SQAP analogues containing base-sensitive functional groups such as carborane. Furthermore, the conversion of a nucleophile (-OH) of **32** to a leaving group (-OMs) enables the introduction of various acyl moieties by S_N2 reaction to give **35**, which corresponds to **27** and **28**. This novel synthesis route, as presented in **Figure 14**, would be useful for preparing a wide variety of SQAP derivatives.

The design and synthesis of 2-boryl-1,2-dideoxy-D-glucose derivatives **29a–e** were also carried out (**Figure 13b**) [36]. It is well known that cancer cells exhibit high glucose consumption and upregulation of glucose transporters (GLUTs) for rapid growth and proliferation, a process that is known as the Warburg effect [37]. It was also reported that hydrogen bonding interactions between the hydroxy groups of D-glucose and amino acid residues of GLUT trigger the intracellular uptake of glucose, and that the modification of D-glucose with bulky moieties at the C2 and C6 positions is tolerated [38]. In clinical applications, for instance, the D-glucose analogue, 2-deoxy-2-[^{18}F]fluoro-D-glucose, has been used for the diagnosis of cancer by means of positron emission tomography (PET) based on the aforementioned issues [39].

We therefore performed the regio- and stereoselective hydroboration of D-glucal **36** at the C1-C2 double bond, esterification with a diol, and deprotection of the hydroxy groups to provide **29a–e** via the intermediate **37** (**Figure 15**). Although hydroboration is one of traditional methods for the conversion of alkenes into alcohols such as **38** after the treatment of a boryl intermediate such as **37** with H_2O_2/NaOH, **37** was directly converted into **29**. Further investigations of their biological activity indicated that these sugar derivatives exhibit the moderate intracellular uptake against cancer cell lines through GLUT1, while their BNCT activity was not satisfying.

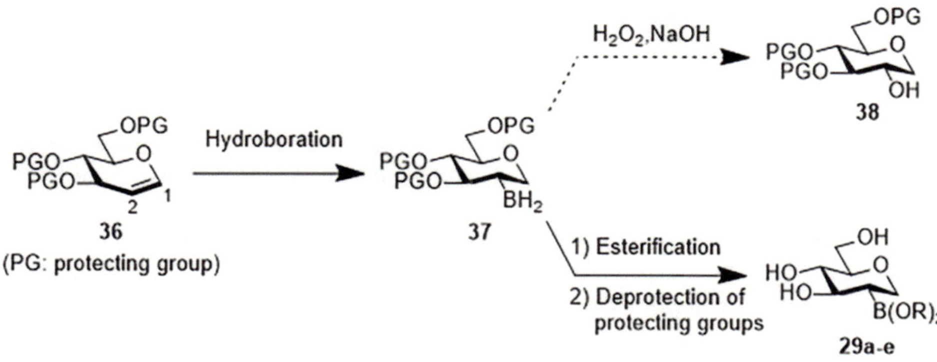

Figure 15.
Synthesis of 2-boryl-1,2-dideoxy-D-glucose derivatives **29a–e** *via the hydroboration of the protected D-glucal* **36**.

3.3 Design and synthesis of boron-containing macrocyclic polyamines for BNCT

It is known that natural polyamines play multiple roles in cellular functions, including gene expression and the stabilization of chromatin structure, and that the activated polyamine transport system and biosynthesis in cancer cells are related to the increase in polyamine concentrations and proliferation activity [40, 41]. Therefore, it is expected that polyamines would be desirable scaffolds for cancer selective and DNA-targeting boron delivery agents [42, 43].

Kimura and coworkers reported that Zn^{2+}–cyclen complexes **39** selectively recognize thymidine (dT) units in DNA to form a stable complex **40** in aqueous solution at neutral

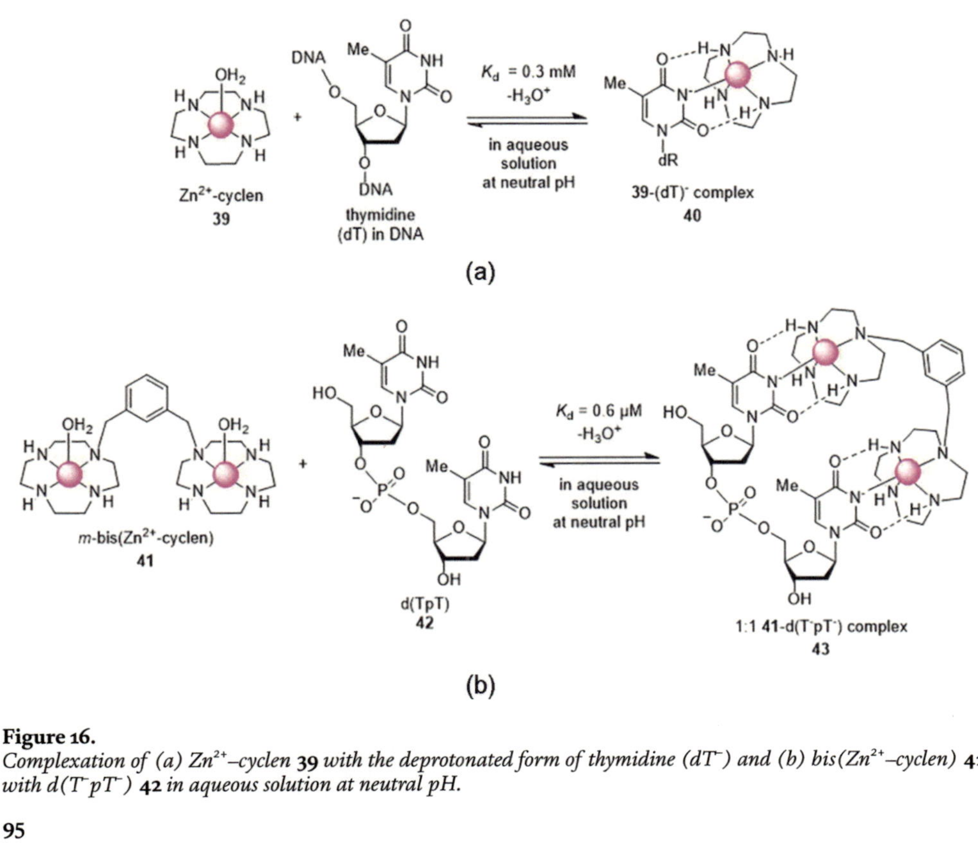

Figure 16.
Complexation of (a) Zn^{2+}*–cyclen* **39** *with the deprotonated form of thymidine* (dT^-) *and (b) bis(*Zn^{2+}*–cyclen)* **41** *with* $d(T^-pT^-)$ **42** *in aqueous solution at neutral pH.*

pH by coordination bonding between the deprotonated imide part of dT (dT$^-$) and Zn^{2+} and by hydrogen bonding between the NH of cyclen and the imide oxygens of dT$^-$ (**Figure 16a**) [44–47]. In addition, the bis(Zn^{2+}–cyclen) complexes **41** strongly bind two adjacent thymidine (thymidyl(3′–5′)thymidine, d(TpT)) **42**, yielding a very stable 1:1 complex **43** (**Figure 16b**) [48–51]. The dissociation constants (K_d) were reported to be 0.3 mM for **40** (1:1 complex of dT$^-$ with **39**) and 0.6 μM for **43** (1:1 complex of d(T$^-$pT$^-$) with **41**), respectively, at physiological pH in aqueous solution [52–54].

In this context, we designed and synthesized some novel DNA-targeting BNCT agents containing macrocyclic polyamine scaffolds such as [9]aneN_3, [12]aneN_4, and [15]aneN_5 and their Zn^{2+} complexes, which contain phenylboronic acid units, as shown in **Figures 17** and **18** [55, 56]. It was assumed that these boron-containing macrocyclic polyamine monomers **44–49** (L^6–L^{12}) and their Zn^{2+} complexes **50–52** (ZnL^6–ZnL^{12}) would be efficiently transferred into cancer cells and that thermal neutron irradiation would induce effective DNA damage in cancer cells due the ^{10}B atoms being located in close proximity to DNA molecules (**Figure 17**). We also expected that the interaction of homo- and heterodimer of macrocyclic polyamines **53–62** (L^{13}–L^{22}) and their corresponding monozinc(II) complexes **63–68** (ZnL^{13}–ZnL^{21}) and dizinc(II) complexes **69–78** (Zn_2L^{13}–Zn_2L^{22}) with DNA would be stronger than that of monomeric polyamines, resulting in efficient DNA damage upon thermal neutron irradiation (**Figure 18**). These mono- and

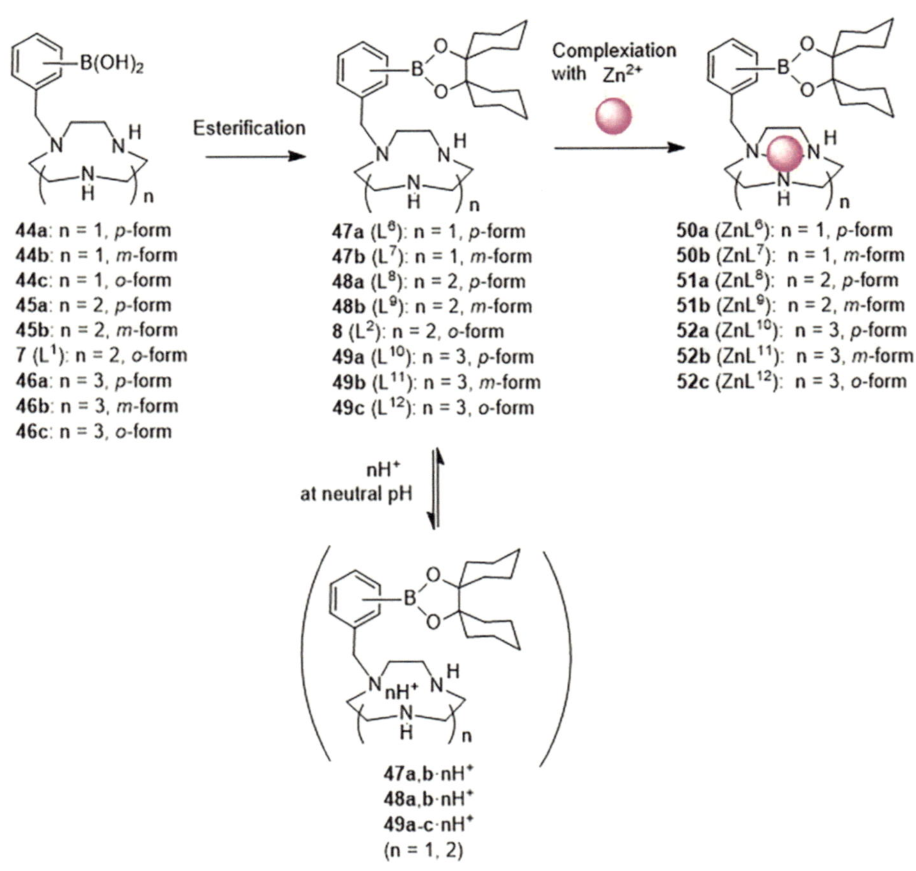

Figure 17.
Structures of B-containing macrocyclic polyamine monomers and their Zn^{2+} complexes.

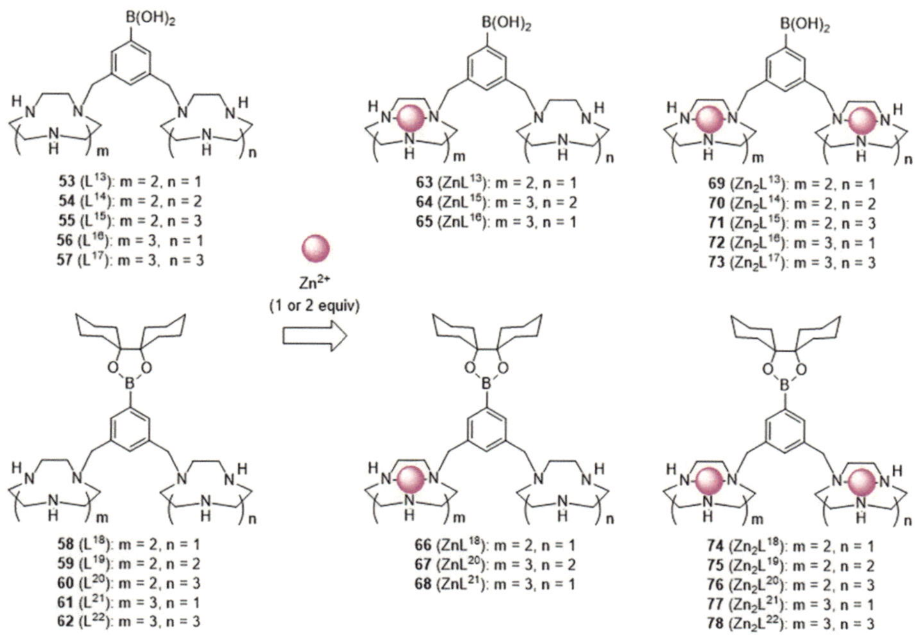

Figure 18.
Structures of B-containing macrocyclic polyamine dimers **53–62** *(L^{13}–L^{22}) and their Zn^{2+} complexes* **63–78** *(ZnL^{13}–ZnL^{21} and Zn_2L^{13}–Zn_2L^{22}).*

dimeric macrocyclic polyamines were first prepared with boron in a natural abundance ratio ($^{10}B/^{11}B$ = 19.9/80.1) to evaluate their cytotoxicity and intracellular uptake in several cancer cell lines, and some of the promising compounds were synthesized in the corresponding ^{10}B-enriched forms for the BNCT experiments. It should also be noted that these compounds possess macrocyclic polyamine units at the *m*- or *p*-position, but not at the *o*-position, of the C–B bonds to avoid the C–B hydrolysis upon metal complexation, as described in **Figures 2** and **3**.

The results of biological studies suggested that the boron-containing macrocyclic polyamine monomers **47b** (L^7), **48b** (L^9), and **49a** (L^{10}) have a weak cytotoxicity against normal cells and are efficiently transferred into cancer cells such as A549 and HeLa S3 cells, possibly via a polyamine transport system. In addition, it was found that ditopic macrocyclic polyamines possess much less cytotoxicity than that of the monomers and moderate uptake activity into cancer cells. Therefore, some of the more promising compounds were selected and their ^{10}B-enriched forms (>99% of ^{10}B) were prepared for BNCT experiments.

In vitro neutron irradiation experiments using A549 cells in the presence of the ^{10}B-enriched **^{10}B-47b** (L^7), **^{10}B-48b** (L^9), and **^{10}B-49a** (L^{10}) were performed at the Institute for Integrated Radiation and Nuclear Science, Kyoto University, and the BNCT effect of these drugs was evaluated by colony formation assays. It was found that **^{10}B-47b** (L^7), **^{10}B-48b** (L^9), and **^{10}B-49a** (L^{10}) showed higher cytotoxic effects than ^{10}B-BSH **23** and ^{10}B-BPA **24** and that the BNCT effect of ^{10}B-enriched dimers is nearly the same as ^{10}B-BPA (**Figure 19**). The BNCT effect of **^{10}B-47b** (L^7) and **^{10}B-50b** (ZnL^7) is almost identical and that of **^{10}B-50b** (ZnL^7) is even better, although the intracellular uptake of the Zn^{2+} complexes is generally lower than that of the corresponding Zn^{2+}-free ligands. It is possibly due to weak complexation of the 9-membered ring of **^{10}B-47b** (L^7) with Zn^{2+}. In addition, 12- and 15-membered macrocycles

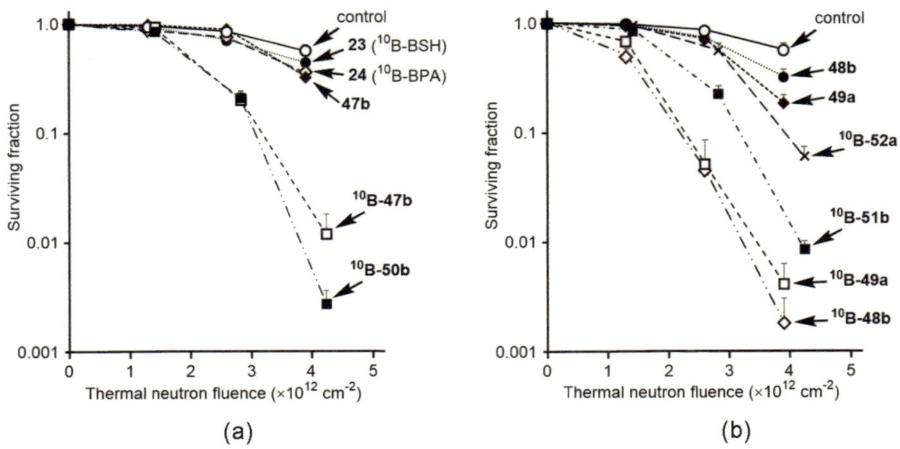

Figure 19.
BNCT effect of macrocyclic polyamine monomers **23**, **24**, **47b**, **^{10}B-47b**, **48b**, **^{10}B-48b**, **49a**, **^{10}B-49a**, **^{10}B-50b**, **^{10}B-51b**, *and* **^{10}B-52a** *(30 μM) against A549 cells was examined by a colony formation assay: (a) control (in the absence of a boron compound) (○),* **23** *(•),* **24** *(◇),* **47b** *(◆),* **^{10}B-47b** *(□), and* **^{10}B-50b** *(■). (b) Control (○),* **48b** *(•),* **^{10}B-48b** *(◇),* **49a** *(◆), and* **^{10}B-49a** *(□), and* **^{10}B-51b** *(■), and* **^{10}B-52a** *(×). After treatment with the boron compound for 24 h, the cells were irradiated with thermal neutrons for 0, 15, 30, and 45 min and then incubated without neutron irradiation for 7 days.*

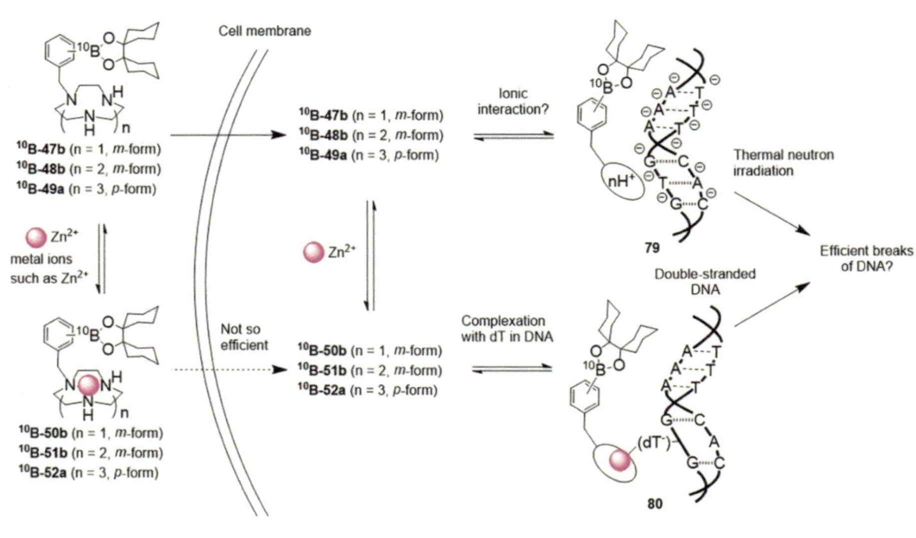

Figure 20.
Proposed scheme for BNCT effect of **^{10}B-47b**, **^{10}B-48b**, **^{10}B-49a** *and their Zn^{2+} complexes.*

^{10}B-48b and **^{10}B-49a** effectively inhibited the proliferation of cancer cells upon irradiation with thermal neutrons, while their intracellular uptake was lower than that of the [9]aneN_3-type **47b**.

According to the results of biological evaluations and DNA interaction studies using double-stranded calf-thymus DNA, it was concluded that metal-free monomers would be efficiently taken up by cancer cells and then form complexes with intracellular Zn^{2+}. Both the cationic metal-free macrocycles and their Zn^{2+} complexes would bind to DNA via electrostatic interactions between cationic macrocyclic polyamine moieties and anionic double-stranded DNA (**79** in **Figure 20**), or via the selective recognition of Zn^{2+}-complexes such as **^{10}B-51b** with dT^- units in DNA as depicted in **Figure 16** (and **80** in **Figure 20**), resulting in effective DNA damage upon thermal neutron irradiation (**Figure 20**). These findings suggest that ^{10}B delivery agents equipped with monomeric [12]aneN_4- and [15]aneN_5-type macrocycles are preferable for use in BNCT.

4. Conclusion

In this review, we summarize the current state of knowledge regarding the design and synthesis of ^{10}B and/or ^{11}B containing agents for biomedical applications such as ^{11}B NMR probes and BNCT agents. We developed the d-block metal ion probes based on changes in ^{11}B NMR signals due to the hydrolysis of C–B bond in **7** (L^1) and **8** (L^2) and the decomposition of *o*-carborane moieties in derivatives such as **14** and **16** (L^5) upon complexation with metal ions in aqueous solution at physiological pH. Some novel BNCT agents based on sugar and macrocyclic polyamine scaffolds were also designed and synthesized. The findings indicate that ^{10}B-enriched monomeric macrocyclic polyamines **^{10}B-48b** (L^9) and **^{10}B-49a** (L^{10}) exhibit potent BNCT activity upon thermal neutron irradiation, possibly due to interaction with DNA, resulting in the efficient damage of DNA molecules that are in close proximity to the boron compounds.

We believe that this review provides useful information for the future design and synthesis of novel boron-containing compounds and their applications for the treatment and diagnosis of cancer and other diseases, as well as in related research fields.

Acknowledgements

We wish to thank our collaborators and coworkers for their contributions to work described in this review. We appreciate Dr. Motoo Shiro (Rigaku Co. Ltd.), Prof. Reiko Kuroda (Chubu University), and Dr. Yasuyuki Yamada (Nagoya University) for their great assistance and helpful discussion. Financial supports from the Ministry of Education, Culture, Sports, Science and Technology (MEXT) of Japan, the Uehara Memorial Foundation, the Tokyo Ohka Foundation for the Promotion of Science and Technology, Kanagawa, Japan, the Tokyo Biochemical Research Foundation, Tokyo, Japan, Japan Society for the Promotion of Science (JSPS), and Tokyo University of Science are gratefully acknowledged.

Conflicts of interest

The authors declare no conflict of interest.

Author details

Shin Aoki[1,2,3]*, Hiroki Ueda[1,4], Tomohiro Tanaka[1], Taiki Itoh[1], Minoru Suzuki[4] and Yoshinori Sakurai[4]

1 Faculty of Pharmaceutical Sciences, Tokyo University of Science, Noda, Chiba, Japan

2 Research Institute for Science and Technology, Tokyo University of Science, Noda, Chiba, Japan

3 Research Institute for Biomedical Sciences, Tokyo University of Science, Noda, Chiba, Japan

4 Institute for Integrated Radiation and Nuclear Science, Kyoto University, Osaka, Japan

*Address all correspondence to: shinaoki@rs.tus.ac.jp

References

[1] Heřmánek S. ^{11}B NMR spectra of boranes, main-group heteroboranes, and substituted derivatives. factors influencing chemical shifts of skeletal atoms. Chemical Reviews. 1992;**92**:325-362

[2] Rinard PM. Neutron interactions with matter. In: Reilly D, Ensslin N, Smith H, Kreiner S, editors. Passive Nondestructive Assay of Nuclear Materials. Washington: Nuclear Regulatory Commission; 1991. pp. 357-377

[3] Morin C. The chemistry of boron analogues of biomolecules. Tetrahedron. 1994;**50**:12521-12569. DOI: 10.1016/S0040-4020(01)89389-3

[4] Locher GL. Biological effects and therapeutic possibilities of neutrons. The American Journal of Roentgenology and Radium Therapy. 1936;**36**:1-13

[5] Soloway AH, Tjarks W, Barnum BA, Rong FG, Barth RF, Codogni IM, et al. The chemistry of neutron capture therapy. Chemical Reviews. 1998;**98**:1515-1562. DOI: 10.1021/cr941195u

[6] Salt C, Lennox AJ, Takagaki M, Maguire JA, Hosmane NS. Boron and gadolinium neutron capture therapy. Russian Chemical Bulletin. 2004;**53**:1871-1888. DOI: 10.1007/s11172-005-0045-6

[7] Barth RF, Coderre JA, Vicente MGH, Blue TE. Boron neutron capture therapy of cancer: Current status and future prospects. Clinical Cancer Research. 2005;**11**:3987-4002

[8] Suzuki M. Boron neutron capture therapy (BNCT): A unique role in radiotherapy with a view to entering the accelerator-based BNCT era. International Journal of Clinical Oncology. 2020;**25**:43-50. DOI: 10.1007/s10147-019-01480-4

[9] Dai Q, Yang Q, Bao X, Chen J, Han M, Wei Q. The development of boron analysis and imaging in boron neutron capture therapy (BNCT). Molecular Pharmaceutics. 2022;**19**:363-377. DOI: 10.1021/acs.molpharmaceut.1c00810

[10] McRae R, Bagchi P, Sumalekshmy S, Fahrni CJ. In situ imaging of metals in cells and tissues. Chemical Reviews. 2009;**109**:4780-4827. DOI: 10.1021/cr900223a

[11] Que EL, Domaille DW, Chang CJ. Metals in neurobiology: Probing their chemistry and biology with molecular imaging. Chemical Reviews. 2008;**108**:1517-1549. DOI: 10.1021/cr078203u

[12] Zhu H, Fan J, Wang B, Peng X. Fluorescent, MRI, and colorimetric chemical sensors for the first-row d-block metal ions. Chemical Society Reviews. 2015;**44**:4337-4366. DOI: 10.1039/c4cs00285g

[13] Hingorani DV, Bernstein AS, Pagel MD. A review of responsive MRI contrast agents: 2005-2014. Contrast Media & Molecular Imaging. 2015;**10**:245-265. DOI: 10.1002/cmmi.1629

[14] Caravan P, Ellison JJ, McMurry TJ, Lauffer RB. Gadlinium(III) chelates as MRI contrast agents: Structure, dynamics, and applications. Chemical Reviews. 1999;**99**:2293-2352

[15] Xu Z, Liu C, Zhao S, Chen S, Zhao Y. Molecular sensors for NMR-based detection. Chemical Reviews.

2019;**119**:195-230. DOI: 10.1021/acs.chemrev.8b00202

[16] Kodama M, Kimura E. Equilibria and kinetics of complex formation between zinc(II), lead(II), andcadmium(II), and 12-, 13-, 14-, and 15-memberd macrocyclic tetraamines. Journal of the Chemical Society Dalton Transactions. 1977:2269-2276. DOI: 10.1039/DT9770002269

[17] Kodama M, Kimura E. Equilibria of Complex formation between several bivalent metal ions and macrocyclic tri- and penta-amines. Journal of the Chemical Society Dalton Transactions. 1978:1081-1085. DOI: 10.1039/DT9780001081

[18] Kimura E. Macrocyclic polyamines with intelligent functions. Tetrahedron. 1992;**48**:6175-6217. DOI: 10.1016/S0040-4020(01)88212-0

[19] Kimura E. Model studies for molecular recognition of carbonic anhydrase and carboxypeptidase. Accounts of Chemical Research. 2001;**34**:171-179. DOI: 10.1021/ar000001w

[20] Itoh S, Sonoike S, Kitamura M, Aoki S. Design and synthesis of chiral Zn^{2+} complexes mimicking natural aldolases for catalytic C–C bond forming reactions in aqueous solution. International Journal of Molecular Sciences. 2014;**15**:2087-2118. DOI: 10.3390/ijms15022087

[21] Aoki S, Zulkefeli M, Kitamura M, Hisamatsu Y. Supramolecular host and catalysts formed by the synergistic molecular assembly of multinuclear zinc(II) complexes in aqueous solution. In: Nabeshima T, editor. Synergy in Supramolecular Chemistry. Boca Raton: CRC; 2015. pp. 33-56. DOI: 10.1201/b17940

[22] Kimura E, Koike T, Aoki S. Evolution of Zn^{II}–macrocyclic polyamines to biological probes and supramolecular assembly. In: Izatt RM, editor. Macrocyclic and Supramolecular Chemistry: How Izatt-Christensen Award Winners Shaped the Field. Hoboken: John Wiley & Sons; 2016. pp. 417-445. DOI: 10.1002/9781119053859.ch21

[23] Aoki S, Rahman AB, Hisamatsu Y, Miyazawa Y, Zulkefeli M, Saga Y, et al. Development of metallosupramolecular phosphatases based on the combinatorial self-assembly of metal complexes and organic building blocks for the catalytic hydrolysis of phosphate monoesters. Result in Chemistry. 2021;**3**:100133. DOI: 10.1016/j.rechem.2021.100133

[24] Bendel P. Biomedical applications of ^{10}B and ^{11}B NMR. NMR in Biomedicine. 2005;**18**:74-82. DOI: 10.1002/nbm.886

[25] Bendel P, Margalit R, Koudinova N, Salomon Y. Noninvasive quantitative in vivo mapping and metabolism of boronophenylalanine (BPA) by nuclear magnetic resonance (NMR) spectroscopy and imaging. Radiation Research. 2005;**164**:680-687. DOI: 10.1667/RR3450.1

[26] Kitamura M, Suzuki T, Abe R, Ueno T, Aoki S. ^{11}B NMR Sensing of d-block metal ions in vitro and in cells based on a carbon–boron bond cleavage of phenylboronic acid-pendant cyclen (cyclen = 1,4,7,10-tetraazacyclododecane). Inorganic Chemistry. 2011;**50**:11568-11580. DOI: 10.1021/ic201507q

[27] Ohshima R, Kitamura M, Morita A, Shiro M, Yamada Y, Ikekita M, et al. Design and synthesis of fluorescent probe for Zn^{2+}, 5,7-bis(*N*,*N*-dimethylaminosulfonyl)-8-hydroxyquinoline-pendant

1,4,7,10-tetraazacyclododecane and Zn^{2+}-dependent hydrolytic and Zn^{2+}-independent photochemical reactivation of its benzenesulfonyl-caged derivatives. Inorganic Chemistry. 2010;**49**:888-899. DOI: 10.1021/ic901279t

[28] Hawthorne MF. The role of chemistry in the development of boron neutron capture therapy of cancer. Angewandte Chemie, International Edition. 1993;**32**:950-984. DOI: 10.1002/anie.199309501

[29] Tanaka T, Nishiura Y, Araki R, Saido T, Abe R, Aoki S. ^{11}B NMR probes of copper(II): Finding and implications of the Cu^{2+}-promoted decomposition of *ortho*-carborane derivatives. European Journal of Inorganic Chemistry. 2016;**2016**(12):1819-1834. DOI: 10.1002/ejic.201600117

[30] Tanaka T, Araki R, Saido T, Abe R, Aoki S. ^{11}B NMR/MRI sensing of copper(II) ions in vitro by the decomposition of a hybrid compound of a *nido-o*-carborane and a metal chelator. European Journal of Inorganic Chemistry. 2016;**2016**(20):3330-3337. DOI: 10.1002/ejic.201600346

[31] Cerecetto H, Couto M. Medicinal chemistry of boron-bearing compounds for BNCT-glioma treatment: Current challenges and perspectives. In: Omerhodžić I, Arnautović K, editors. Glioma – Contemporary Diagnostic and Therapeutic Approaches. London, UK: IntechOpen; 2018. pp. 205-230. DOI: 10.5772/intechopen.76369

[32] Hu K, Yang Z, Zhang L, Xie L, Wang L, Xu H, et al. Boron agents for neutron capture therapy. Coordination Chemistry Reviews. 2020;**405**:213139. DOI: 10.1016/j.ccr.2019.213139

[33] Ohta K, Mizushina Y, Yamazaki T, Hanashima S, Sugawara F, Sakaguchi K. Specific interaction between an oligosaccharide on the tumor cell surface and the novel antitumor agents, sulfoquinovosylacylglycerols. Biochemical and Biophysical Research Communications. 2001;**288**:893-900. DOI: 10.1006/bbrc.2001.5852

[34] Brahmi MM, Portmann C, D'Ambrosio D, Woods TM, Banfi D, Reichenbach P, et al. Telomerase inhibitors from cyanobacteria: Isolation and synthesis of sulfoquinovosyl diacylglycerols from Microcystis aeruginosa PCC 7806. Chemistry--A European Journal. 2013;**19**:4596-4601. DOI: 10.1002/chem.201203296

[35] Tanaka T, Sawamoto Y, Aoki S. Concise and versatile synthesis of sulfoquinovosyl acyl glycerol derivatives for biological applications. Chemical & Pharmaceutical Bulletin. 2017;**65**:566-572. DOI: 10.1248/cpb.c17-00135

[36] Itoh T, Tamura K, Ueda H, Tanaka T, Satoh K, Kuroda R, et al. Design and synthesis of boron containing monosaccharides by the hydroboration of D-glucal for use in boron neutron capture therapy (BNCT). Bioorganic & Medicinal Chemistry. 2018;**26**:5922-5933. DOI: 10.1016/j.bmc.2018.10.041

[37] Heiden MGV, Cantley LC, Thompson CB. Understanding the Warburg effect: The metabolic requirements of cell proliferation. Science. 2009;**324**:1029-1033. DOI: 10.1126/science.1160809

[38] Calvaresi EC, Hergenrother PJ. Glucose conjugation for the specific targeting and treatment of cancer. Chemical Science. 2013;**4**:2319-2333. DOI: 10.1039/C3SC22205E

[39] Patching SG. Role of facilitative glucose transporter GLUT1 in [18F]FDG positron emission tomography (PET)

imaging of human diseases. Journal of Diagnostic Imaging in Therapy. 2015;**2**:30-102

[40] Nowotarski SL, Woster PM, Casero RA. Polyamines and cancer: Implications for chemoprevention and chemotherapy. Expert Reviews in Molecular Medicine. 2013;**15**:e3. DOI: 10.1017/erm.2013.3

[41] Murray-Stewart TR, Woster PM, Casero RA. Targeting polyamine metabolism for cancer therapy and prevention. The Biochemical Journal. 2016;**473**:2937-2953. DOI: 10.1042/BCJ20160383

[42] Hosmane NS, Maguire JA, Zhu Y, Takagaki M. Boron and Gadolinium Neutron Capture Therapy for Cancer Treatment. Singapore: World Scientific Publishing; 2012. p. 272

[43] Zhuo JC, Cai J, Soloway AH, Barth RF, Adams DM, Ji W, et al. Synthesis and biological evaluation of boron-containing polyamines as potential agents for neutron capture therapy of brain tumors. Journal of Medicinal Chemistry. 1999;**42**:1282-1292. DOI: 10.1021/jm960787x

[44] Shionoya M, Kimura E, Shiro M. A new ternary zinc(II) complex with [12] aneN_4 (= 1,4,7,10-tetraazacyclododecane) and AZT (= 3'-azido-3'-deoxythymine). highly selective recognition of thymidine and its related nucleosides by a zinc(II) macrocyclic tetraamine complex with novel complementary associations. Journal of the American Chemical Society. 1993;**115**:6730-6737

[45] Kimura E, Ikeda T, Aoki S, Shionoya M. Macrocylic zinc(II) complexes for selective recognition of nucleobases in single- and double-stranded polynucleotides. Journal of Biological Inorganic Chemistry. 1998;**3**:259-267. DOI: 10.1007/s007750050230

[46] Kikuta E, Murata M, Katsube N, Koike T, Kimura E. Novel recognition of thymine base in double-stranded DNA by zinc(II)–macrocyclic tetraamine complexes appended with aromatic groups. Journal of the American Chemical Society. 1999;**121**:5426-5436. DOI: 10.1021/ja983884j

[47] Kimura E, Kikuta E. Why zinc in zinc enzymes? From biological roles to DNA based-selective recognition. Journal of Biological Inorganic Chemistry. 2000;**5**:139-155. DOI: 10.1007/s007750050359

[48] Aoki S, Sugimura C, Kimura E. Efficient inhibition of photo[2 + 2] cycloaddition of thymidilyl(3′–5′) thymidine and promotion of photosplitting of the cis-syn- cyclobutane thymine dimer by dimeric zinc(II)–cyclen complexes containing *m*- and *p*-xylyl spacers. Journal of the American Chemical Society. 1998;**120**:10094-10102. DOI: 10.1021/ja981788c

[49] Kimura E, Kikuchi M, Kitamura H, Koike T. Selective and efficient recognition of thymidylylthymidine (TpT) by bis(Zn^{II}–cyclen) and thymidylylthymidylylthymidine (TpTpT) by tris(Zn^{II}–cyclen) at neutral pH in aqueous solution. Chemistry--A European Journal. 1999;**5**:3113-3123. DOI: 10.1002/(SICI)1521-3765(19991105)5:11<3113::AID-CHEM3113>3.0.CO;2-L

[50] Aoki S, Kimura E. Highly selective recognition of thymidine mono- and diphosphate nucleotides in aqueous solution by ditopic receptors zinc(II)–bis(cyclen) complexes (cyclen = 1,4,7,10-tetraazacyclododecane). Journal of the American Chemical Society. 2000;**122**:4542-4548. DOI: 10.1021/ja994537s

[51] Kikuta E, Aoki S, Kimura E. A new type of potent inhibitors of HIV-1 TAR RNA–Tat peptide binding by zinc(II)–macrocyclic tetraamine complexes. Journal of the American Chemical Society. 2001;**123**:7911-7912. DOI: 10.1021/ja0108335

[52] Kimura E, Kikuta E. Macrocyclic zinc(II) complexes for selective recognition of nucleobases in single- and double-stranded polynucleotides. Progress in Reaction Kinetics and Mechanism. 2000;**25**:1-64. DOI: 10.3184/007967400103165119

[53] Aoki S, Kimura E. Zinc–nucleic acid interaction. Chemical Reviews. 2004;**104**:769-788. DOI: 10.1021/cr020617u

[54] Kimura E. Evolution of macrocyclic polyamines from molecular science to supramolecular science. Bulletin of Japan Society of Coordination Chemistry. 2012;**59**:26-47. DOI: 10.4019/bjscc.59.26

[55] Ueda H, Suzuki M, Kuroda R, Tanaka T, Aoki S. Design, synthesis, and biological evaluation of boron-containing macrocyclic polyamines and their zinc(II) complexes for boron neutron capture therapy. Journal of Medicinal Chemistry. 2021;**64**:8523-8544. DOI: 10.1021/acs.jmedchem.1c00445

[56] Ueda H, Suzuki M, Sakurai Y, Tanaka T, Aoki S. Design, synthesis and biological evaluation of boron-containing macrocyclic polyamine dimers and their zinc(II) complexes for boron neutron capture therapy. European Journal of Inorganic Chemistry. 2022;**2022**:e202100949. DOI: 10.1002/ejic.202100949

Chapter 6

Boron Clusters in Biomedical Applications: A Theoretical Viewpoint

Ehsan Shakerzadeh, Elham Tahmasebi, Long Van Duong and Minh Tho Nguyen

Abstract

In this chapter, we presented an analysis of the recent advances in the applications of boron clusters in biomedical fields such as the development of biosensors and drug delivery systems on the basis of quantum chemical calculations. Biosensors play an essential role in many sectors, e.g., law enforcement agencies for sensing illicit drugs, medical communities for detecting overdosed medications from human and animal bodies, etc. The drug delivery systems have theoretically been proposed for many years and subsequently implemented by experiments to deliver the drug to the targeted sites by reducing the harmful side effects significantly. Boron clusters form a rich and colorful family of atomic clusters due to their unconventional structures and bonding phenomena. Boron clusters and their complexes have various biological activities such as the drug delivery, imaging for diagnosis, treatment of cancer, and probe of protein-biomolecular interactions. For all of these reactivities, the interaction mechanisms and the corresponding energetics between biomaterials and boron clusters are of essential importance as a basic step in the understanding, and thereby design of relevant materials. During the past few years, attempts have been made to probe the nature of these interactions using quantum chemical calculations mainly with density functional theory (DFT) methods. This chapter provides a summary of the theoretical viewpoint on this issue.

Keywords: boron clusters, drug delivery systems, biosensors, quantum chemical computations

1. Introduction

Nowadays, nanomaterials have been applied in most major scientific and industrial fields [1–5]. Such wide ranges of applications are possible owing to the opportuneness of the extremely different classes of nanomaterials with various novel properties. Noticeably, the biocompatibility of the nanomaterials is a great issue for the scientists to use them in the biomedical applications including, among others, biosensors and drug delivery systems.

A biosensor is a device that can produce a measurable signal proportional to the concentration of the biological analyte target [6, 7]. Biosensors are one of the most widely studied topics due to their contributions to development of innovative medicines, which could be applied as adapted drugs or highly sensitive detectors of disease markers [8–15].

Biosensors become new inventions that are hopeful to help an effective diagnosis in the current COVID-19 pandemic and also to remove experimental drugs during the human trials when they show any unwanted adverse effect [16–18]. Generally, a given biosensor has three components including a biological element, a transducer, and a detector [19]. The biological element leads to a detection of the analyte and a generation of a response. This response is thereafter transformed into a detectable signal through a transducer, which is often the most challenging part. Consequently, the generated signal is intensified and processed via an amplifier for exhibiting it by an electronic display device. **Figure 1** schematically illustrates the various steps of the signal processing in a biosensor.

Nanomedicine emerges as a revolutionary medical technology, particularly in the cancer therapy. Recently, much effort has been devoted to the study of nanostructures for applications in nanomedicine domain owing to their particular role in cancer therapies [20–26]. Undoubtedly, the most challenging task in cancer therapy is the finding of a suitable drug delivery system. As the design of efficient and promising drug delivery systems could be developed on the basis of nanostructures, a survey of the relevant prospective drug delivery agents constitutes a primordial subject [27–29].

A key requirement for a drug delivery system is that the delivery of the drug to the targeted sites needs to be associated with a considerable decrease in adverse effects. It is worth mentioning that the experimental research in this field is rather long and expensive, and thereby computational studies can effectively help experimentalists in the design of nanocarriers [30–45]. In this regard, the nature of the interactions between drugs and nanostructures emerges as an essential step. **Figure 2** represents a schematic boron-based drug delivery system.

Boron is an effective element in a wide range of fields. The history of boron chemistry started from the isolation of a series of simple boranes by Stock and his co-workers [46]. In the last two decades, several types of low-dimensional boron nanomaterials such as nanoclusters, nanowires, nanotubes, nanobelts, nanoribbons, nanosheets, and monolayer crystalline sheets have been experimentally synthesized and characterized [47–57]. These boron-based nanomaterials exhibit different bonding patterns from those of bulk boron crystals that exist as the α-, β-, γ-rhombohedral, and α-tetragonal forms. Accordingly, their resulting unique physical and chemical properties are fascinating from a standpoint of materials science. Noticeably, boron-based nanomaterials, such as clusters, can be used as superatoms or building blocks of other nanostructures with novel functionalities and properties.

Of the various types of boron-based nanomaterials, the pure boron clusters (BC) represent a distinctive category of structures owing to their unconventional structures and bonding patterns. During the past decades, boron-based compounds at the

Figure 1.
Steps of signal processing in a biosensor.

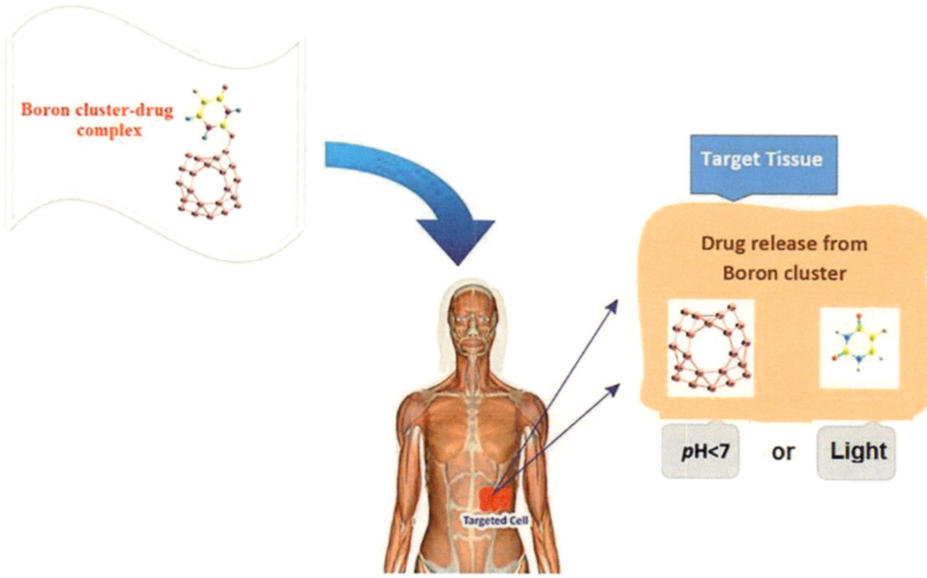

Figure 2.
A schematic boron-based drug delivery system.

nanoscale have been the subject of a large number of theoretical and experimental studies. These systems have intriguing features with different structures such as planar, quasi-planar, ribbon, bowl, cage, teetotum, tubular drum-like forms, multiple ring tubes, and fullerenes [58–72]. This arises from the fact that the boron atom with electron deficiency can take part in both localized and delocalized electronic systems in many geometric shapes. In other words, the most attractive nature of boron skeletons is due to the electron deficiency of the boron atom, leading to a rich bonding capacity.

The neutral B_n clusters with the size of smaller than 20 atoms prefer a planar or quasi-planar structure, except for B_{14} which has a fullerene-type [73]. The B_{40}, $B_{32}C_4$, and $B_{32}Si_4$ fullerene-like clusters together with the B_{30} and B_{36} bowls have attracted some interest in biomedical applications. The schematic structures of B_{40}, $B_{32}C_4$, $B_{32}Si_4$, and B_{36} are provided in **Figure 3**.

Tai et al. [74] reported a computational study on the structural, electronic properties, and chemical bonding of the bowl-like B_{30} global-minimum cluster, exhibiting a disk-aromaticity [11]. Similarly, the B_{36} was theoretically predicted to have a bowl shape stabilized by a disk aromaticity [75]. Piazza et al. [76] subsequently reported an experimental identification of the neutral B_{36} from the photoelectron spectrum of the B_{36}^- anion, confirming a highly stable quasi-planar boron cluster with a central hexagonal hole, providing the first experimental evidence that single-atom layer boron sheets with hexagonal vacancies are potentially viable. The neutral B_{36} is in fact the smallest boron cluster exhibiting a sixfold symmetry and a hexagonal hole, and it can be viewed as a potential basis for extended two-dimensional boron sheets. Recently, it was revealed that the B_{36}^{4-} cluster has a six-membered hole, but the presence of four extra electrons renders the considered system difficult to be synthesized [77]. Thus, the use of carbon or silicon atoms instead of boron anion to neutralize the extra electrons in the carbon or silicon-doped cluster (C_4B_{32} and Si_4B_{32}) has been suggested and comprehensively studied [78].

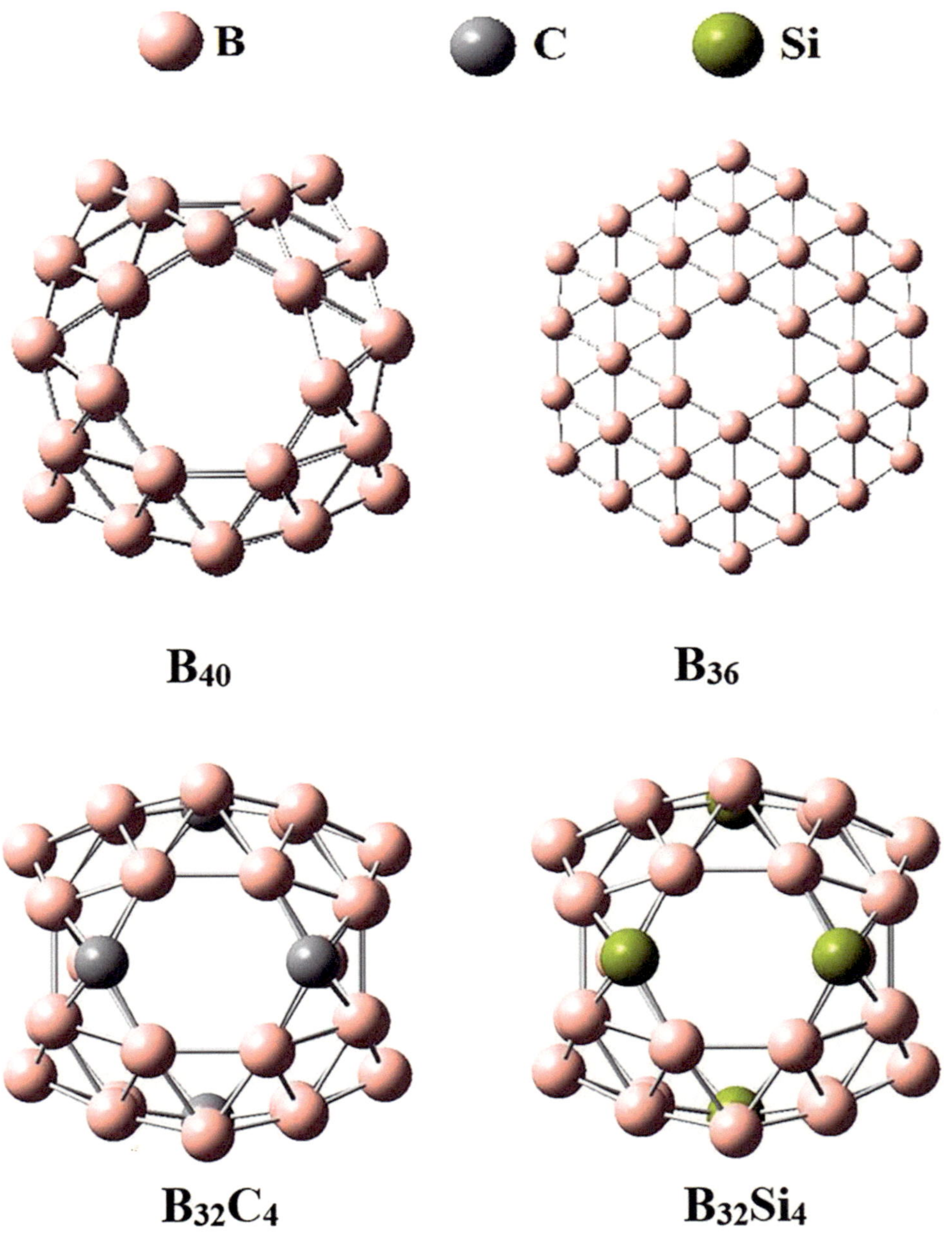

Figure 3.
Shapes of the B_{40}, $B_{32}C_4$, $B_{32}Si_4$ fullerenes, and B_{36} bowl clusters.

The fullerene B_{40} was also predicted by computations [75] and subsequently prepared [79] by utilizing a laser vaporization supersonic source and identified via photoelectron spectroscopy (PES). The B_{40} fullerene with a D_{2d} symmetry consists of four heptagonal rings and two hexagonal rings. With exceptional properties, it has been subjected to many theoretical studies due to its potential applications in molecular devices [80–85]. It is noteworthy that its electronic and reactivity features could be tuned via metal encapsulation or substitution.

Boron neutron capture therapy (BNCT) for cancer treatment remains the main biomedical application of boron-based compounds [86, 87]. Boron compounds have thus facilitated the mission of BNCT. Furthermore, novel biological activities of boron cages and their complexes have been reported [88, 89].

The drugs that are commonly explored for anticancer treatment include 5-fluorouracil (FU), metronidazole (ML), hydroxyurea (HU), nitrosourea (NU), 6-thioguanine (TG), melphalan (MP), and cisplatin and nedaplatin (cf. **Figure 4**). Some nitrosoureas have been used in chemotherapy for treatment of brain tumors, breast carcinoma, lymphomas, and leukemia. The MP drug is conventionally applied for the treatment of specific cancers such as multiple myeloma, ovarian cancer, and breast cancer. FU also has multiple applications and is one of the most beneficial drugs to date to treat breast, head, neck, anal, stomach, colon, and skin cancers [24]. Cisplatin, which is one of the most common anticancer chemotherapy drugs, is particularly effective in treatment of testis, ovary, esophageal, bladder, non-small-cell lung cancers, and head and neck malignancies [90, 91]. Nedaplatin is also an antineoplastic drug which is used for cancer chemotherapy with the

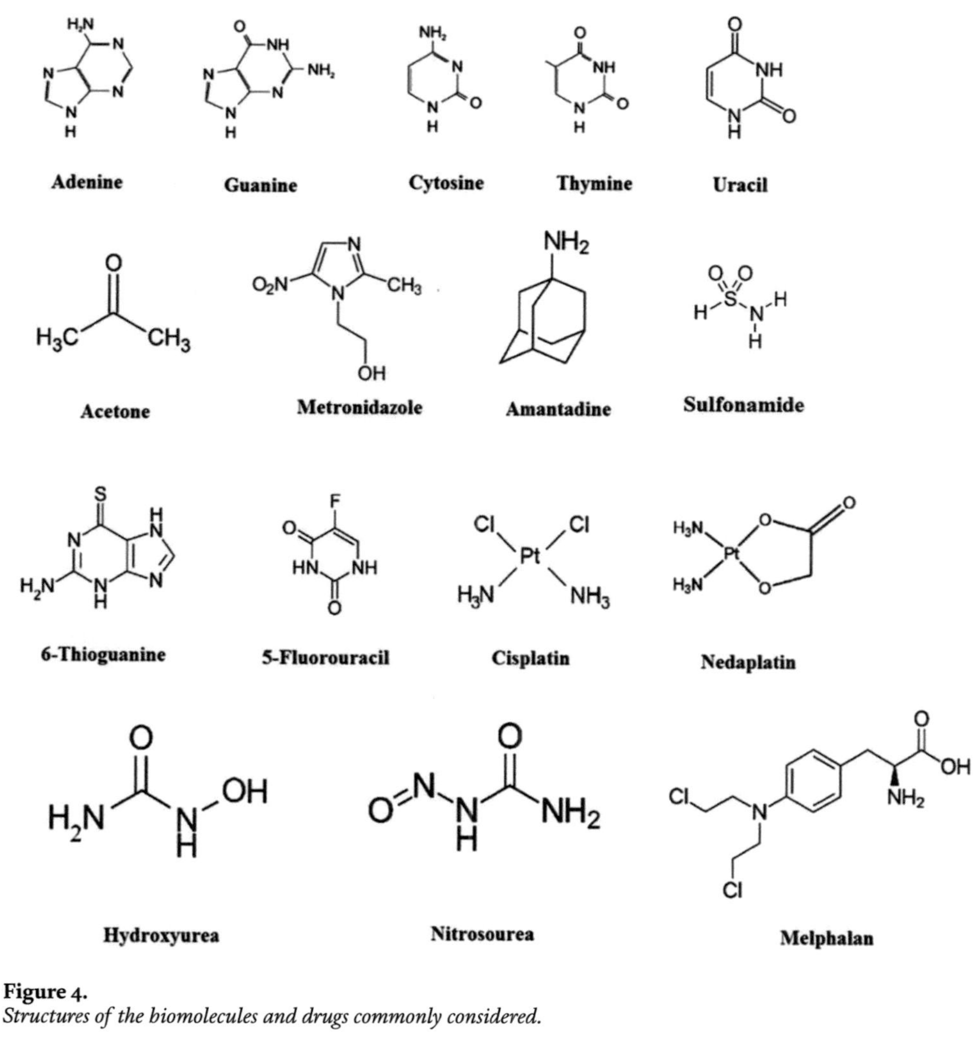

Figure 4.
Structures of the biomolecules and drugs commonly considered.

purpose to decrease the inherent toxicities induced by cisplatin [92]. However, the long-term use of such drugs may lead to some secondary tumors such as leukemia [93]. Hence, improvement of the efficacy and reduction of the toxicity of these drugs is of great importance. Of the diverse strategies recently put forward, the drug delivery is one of the most widely used techniques to improve the therapeutic efficiency and targeting of various drugs. In this context, the design of boron-based drug delivery systems appears to be an important issue for the beneficial usage of boron clusters.

The main contribution of this chapter is to scrutinize the functionality of calculated predictions for boron clusters to be considered as prospective biosensors or drug delivery systems. The theoretical methodologies will first be presented. A brief discussion on the various features of promising biosensors or drug delivery systems that should further be investigated for biomedical applications.

2. Methodology

In this section, a brief discussion is presented on the various features of the biosensors and drug delivery systems, that can be predicted using quantum chemical methods. Density functional theory (DFT) ranks as the most widely used quantum mechanical method and plays an increasingly larger role in a number of disciplines besides chemistry, such as physics, materials, biology, and pharmacy [94–103]. While DFT computations have long been used to complement experimental investigations, the approach has emerged as an indispensable and powerful tool for predictions of different fields.

A general theoretical approach to this topic boils down to an assessment of the interactions between the materials and the biomolecules or drugs considered. It simply leads to an examination of the structures and properties of the interacting complexes. This requires a determination of all possible configurations of the complexes by carrying out systematic geometry optimizations and making use of appropriate DFT methods. The nature of local energy minima corresponding to various configurations needs to be verified through an analysis of their vibrational frequencies. In order to assess the capability of a boron cluster for detection of a biomolecule or drug delivery system, the structural, energetic, and electronic properties can simply be computed for the relaxed favorable geometries. These properties can provide us with valuable information for biomedical applications. All the mentioned calculations can be performed in both vacuum and aqueous media. It is essential to evaluate these parameters in aqueous medium since these systems are anticipated to act in human body. The polarizable continuum model (PCM) and the conductor-like screening model (COSMO) are common continuum models for the treatment of the solvent effects. The key factors of the properties mentioned above are described as follows.

2.1 Structural parameters

From an optimized geometry, the bond lengths and bond angles between the constituent atoms in the complexes can be determined. These are simple but essential parameters determining the nature of the interaction between the drug molecules and respective adsorbents.

2.2 Energetic properties

The interaction energy (E_{int}) of a biomolecule or drug with a boron cluster is the key parameter that should be determined in order to emphasize the nature of the interaction. The interaction energy is usually computed as in Eq. (1):

$$E_{\text{int}} = E_{BC-\text{adsorbent}} - E_{\text{BC}} - E_{\text{adsorbate}} \tag{1}$$

where $E_{\text{BC–adsorbent}}$ denotes the total energy of the adduct formed upon interaction between the boron cluster with the corresponding drug or biomolecule. The E_{BC} and $E_{\text{adsorbent}}$ terms correspond to the total energies of the isolated boron cluster and the drug or biomolecule, respectively. These energies are usually calculated using DFT methods. The efficiency of DFT methods, namely the functionals, for evaluating interaction energies was documented previously. The reported results show a good performance with the root-mean-square deviation of 0.05 kcal/mol and the mean absolute deviation of 0.07–0.13 kcal/mol against the benchmark energies of N-methylacetamide-water complex obtained at the CCSD (T)/CBS-aTQ complete basis set limit level [104]. N-methylacetamide is the simplest model for the peptide linkage in peptides and proteins.

As for a convention, a negative interaction energy indicates that the obtained complex is thermodynamically stable, while a positive adsorption energy refers to a local minimum where the interaction is prevented by an energy barrier connecting it with the global minimum. The interaction energy can provide us with meaningful insights to distinguish between a chemisorption and a physisorption process.

The recovery time (τ) is one of the important factors for biomedical applications. It can be used for estimation of the drug desorption from the cluster surface or the sensor refreshing, which can be occurred by exposing to light. Based on the conventional transition state theory, the recovery time can be computed using the Arrhenius-type Eq. (2):

$$\tau = \nu_0^{-1} \exp\left(\frac{-E_{\text{int}}}{kT}\right) \tag{2}$$

where ν_0, T, and k terms stand for the attempt frequency, the temperature of the system, and the Boltzmann constant, respectively. A larger interaction energy inherently leads to a longer recovery time, which is not a good factor for drug release or for a biosensor refreshing. Thus, the adsorption process energy should be neither chemisorption nor physisorption; it should be in a semi-chemisorption to provide an efficient recovery time. Accordingly, an interaction characterized by *a large interaction energy is not always favorable for biomedical applications.*

It is possible to investigate the thermodynamical nature of the interaction, through the change in the Gibbs energy using Eq. (3):

$$\begin{aligned} \Delta G &= G_{BC-\text{adsorbent}} - G_{\text{BC}} - G_{\text{adsorbate}} \\ &= \left(H_{BC-\text{adsorbent}} - H_{\text{BC}} - H_{\text{adsorbate}}\right) - T\left(S_{BC-\text{adsorbent}} - S_{\text{BC}} - S_{\text{adsorbate}}\right) \end{aligned} \tag{3}$$

where G represents the sum of electronic and thermal free energies. H stands for the sum of electronic and thermal enthalpies. S and T refer to entropy and temperature, respectively. Computations of the S and G quantities are carried out using the electronic, rotational, and vibrational parameters associated with the equilibrium structures according to the well-known thermochemical equations. A negative change in the Gibbs energy (free energy) represents a spontaneous interaction between the adsorbate molecule and adsorbent, which is desired in both drug delivery and drug sensor devices.

It is worth mentioning that a drug release from a carrier in the target cell is the most vital step in a drug delivery process. Owing to the excessive lactic production, a cancer cell is generally more acidic than normal cells (pH < 7) [105]. Thus, it is crucial to examine the performance of anticancer drug delivery systems in a low pH cancerous cell region for a better evaluation of the drug release performance of the nanostructure in the targeted region. This is well-known as the pH-dependent drug release mechanism [106].

Furthermore, the photochemical mechanism of light-triggered release from nanocarriers is also well known. Distinct wavelengths including ultraviolet (UV, 200–400 nm), visible (400–750 nm), and near-infrared (NIR, 780–1700 nm) lights, can be utilized to activate the light responsiveness [107]. Although the UV light is a relatively poor candidate due to its limited tissue penetration capacities and potentially carcinogenic effects under prolonged exposure, the NIR light has the advantages of lower phototoxicity, improved penetration depth in biological tissues, and reduced background signal. Thus, it is more suitable for biological applications. The NIR light is regarded as a transparent therapeutic window for light-activated delivery system in vivo due to its deep tissue penetration and minimum cellular damage [108]. The recovery time (Eq. (2)) could provide a theoretical estimation for light controlled release mechanism.

2.3 Electronic properties

The electronic properties investigation is usually performed using the HOMO-LUMO gap as a quantum descriptor, to establish correlation in various chemical and biochemical systems. The HOMO-LUMO gap E_g values are considered to explore the electronic properties and reactivities of the complexes formed upon interaction. This parameter is simply calculated by the following operational Eq. (4):

$$E_g = \varepsilon_{LUMO} - \varepsilon_{HOMO} \tag{4}$$

where ε_{HOMO} and ε_{LUMO} are the energies of the highest occupied molecular orbital (HOMO) and the lowest unoccupied molecular orbital (LUMO), respectively.

The electrical conductivity is exponentially related to the energy gap in a semiconductor material as follows (5):

$$\sigma = AT\frac{3}{2}e\left(-\frac{E_g}{2kT}\right) \tag{5}$$

where A (electrons/m^{3}K$^{3/2}$) is a pre-factor constant, k is the Boltzmann's constant, and T is the absolute temperature. This equation has frequently been used and previously demonstrated to yield results in agreement with experiment. The change in energy gap is a proper pointer for identification of the presence and attachment of a drug or biomolecule to a substrate.

Furthermore, the charge transfer between the adsorbate molecules and the adsorbent is generally performed through the natural bond orbital (NBO) or Hirshfeld population analyses. The amount of charge transfer plays an essential role in the development of a biosensor device. It helps determine the capability of a boron cluster in generating a detectable electrochemical signal on the presence of a biomolecule or a drug.

The electronic dipole moment is also an important issue for design of nanocarriers. The dissolvability of a nanostructure into a polar medium, such as an aqueous solution, can be explored using the dipole moment (μ). It plays a vital part in the design of a drug delivery device. A dipole moment enhancement is necessary for their solubility in a polar solvent. An increase in the hydrophilicity upon formation of the complex is a valuable factor for the efficient drug delivery system.

In summary, the structural, energetic, and electronic parameters necessary for the design of relevant materials are the basic molecular properties that can easily be determined using simple quantum chemical computations.

3. Boron clusters for biomedical applications

In this section, we discuss the studies reported on boron clusters in two separate categories, biosensors and drug delivery systems.

3.1 Biosensor applications

Kaur and Kumar [109] proposed a B_{40}-based biomarker for DNA sequencing from the results of DFT calculations using the Perdew-Burke-Ernzerhof (PBE) functional along with a double-zeta polarized basis set (DZP). These authors reported that all nucleobases are adsorbed on the surface of the B_{40} fullerene with the interaction energies of −18, −15, −16, and −23 kcal/mol for adenine, thymine, cytosine, and guanine, respectively. No complex between the nucleobases and B_{40} was visualized. The analysis of transmission spectra, density of states, and eigenstates of the HOMO and LUMO revealed that all molecular junctions show transmission dominated by the HOMO. The highest energy gap was found in the adenine molecular junction, and this molecule gives the least value of current in comparison to the other molecular junctions.

Thus, by analysis of differential conductance curves for all the nucleobase-B_{40} junctions, it is deduced that the values of conductance are different from each other for all the junctions considered. This implies that B_{40} can appropriately be used as a biomarker for DNA sequencing applications, in predicting the sequence of nucleobases in a DNA strand. As another direct application, B_{40} can thus be employed as a multipurpose sensor for detection of the DNA nucleobases.

Kaur and coworkers explored in 2022 [110] the interaction of uracil on B_{40} utilizing DFT (PBE/DZP) and nonequilibrium Green's function regime computations. The physisorption phenomenon of the uracil molecule on the B_{40} surface is found, with an interaction distance of 2.38 Å and an interaction energy of −19 kcal/mol. No orbital overlapping exists between uracil and B_{40} moiety according to an electron density analysis. The HOMO–LUMO energy gap of B_{40} decreases upon adsorption of uracil.

Although these authors suggested B_{40} as an effective biomarker to detect the presence of uracil molecule and thereby the mutations and cancerous tumors, the nature of the interaction is not well understood yet.

Rastgou et al. [111] examined the sensing ability of the quasi-planar B_{36} toward DNA nucleobases that might be used in a DNA sequencing device. The interaction energies for the most stable configuration of each complex were computed to be −57, −43, −38, and −10 kcal/mol for adenine-B_{36}, guanine -B_{36}, cytosine-B_{36}, and thymine-B_{36}, respectively. It was found from DFT calculations using the B97D/6-31G(d) method that the cytosine interacts more considerably with the edge of B_{36} than other nucleobases, resulting in a large decrease in the energy gap, by 96% with respect to the isolated cluster. Such a decrease in the energy gap was observed at 36, 20, and 15% for thymine, adenine, and guanine, respectively. As a result, a change in conductivity could allow cytosine, followed by thymine, adenine, and guanine to be detected.

In particular, acetone ($CH_3C{=}OCH_3$) in the human breath exhaust is one of the commonly considered biomarkers for type-I as well as type-II diabetes. Yong and coworkers [112] studied in 2018 the potential capability of B_{40} and the doped M@B_{40} (M = Li and Ba) as acetone gas sensors using DFT calculations at the PBE/DZP level. The @ symbol stands hereafter for an encapsulation. The acetone molecule can easily adsorb on the B_{40}, Li@B_{40}, and Ba@B_{40} clusters with interaction energies of −16, −19, and − 8 kcal/mol, respectively. The recovery times were computed at 9.2 seconds for Li@B_{40} and 1.2 seconds for Ba@B_{40}. Such a recovery time can be considered to be relatively long, as compared to a spectroscopic signal at the order of a microsecond, but it could be suitable for a sensor. The HOMO-LUMO gaps of M@B_{40} again decrease upon acetone adsorption. Accordingly, the change in eclectic conductance of Li@ B_{40} or Ba@B_{40} before and after the adsorption of acetone would be very distinctive, exhibiting the high sensitivity of M@B_{40} for sensing acetone. Thus, the B_{40} and M@ B_{40} were introduced as highly sensitive molecular sensors for acetone detection, but the recovery time is relatively long at the order of a second.

The quasi-planar B_{36} was further explored for prospective sensing of the metronidazole (ML, cf. **Figure 4**) drug, which is an antibiotic drug with widespread usage but can cause unwanted hazardous effects on the human body. DFT calculations at the B3LYP-D3/6-31G(d) level demonstrated that ML interacts more strongly with B_{36} by its edge with an adsorption energy as high as −22 and −21 kcal/mol in both gaseous and aqueous phases, respectively. The change in Gibbs energy of −19 kcal/mol implies spontaneous adsorption. The decrease of 64% in the energy gap upon complexation is considerable, resulting in a substantial increase in the conductivity of the structure. The recovery time of the sensor was further found to be as 1.5 s for the most stable adsorption complex at room temperature. Again, such a time is rather long, but these results could be used to develop a boron-based sensor to detect the ML drug [113] in more appropriate time.

3.2 Drug delivery application

Solimannejad and coworkers investigated in 2018 [114] the possible complexes generated from the interaction between the amantadine drug (cf. **Figure 4**) and the bowl-like B_{30} using the DFT ωB97XD/6-31G (d, p) method in both gaseous and aqueous media. Amantadine drug has been used to treat the Parkinson's disease, influenza, or hepatitis for many years, even though in some cases it can cause some impairment of corneal endothelial function or corneal edema. The strongest interaction occurs between an edge boron atom of the B_{30} and an N atom of amantadine with binding energy −46 and −53 kcal/mol in both gaseous and aqueous phases, respectively. The energy gap of the complex is remarkably reduced in both phases, with respect to the separated B_{30}. Thus B_{30} is quite sensitive to the presence of amantadine drug molecule, in such a way that it may be used in the sensor technology and possible drug delivery for amantadine for medicinal applications.

The interaction of fluorouracil (FU) with the quasi-planar B_{36} cluster was studied in 2017 [115] using the hybrid TPSSh functional with the 6–31 + G (d) basis set. The FU drug failed to generate any noticeable signal owing to the very weak interaction of this drug with the concave and convex surface of B_{36} ranging from −2 to −5 kcal/mol. Meanwhile, the FU drug remarkably interacts at its O atom site on the edge of the B_{36} with interaction energy of −24 and −27 kcal/mol in the gaseous and aqueous media, respectively (cf. **Figure 5**). The FU drug can also be detected by the B_{36} cluster with a noticeable signal owing to a significant decrease of 47% in the energy gap with respect to the free cluster. The dipole moment of FU-B_{36} complex was also observed as high as 17 and 36 Debye in the gas and water media, respectively, which indicates a large increase of the solubility in a polar medium.

Kamalinahad et al. performed in 2020 [116] a study on the interactions between sulfonamide (cf. **Figure 4**) and the B_{36} nanocluster through M06-2X/6-31G (d, p) computations. As a functional group, sulfonamide exists in several classes of drugs. Sulfonamide remarkably tends to adsorb via its oxygen atoms at the edge of B_{36}, alike FU drug, with interaction energy of −15 kcal/mol in both gaseous and aqueous

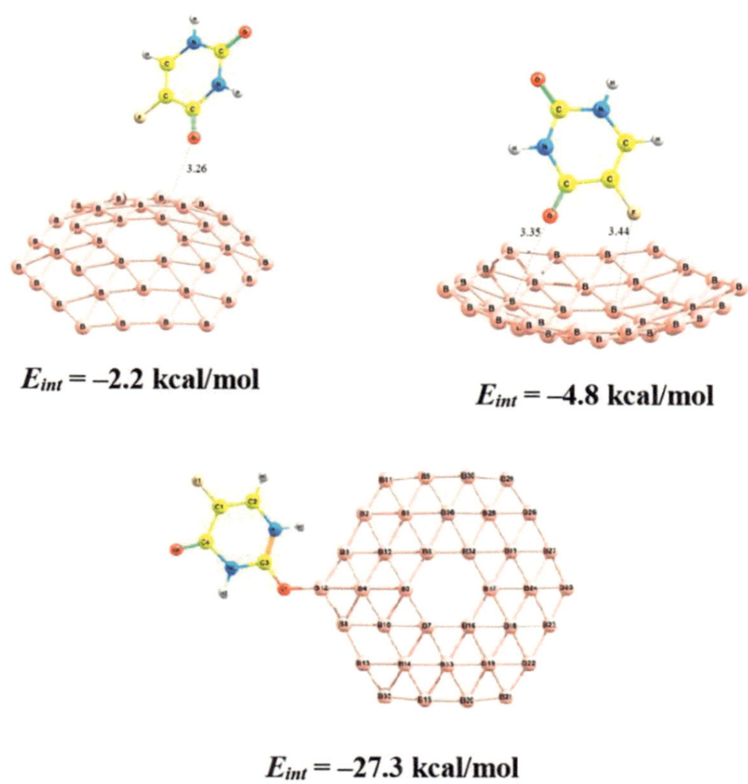

Figure 5.
Configurations of the interaction in the FU-B_{36} complex. Values given are the interaction energies obtained by TPSSh/6-31G (d.p) computations.

media. The results illustrate that the edge B atoms are more reactive than the inner atoms toward the sulfonamide molecule leading to some large changes in its electronic features. The dipole moment of the complex increases to 13 Debye with respect to 4 Debye for the bare B_{36} cluster in aqueous medium. The high polarity together with appreciable adsorption energy suggested that these systems could be a vehicle for drug delivery.

Zheng et al. in 2020 [117] reported DFT computations at the PBE0/6–31 + G (d) level on the pristine and amino acid-functionalized C_4B_{32} fullerene as drug delivery agents for hydroxyurea (HU, cf. **Figure 4**) anticancer drug. These authors found that an alanine functionalization can significantly enhance the tendency of the carbon-doped C_4B_{32} cluster to the adsorption of HU. In this regard, the drug adsorption on the B atom of the clusters is more favorable than on the others. Indeed, the adsorption of HU drug on the cage part of the ala-C_4B_{32} isomers is stronger than that adsorbed on the alanine within a range of −16 to −19 kcal/mol in gas phase. Also, more negative adsorption occurs in aqueous medium, ranging from −20 to −23 kcal/mol, whose solubility can modify their interactions with the HU drug. The interactions between the HU drug and the clusters in the acidic condition become weak, and thereby the drug can faster be released from the carrier.

Yunyu and Jameh-Bozorghi [118] reported a DFT study at the PBE0/6–31 + G (d) level on the endohedral fullerenes Li@C_4B_{32} and Li@Si_4B_{32} as materials for drug delivery applications of the 6-thioguanine (TG) anticancer drug. These authors suggested the pristine and Li-encapsulation C_4B_{32} and Si_4B_{32} clusters as suitable for drug delivery applications. Calculated interaction energies were found to be −42, −56, −38, and −43 kcal/mol for the TG/C_4B_{32}, TG/Li@C_4B_{32}, TG/Si_4B_{32}, and TG/Li@Si_4B_{32} complexes, respectively. Such interaction energies are however quite large.

In fact, the strongest feature of the studied complexes bonding was found for TG/Li@C_4B_{32} with the maximum positive charge on B atoms, and the system with LUMOs orbitals distributed on B atoms that has been predicted as the most favorable site for the nucleophilic agents. Moreover, their computed ultraviolet–visible spectra reveal that the electronic spectra of the drug/cluster complexes exhibit a red shift toward higher wave lengths (lower transition energies). Furthermore, the interaction of TG with the clusters leads to narrower E_g values resulting again in an increase in conductivity. The effect of pH on the TG/Li@C_4B_{32} pointed out that the interaction energy in the acidic environment tends to decrease from 56 to 30 kcal/mol. Hence, the interactions between the drug and Li@C_4B_{32} become again weaker in an acidic medium.

The alkali metal encapsulated fullerenes M@C_4B_{32} with M = Li, Na, and K were considered as drug carrier agents for nitrosourea (NU) anticancer drug (cf. **Figure 4**) on the basis of calculations carried out using the PBE0/6–31 + G (d) approach [119]. A comparison between the interaction energies reveals that a potassium encapsulation inside C_4B_{32} can considerably enhance the tendency of cluster for adsorption of NU drug with an interaction energy of −37 kcal/mol. In this case, the interaction energy tends to increase to −41 kcal/mol in aqueous medium, and thereby the K@C_4B_{32} cluster can increase its solubility and modify its interaction with the NU drug. The *p*H-dependent mechanism for drug release was also explored in which the proton (H^+) species attached to the NU. Results showed that the interaction between the NU drug and the K@C_4B_{32} in an acidic environment is weaker with an interaction energy of −20 kcal/mol. Hence, the NU drug could better be released from a carrier in the targeted cancer cell in an acidic environment.

Furthermore, Luo and Gu [120] explored the ability of C_4B_{32} and Si_4B_{32} together with the Li encapsulated clusters for cisplatin (cf. **Figure 4**), using the PBE0/6–31 + G

(d) level leads to interaction energies of −28, −12, −18, and −11 kcal/mol for the cisplatin/C_4B_{32}, cisplatin/Li@C_4B_{32}, cisplatin/Si_4B_{32}, and cisplatin/Li@Si_4B_{32} complexes, respectively. The interaction distance for the cisplatin/C_4B_{32} is relatively short (1.86 Å) in spite of relatively small interaction energy. Also, a blue shift toward lower wavelengths (larger transition energies) was observed from ultraviolet-visible spectra. Noticeably larger adsorption energies (more negative) are found in the solvent phase.

Sun and coworkers [121] explored the adsorption behavior of FU drug on B_{40} and some derivatives including MB_{39} and M@B_{40} (M = Mg, Al, Si, Mn, Cu, Zn). These authors applied calculations using the B3LYP functional in conjunction with the SDD basis set with effective core potential for Cu, Mn, and Zn atoms and 6-31G(d) basis set for the other atoms. Accordingly, the FU drug prefers to attach to the corner boron atom of the B_{40} through one of its oxygen atoms, resulting in a strong polar covalent B–O bond. The corresponding interaction energy is calculated to be −11 kcal/mol. Additionally, the ΔH and ΔG values for the interaction of FU drug via B_{40} are both negative. Furthermore, they found that FU-B_{40} complex exhibits a much larger dipole moment of 9 Debye than those of 6 and 0 Debye for 5-FU and B_{40}, respectively, resulting in an increase in polarity for the whole system, and thus, enhancing the solubility of the resulting FU-B_{40} in an aqueous medium.

The drug release was also studied through a pH-dependent mechanism approach. The influence of pH on the FU-B40 complex was further examined by approaching a proton to the O atom of FU in complex. As seen in **Figure 6**, the distance between the O and B atoms greatly increases from 1.55 to 4.05 Å during the structural optimization. As a result, the interaction energy of FU-B_{40} severely decreases from −11 to −5 kcal/mol in the acidic environment, reflecting that the interaction between FU and B_{40} cluster is distinctly weakened under the attack of a single proton. Therefore, the FU drug can be released from the B_{40} carrier within the targeted tumor tissue where the medium is more acidic.

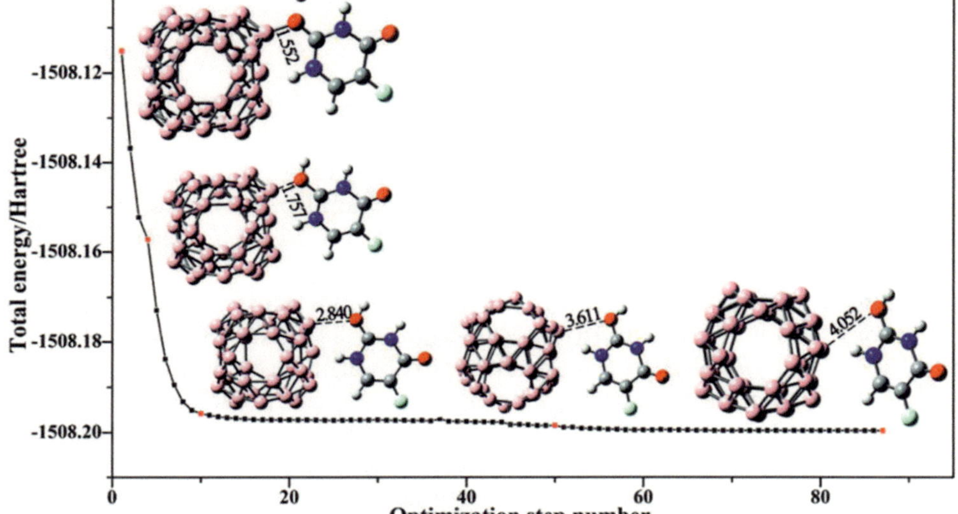

Figure 6.
Optimization process for the protonation of FU drug adsorbed on B_{40} cluster. The distances (in Å) between B and O atoms are also given. Figure reprinted with permission from ref. [121].

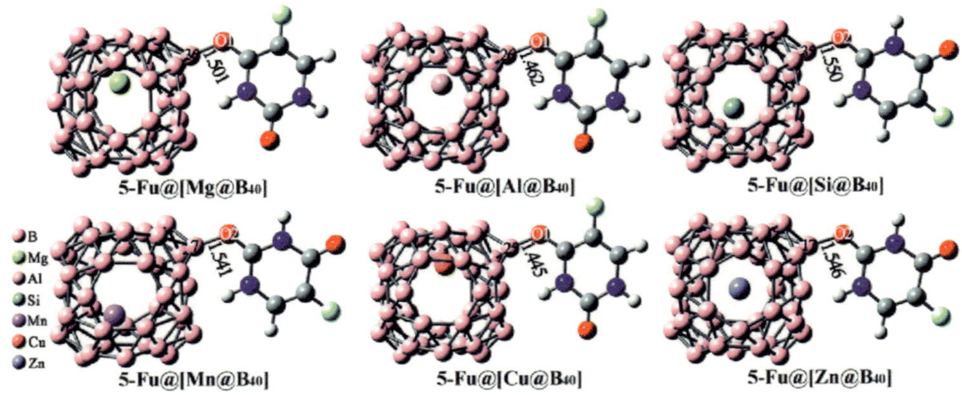

Figure 7.
Optimized geometries of the most stable FU-[M@B$_{40}$] with M = Mg, Al, Si, Mn, Cu, and Zn complexes. The lengths of the newly formed bonds (in Å) are also given. Figure is reprinted with permission from ref. [121].

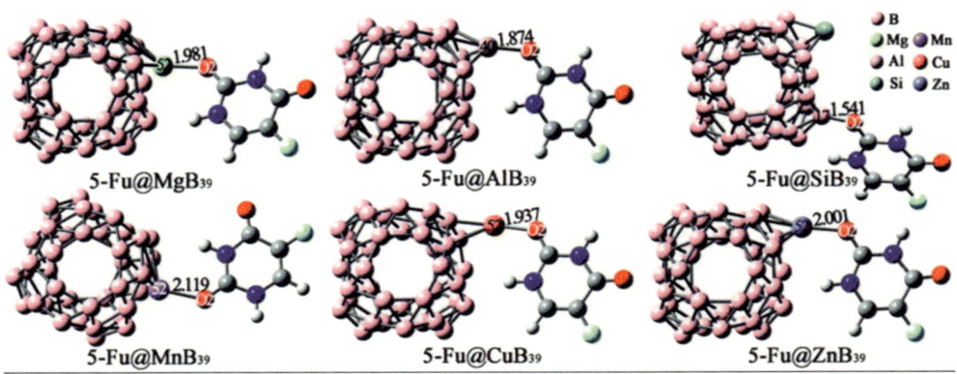

Figure 8.
Optimized geometries of the most stable FU-B$_{39}$M (M = Mg, Al, Si, Mn, Cu, and Zn) complexes. The lengths of newly formed bonds (in Å) are also given. Figure is reprinted with permission from ref. [121].

Additionally, the substituent and encapsulation effects of Mg, Al, Si, Mn, Cu, and Zn atoms on the drug delivery performance of B_{40} have been also explored. The FU oxygen atom tends to combine with MB_{39} or $M@B_{40}$ cages, which are depicted in **Figures** 7 and **8**, respectively. Interaction energies vary in the ordering (values in kcal/mol) –16 (FU-[Al@B_{40}]) –16 (FU-[Mg@B_{40}]) > –15 (FU-[Cu@B_{40}]) > –13 (FU-[Mn@B_{40}]) > –12 (FU-[Zn@B_{40}]) > –12 (FU-[Si@B_{40}]).

Meanwhile, the variation of interaction energies for the substituted complex is in the ordering of –30 (FU-B_{39}Al) –22 (FU-B_{39}Mg) > –13 (FU-B_{39}Cu]) > –12 (FU-B_{39}Zn) > –12 (FU-B_{39}Mn) > –9 kcal/mol (FU-B_{39}Si). The absorption of FU on B_{39}M or M@B_{40} cages is more favorable than pristine B_{40} except for SiB_{39}. Therefore, the encapsulation and substitution of impurities can be regarded as an efficient approach to control and/or tune-up the interaction between the FU and B_{40}.

Sun and coworkers [122] explored the potential application of all-boron fullerene B_{40} as a drug carrier for anti-cancer nitrosourea (NU, cf. **Figure 4**) by means of PBE0/6-31G (d, p) computations. The NU drug tends to combine with a corner B atom of the B_{40} cage via its oxygen and nitrogen atoms with a moderate adsorption energy of –25 kcal/mol. The E_g value is decreased remarkably following adsorption of NU drug because this raises its HOMO and reduces its LUMO level. However, a long recovery time of 52 seconds was predicted for the NU desorption process at 310 K, indicating quite long and difficult desorption of NU from B_{40} at human body temperature.

Moreover, B_{40} with a high drug loading capacity can simultaneously carry up to five NU drug molecules. Additionally, the substituent effect of C, N, Al, and Ga atoms on the drug delivery performance of this B_{40} cluster was investigated. The interaction energies vary in the sequence of −68 kcal/mol (NU-B_{39}C) > −37 (NU-B_{39}Al) > −19 (NU-B_{39}Ga) > −18 (NU-B_{39}N). Also, the dipole moments were greatly enlarged from 1 to 6 Debye of B_{39}M to 15–21 Debye of NU-B_{39}M (M = C, N, Al, and Ga). Therefore, it can be deduced that substitution of one boron atom of B_{40} by an exogenous atom indeed induces an obvious influence on the interaction between B_{40} and NU drug. As a result, the substituent effect of foreign atoms can be employed to modulate or tune up the drug adsorption performance of B_{40} cluster.

Interaction between the FU anticancer drug and the B_{40} fullerene was also investigated using the PBE-D/DZP level in both the gaseous and aqueous phases [123]. Results indicate that the FU molecule remarkably adsorbs on the top of B_{40} through its oxygen atom with moderate interaction energy of −24 kcal/mol (cf. **Figure 9**). The energy gap value of the FU-B_{40} complexes is relatively decreased by 21% as compared to the isolated B_{40} fullerene. The HOMO-LUMO gap of B_{40} amounts to 1.8 eV, which is reduced to 1.4 eV in FU-B_{40}. Thus, the adsorption of the FU molecule can be identified from electronic response, resulting from the decrease of electric conductivity. Furthermore, the FU molecule bears a Hirshfeld charge of 0.35 a.u. in complex, resulting in a charge-transfer complex, in which the charge is effectively transferred from the FU molecule to the B_{40} fullerene.

The capacity of B_{40} for carrying the FU drug was explored. All the six holes of B_{40} interact with FU molecules and the corresponding 6FU-B_{40} complexes in both gas phase and aqueous solution are achieved. The interaction energy was estimated to be −13 kcal/mol per FU drug molecule in both phases. Moreover, the energy gap is

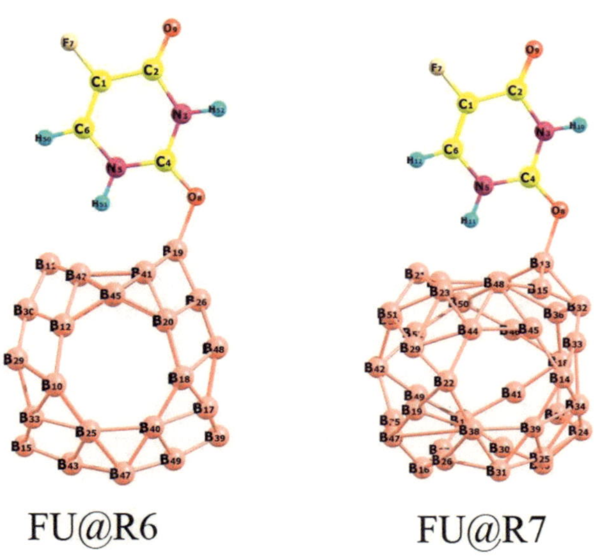

Figure 9.
Optimized B_{40}-FU complexes. Figure is reprinted with permission from Ref. [123].

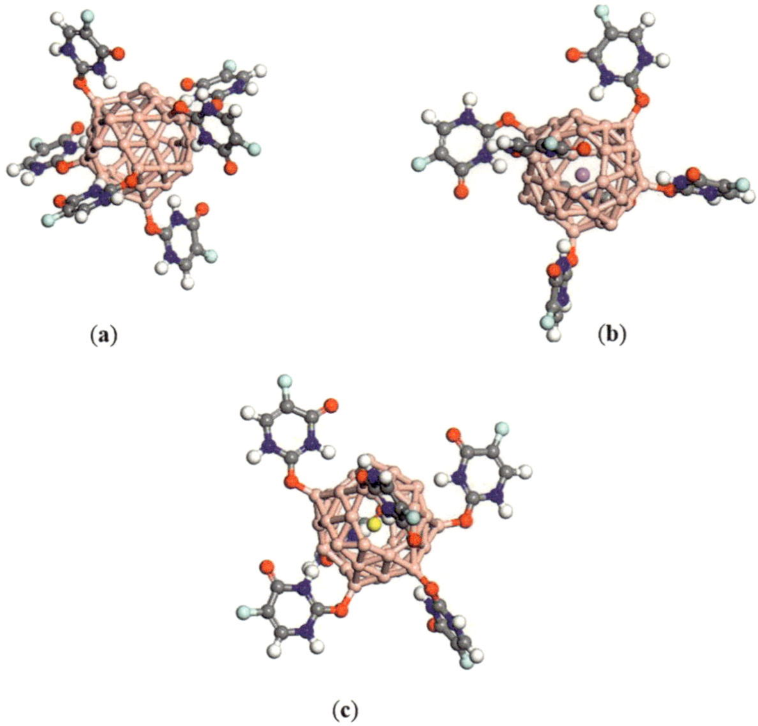

Figure 10.
Optimized geometries of (a) 6FU-B_{40}, (b) Na@B_{40}-6FU, and (c) Ca@B_{40}-6FU complexes. Figure reprinted with permission from ref. [123].

distinctly decreased for this complex. The 6FU-B_{40} system has E_g = 0.35 eV in the gas phase and E_g = 0.71 eV in the aqueous phase.

The effect of Na and Ca encapsulation inside the B_{40} cluster on the FU adsorption behavior was also examined (cf. **Figure 10**). The interaction energy per FU molecule becomes now about −31 kcal/mol for 6FU-Na@B_{40} and 6FU-Ca@B_{40} systems in solution. Noticeably, the dipole moments enhance for the studied complexes in both phases. Further studies are needed to evaluate, in particular the recovery times, as to whether these fullerenes might behave as innovative boron-based candidates as drug delivery systems.

DFT (PBE-D/DZP) calculations were performed to investigate the interaction between the melphalan (MP; cf. **Figure 4**) as a chemotherapy medication and the bare as well as Na and Ca endohedral encapsulated B_{40} fullerenes (M@B_{40} with M = Na and Ca) [124]. The interaction energy of one MP drug with B_{40} was computed to be −15 kcal/mol. This interaction is a charge-transfer type occurring from the drug to the fullerene. The simultaneous adsorption of six MP molecules onto the fullerenes was also studied. An interaction energy of −4 kcal/mol per MP is obtained for 6MP-B_{40} system. Thus, it is deduced that the bare B_{40} fullerene suffers from a low adsorption energy per MP molecule in gas phase when it is fully loaded by MP drugs.

In order to improve the absorbency of B_{40} toward MP drug, the Na and Ca-encapsulated Na@B_{40} and Ca@B_{40} could yield some improvement (cf. **Figure 11**). The interaction energy per MP molecule increases now to −9 kcal/mol for the encapsulated fullerene in gas phase. Also, the dipole moment is enhanced in both gaseous

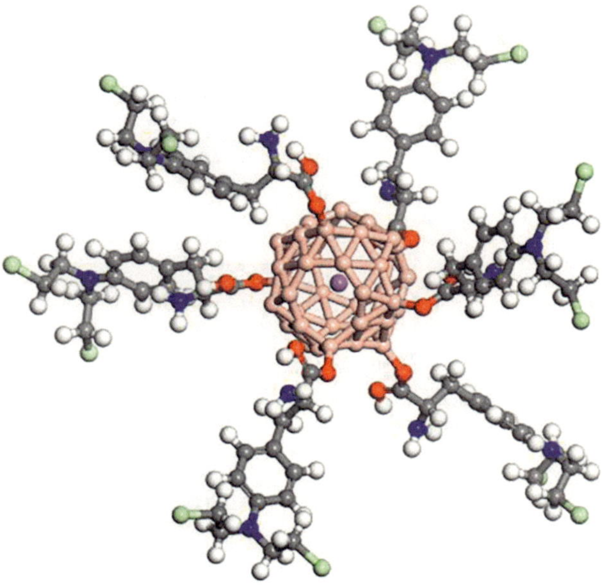

Figure 11.
Optimized geometries of the M@B_{40}-6MP complexes.

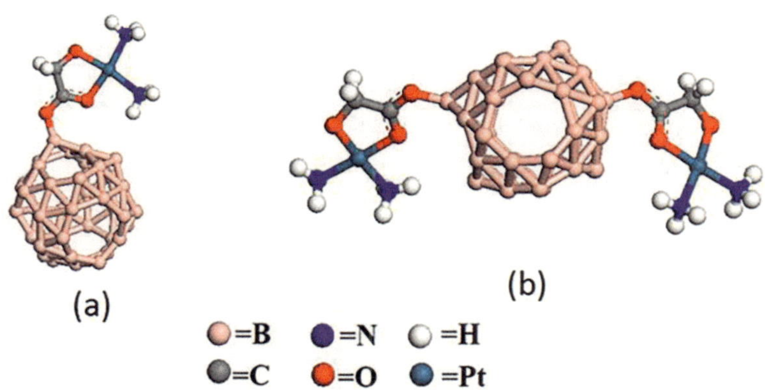

Figure 12.
Optimized geometries of (a) NedaPt-B_{40} and (b) 2NedaPt-B_{40} complexes. Figure reprinted with permission from Ref. [125].

and aqueous phases for the resulting complexes, which is a crucial factor for the design of a drug carrier. The release of the MP drug from the carrier surface could be occurred through a *p*H-dependent mechanism.

The interaction between the nedaplatin anticancer drug (cf. **Figure 4**) with the B_{40} fullerene was also explored using PBE-D/DZP calculations in both vacuum and water mediums [125]. The nedaplatin molecule remarkably tends to adsorb on the top of B_{40} through its oxygen atom with an interaction energy of −18 kcal/mol. The adsorption of two nedaplatin molecules onto the fullerene holes has occurred with an interaction energy of −14 kcal/mol per drug molecule (cf. **Figure 12**).

Furthermore, reported results illustrate that the Li and Na encapsulation into B_{40} greatly increases the adsorption of nedaplatin in both the gaseous and solution

Adsorbate	Substrate	Level of theory	E_{int} (kcal/mol)	Ref.
Adenine	B_{40}	PBE/DZP	–18	[111]
Thymine	B_{40}	PBE/DZP	–15	[111]
Cytosine	B_{40}	PBE/DZP	–16	[111]
Guanine	B_{40}	PBE/DZP	–23	[111]
Uracil	B_{40}	PBE/DZP	–19	[110]
Adenine	B_{36}	B97D/6-31G(d)	–5	[109]
Thymine	B_{36}	B97D/6-31G(d)	–10	[109]
Cytosine	B_{36}	B97D/6-31G(d)	–38	[109]
Guanine	B_{36}	B97D/6-31G(d)	–43	[109]
Acetone	B_{40}	PBE/DNP	–15	[112]
Acetone	Li@B_{40}	PBE/DNP	–19	[112]
Acetone	Ba@B_{40}	PBE/DNP	–8	[112]
Metronidazole	B_{36}	B3LYP-D3/6-31G(d)	–22	[113]
Amantadine	B_{30}	ωB97XD/6-31G(d,p)	–46	[114]
Sulfonamide	B_{36}	M06-2X/6-31G(d,p)	–15	[116]
5-Fluorouracil	B_{36}	TPPSh/6-31 + G(d)	–24	[115]
Hydroxyurea	alanine-C_4B_{32}	PBE0/6-31 + G(d)	–19	[117]
6-thioguanine	C_4B_{32}	PBE0/6-31 + G(d))	–42	[118]
6-thioguanine	Si_4B_{32}	PBE0/6-31 + G(d))	–56	[118]
6-thioguanine	Li@C_4B_{32}	PBE0/6-31 + G(d))	–38	[118]
6-thioguanine	Li@Si_4B_{32}	PBE0/6-31 + G(d))	–43	[118]
Nitrosourea	C_4B_{32}	PBE0/6-31 + G(d))	–37	[119]
Cisplatin	C_4B_{32}	PBE0/6-31 + G(d))	–28	[120]
Cisplatin	Si_4B_{32}	PBE0/6-31 + G(d))	–18	[120]
Cisplatin	Li@C_4B_{32}	PBE0/6-31 + G(d))	–12	[120]
Cisplatin	Li@Si_4B_32	PBE0/6-31 + G(d))	–11	[120]
5-Fluorouracil	B_{40}	B3LYP/6-31G(d)	–11	[121]
5-Fluorouracil	Al@B_{40}	B3LYP/6-31G(d)	–15	[121]
5-Fluorouracil	Mg@B_{40}	B3LYP/6-31G(d)	–15	[121]
5-Fluorouracil	Cu@B_{40}	B3LYP/6-31G(d)	–15	[121]
5-Fluorouracil	Mn@B_{40}	B3LYP/6-31G(d)	–13	[121]
5-Fluorouracil	Zn@B_{40}	B3LYP/6-31G(d)	–12	[121]
5-Fluorouracil	Si@B_{40}	B3LYP/6-31G(d)	–12	[121]
5-Fluorouracil	$B_{39}Al$	B3LYP/6-31G(d)	–29	[121]
5-Fluorouracil	$B_{39}Mg$	B3LYP/6-31G(d)	–22	[121]
5-Fluorouracil	$B_{39}Cu$	B3LYP/6-31G(d)	–13	[121]
5-Fluorouracil	$B_{39}Zn$	B3LYP/6-31G(d)	–12	[121]
5-Fluorouracil	$B_{39}Mn$	B3LYP/6-31G(d)	–12	[121]

5-Fluorouracil	$B_{39}Si$	B3LYP/6-31G(d)	−9	[121]
Nitrosourea	B_{40}	PBE0/6-31G(d,p)	−25	[122]
Nitrosourea	$B_{39}C$	PBE0/6-31G(d,p)	−36	[122]
Nitrosourea	$B_{39}Al$	PBE0/6-31G(d,p)	−19	[122]
Nitrosourea	$B_{39}Ga$	PBE0/6-31G(d,p)	−18	[122]
5-Fluorouracil	B_{40}	PBE-D/DNP	−13	[123]
5-Fluorouracil	$Na@B_{40}$	PBE-D/DNP	−16	[123]
5-Fluorouracil	$Ca@B_{40}$	PBE-D/DNP	−17	[123]
Melphalan	B_{40}	PBE-D/DNP	−3	[124]
Melphalan	$Na@B_{40}$	PBE-D/DNP	−8	[124]
Melphalan	$Ca@B_{40}$	PBE-D/DNP	−9	[124]
Nedaplatin	B_{40}	PBE-D/DNP	−13	[125]
Nedaplatin	$Li@B_{40}$	PBE-D/DNP	−20	[125]
Nedaplatin	$Na@B_{40}$	PBE-D/DNP	−18	[125]

Table 1.
Summary of the interaction energies of boron-based clusters with bio-molecules considered in biomedical applications.

phases. The adsorption energy per nedaplatin molecule is about −28 kcal/mol for both $Li@B_{40}$ and $Na@B_{40}$ fullerene in aqueous solution, which is greater than that of the bare B_{40} fullerene, which is not favorable to be used.

4. Concluding remarks

Table 1 summarizes the reported interaction energies of boron-based clusters with the bio-molecules considered. Interaction energy constitutes the main parameter to evaluate the suitability of a carrier molecule for biomedical applications. Although the number of studies is rather limited, an analysis of these theoretical results suggests that boron-based clusters deserve to be regarded as promising candidates for the bio-sensing and drug delivery-related applications.

A large interaction energy indicates a stable complex, but it invariably causes a longer recovery time, which is not a good factor for drug release or for refreshing a biosensor. Thus, a semi-chemisorption with an effective recovery time (less than 1 second) is more favorable for biomedical applications. Based on Eq. (2), the recovery of a biomedical agent from a typical carrier surface is estimated to be short in the range 0.03–0.06 microsecond for NIR light with the typical interaction energy of −10 kcal/mol at 310 K of the human body. Also, it amounts to 0.3–0.7 second for interaction energies of −20 kcal/mol, which seems to be usable for an appropriate biosensor [107].

The recovery time exponentially increases by an enhancement of interaction energy. According to **Table 1**, most of the reported interaction energies are smaller than −20 kcal/mol, with an expected error margin of $\pm 3\,\text{kcal}/mol$ for DFT computations, which provide suitable recovery times at human body temperature in the range of NIR light.

In spite of the frequent claims of many authors in reported theoretical studies, some systems are not suitable due to a long recovery time (in the order of second or even longer). For example, in the aforementioned condition, B_{36} suffers from long recovery time for sensing cytosine and guanine; similarly, C_4B_{32} suffers from long recovery time for detecting 6-thioguanine, nitrosourea, and cisplatin drugs.

Noticeably, the release mechanism of drug is also a crucial factor, which should be understood for a better design of a drug delivery system, in addition to suitable recovery time. Such studies are yet to be carried out.

In summary, from a theoretical viewpoint, boron-based clusters having some unique structural and electronic properties provide us with great potential in biomedical applications. Quantum chemical calculations can further assist experimental researchers in the understanding of these systems from the molecular insights at low cost but with much detail and substantial accuracy.

Acknowledgements

The work of LVD and MTN is funded by VinGroup (Vietnam) and supported by VinGroup Innovation Foundation (VinIF) under project code VinIF_2020_DA21. LVD is thankful to the Van Lang University.

Conflict of interest

The authors declare no conflict of interest.

Author details

Ehsan Shakerzadeh[1*], Elham Tahmasebi[1], Long Van Duong[2,3]
and Minh Tho Nguyen[4,5*]

1 Chemistry Department, Faculty of Science, Shahid Chamran University of Ahvaz, Ahvaz, Iran

2 Laboratory for Computational Molecular and Materials Sciences, Science and Technology Advanced Institute, Van Lang University, Ho Chi Minh City, Vietnam

3 Faculty of Applied Technology, School of Engineering and Technology, Van Lang University, Ho Chi Minh City, Vietnam

4 Institute for Computational Science and Technology (ICST), Ho Chi Minh City, Vietnam

5 Department of Chemistry, KU Leuven, Leuven, Belgium

*Address all correspondence to: e.shakerzadeh@scu.ac.ir and tho.nm@icst.org.vn

References

[1] Alphandéry E. Ultrasound and nanomaterial: An efficient pair to fight cancer. Journal of Nanobiotechnology. 2022;**20**:139. DOI: 10.1186/s12951-022-01243-w

[2] Jedla MR, Koneru B, Franco A, Rangappa D, Banerjee P. Recent developments in nanomaterials based adsorbents for water purification techniques. Biointerface Research and Applied Chemistry. 2022;**12**:5821-5835. DOI: 10.33263/BRIAC125.58215835

[3] Hu ZT, Chen Y, Fei YF, Loo SL, Chen G, Hu M, et al. An overview of nanomaterial-based novel disinfection technologies for harmful microorganisms: Mechanism, synthesis, devices and application. Science Total Environment. 2022;**837**:155720. DOI: 10.1016/j.scitotenv.2022.155720

[4] Yuan F, Xia Y, Lu Q, Xu Q, Shu Y, Hu X. Recent advances in inorganic functional nanomaterials based flexible electrochemical sensors. Talanta. 2022;**244**:123419. DOI: 10.1016/j.talanta.2022.123419

[5] Rahman H, Hossain MR, Ferdous T. The recent advancement of low-dimensional nanostructured materials for drug delivery and drug sensing application: A brief review. Journal of Molecular Liquids. 2020;**320**:114427. DOI: 10.1016/j.molliq.2020.114427

[6] Zhu X, Chen H, Zhou Y, Wu J, Ramakrishna S, Peng X, et al. Recent advances in biosensors for detection of exosomes. Curr Opin Biomed Eng. 2021;**18**:100280. DOI: 10.1016/j.cobme.2021.100280

[7] Dyussembayev K, Sambasivam P, Bar I, Brownlie JC, Shiddiky MJA, Ford R. Biosensor Technologies for early detection and quantification of plant pathogens. Frontiers in Chemistry. 2021;**9**:636245. DOI: 10.3389/fchem.2021.636245

[8] Eivazzadeh-Keihan R, Bahojb Noruzi E, Chidar E, Jafari M, Davoodi F, Kashtiaray A, et al. Applications of carbon-based conductive nanomaterials in biosensors. Chemical Engineering Journal. 2022;**442**:136183. DOI: 10.1016/j.cej.2022.136183

[9] Machado M, Oliveira AML, Silva GA, Bitoque DB, Ferreira JT, Pinto LA, et al. Graphene biosensors—A molecular approach. Nanomaterials. 2022;**12**:1624. DOI: 10.3390/nano12101624

[10] Pourmadadi M, Dinani HS, Tabar FS, Khassi K, Janfaza S, Tasnim N, et al. Properties and applications of graphene and its derivatives in biosensors for Cancer detection: A comprehensive review. Biosensors. 2022;**12**:269. DOI: 10.3390/bios12050269

[11] Chao L, Liang Y, Hu X, Shi H, Xia T, Zhang H, et al. Recent advances in field effect transistor biosensor technology for cancer detection: A mini review. Journal of Physics D: Applied Physics. 2022;**55**:153001. DOI: 10.1088/1361-6463/ac3f5a

[12] Liu J, Yan L, He S, Hu J. Engineering DNA quadruplexes in DNA nanostructures for biosensor construction. Nano Research. 2022;**15**:3504-3513. DOI: 10.1007/s12274-021-3869-y

[13] Deng Y, Sun Z, Wang L, Wang M, Yang J, Li G. Biosensor-based assay of exosome biomarker for early diagnosis of cancer. Frontiers in Medicine.

2022;**16**:157-175. DOI: 10.1007/s11684-021-0884-z

[14] Yang YJ, Gao ZF. Superwettable biosensor for disease biomarker detection. Frontiers in Bioengineering and Biotechnology. 2022;**10**:872984. DOI: 10.3389/fbioe.2022.872984

[15] Chupradit S, Jasim SA, Bokov D, Mahmoud MZ, Roomi AB, Hachem K, et al. Recent advances in biosensor devices for HER-2 cancer biomarker detection. Analytical Methods. 2022;**14**:1301-1310. DOI: 10.1039/d2ay00111j

[16] Jiang C, Mu X, Du B, Tong Z. A review of electrochemical biosensor application in the detection of the SARS-COV-2. Micro Nano Letters. 2022;**17**:49-58. DOI: 10.1049/mna2.12101

[17] Breshears LE, Nguyen BT, Mata Robles S, Wu L, Yoon JY. Biosensor detection of airborne respiratory viruses such as SARS-CoV-2. SLAS Technology. 2022;**27**:4-17. DOI: 10.1016/j.slast.2021.12.004

[18] El-Sherif DM, Abouzid M, Gaballah MS, Ahmed AA, Adeel M, Sheta SM. New approach in SARS-CoV-2 surveillance using biosensor technology: A review. Environmental Science and Pollution Research. 2022;**29**:1677-1695. DOI: 10.1007/s11356-021-17096-z

[19] Shobana MK. Nanoferrites in biosensors – A review. Materials Science & Engineering, B: Solid-State Materials for Advanced Technology. 2021;**272**:115344. DOI: 10.1016/j.mseb.2021.115344

[20] Aljuffali IA, Fang CL, Chen CH, Fang JY. Nanomedicine as a strategy for natural compound delivery to prevent and treat cancers. Current Pharmaceutical Design. 2016;**22**:4219-4231. DOI: 10.2174/1381612822666160620072539

[21] Edis Z, Wang J, Waqas MK, Ijaz M, Ijaz M. Nanocarriers-mediated drug delivery Systems for Anticancer Agents: An overview and perspectives. International Journal of Nanomedicine. 2021;**16**:1313-1330. DOI: 10.2147/IJN.S289443

[22] Din FU, Aman W, Ullah I, Qureshi OS, Mustapha O, Shafique S, et al. Effective use of nanocarriers as drug delivery systems for the treatment of selected tumors. International Journal of Nanomedicine. 2017;**12**:7291-7309. DOI: 10.2147/IJN.S146315

[23] Alotaibi BS, Buabeid M, Ibrahim NA, Kharaba ZJ, Ijaz M, Noreen S, et al. Potential of nanocarrier-based drug delivery Systems for Brain Targeting: A current review of literature. International Journal of Nanomedicine. 2021;**16**:7517-7533. DOI: 10.2147/IJN.S33365

[24] Entezar-Almahdi E, Mohammadi-Samani S, Tayebi L, Farjadian F. Recent advances in designing 5-fluorouracil delivery systems: A stepping Stone in the safe treatment of colorectal Cancer. International Journal of Nanomedicine. 2020;**15**:5445-5458. DOI: 10.2147/IJN.S257700

[25] Wang LM, Wang YT, Yang WX. Engineered nanomaterials induce alterations in biological barriers: Focus on paracellular permeability. Nanomedicine. 2021;**16**:2725-2741. DOI: 10.2217/nnm-2021-0165

[26] Liu H, Zhang R, Zhang D, Zhang C, Zhang Z, Fu X, et al. Cyclic RGD-decorated liposomal gossypol AT-101 targeting for enhanced antitumor effect. International Journal of Nanomedicine. 2022;**17**:228-244. DOI: 10.2147/IJN.S341824

[27] Li X, Golberg D. Chapter 5 - boron nitride nanotubes as drug carriers. In: Ciofani G, Mattoli VBTBNN in N, editors. Micro and Nano Technologies. Boston: William Andrew Publishing; 2016. p. 79-94. DOI:10.1016/B978-0-323-38945-7.00005-5

[28] Lan HR, Wu ZQ, Zhang LH, Jin KT, Wang SB. Nanotechnology assisted chemotherapy for targeted Cancer treatment: Recent advances and clinical perspectives. Current Topics in Medicinal Chemistry. 2020;**20**:2442-2458. DOI: 10.2174/1568026620666200722110808

[29] Kenchegowda M, Rahamathulla M, Hani U, Begum MY, Guruswamy S, Osmani RAM, et al. Smart nanocarriers as an emerging platform for cancer therapy: A review. Molecules. 2020;**21**:146

[30] Rahimi R, Solimannejad M. First-principles survey on the pristine BC_2N monolayer as a promising vehicle for delivery of β-lapachone anticancer drug. Journal of Molecular Liquids. 2021;**321**:114917. DOI: 10.1016/j.molliq.2020.114917

[31] Sun X, Wan X, Li G, Yu J, Vahabi V. Amantadine antiparkinsonian drug adsorption on the AlN and BN nanoclusters: A computational study. Physics Letters A. 2020;**384**:126128. DOI: 10.1016/j.physleta.2019.126128

[32] Rahimi R, Solimannejad M, Ehsanfar Z. Potential application of XC_3 (X=B, N) nanosheets in drug delivery of hydroxyurea anticancer drug: A comparative DFT study. Molecular Physics. 2022;**120**:6. DOI: 10.1080/00268976.2021.2014587

[33] Zhu H, Zhao C, Cai Q, Fu X, Sheykhahmad FR. Adsorption behavior of 5-aminosalicylic acid drug on the $B_{12}N_{12}$, $AlB_{11}N_{12}$ and $GaB_{11}N_{12}$ nanoclusters: A comparative DFT study. Inorganic Chemistry Communications. 2018;**79**:223-229. DOI: 10.1016/j.inoche.2020.107808

[34] Onsori S, Alipour E. A computational study on the cisplatin drug interaction with boron nitride nanocluster. Journal of Molecular Graphics & Modelling. 2018;**79**:223-229. DOI: 10.1016/j.jmgm.2017.12.007

[35] Balali E, Davatgaran S, Sheikhi M, Shahab S, Kaviani S. Adsorption of doxepin drug on the surface of $B_{12}N_{12}$ and $Al_{12}N_{12}$ nanoclusters: DFT and TD-DFT perspectives. Main Gr Chemistry. 2022;**21**:69-84. DOI: 10.3233/MGC-210083

[36] Ganji MD, Yazdani H, Mirnejad A. $B_{36}N_{36}$ fullerene-like nanocages: A novel material for drug delivery. Physica E Low-Dimensional System Nanostructures. 2010;**42**:2184-2189. DOI: 10.1016/j.physe.2010.04.018

[37] Shabani M, Ghiasi R, Zare K, Fazaeli R. Quantum chemical study of interaction between Titanocene dichloride anticancer drug and $Al_{12}N_{12}$ Nano-cluster. Russian Journal of Inorganic Chemistry. 2020;**65**:1726-1734. DOI: 10.1134/S0036023620110169

[38] Yuan J, Mohamadi A. Study the adsorption process of 5-fluorouracil drug on the pristine and doped graphdiyne nanosheet. Journal of Molecular Modeling. 2021;**27**:32. DOI: 10.1007/s00894-020-04629-5

[39] Abd El-Mageed HR, Abbas HS. Adsorption behavior of mercaptopurine and 6-thioguanine drugs on the $B_{12}N_{12}$, $AlB_{11}N_{12}$ and $GaB_{11}N_{12}$ nanoclusters, a comparative DFT study. Journal of Biomolecular Structure & Dynamics. 2021;**12**:1-20. DOI: 10.1080/07391102.2021.1930163

[40] Yang Y, Ostadhosseini N. A theoretical investigation on the mercaptopurine drug interaction with boron nitride nanocage: Solvent and density functional effect. Physica E Low-Dimensional System Nanostructures. 2021;**125**:114337. DOI: 10.1016/j.physe.2020.114337

[41] Rahimi R, Solimannejad M, Farghadani M. Adsorption of chloroquine and hydroxychloroquine as potential drugs of SARS-CoV-2 infection on BC_3 nanosheets: A DFT study. New Journal of Chemistry. 2021;**45**:17976-17983. DOI: 10.1039/D1NJ03084A

[42] Khezri B, Maskanati M, Ghanemnia N, Shabani Gokeh M, Rezaei S, Chang L. Efficient detection of thioguanine drug using boron nitride nanocage: DFT outlook of solvent effect and AIM analysis. Inorganic Chemistry Communications. 2021;**134**:109015. DOI: 10.1016/j.inoche.2021.109015

[43] Shakerzadeh E. Chapter 4 - theoretical investigations of interactions between boron nitride nanotubes and drugs. In: Ciofani G, Mattoli VBTBNN in N, editors. Micro and Nano Technologies. Boston: William Andrew Publishing; 2016. p. 59-77. DOI:10.1016/B978-0-323-38945-7.00004-3

[44] Golipour-Chobar E, Salimi F, Ebrahimzadeh RG. Boron nitride nanocluster as a carrier for lomustine anticancer drug delivery: DFT and thermodynamics studies. Monatshefte für Chemie. 2020;**151**:309-318. DOI: 10.1007/s00706-020-02564-y

[45] El-Mageed HRA, Mustafa FM, Abdel-Latif MK. Boron nitride nanoclusters, nanoparticles and nanotubes as a drug carrier for isoniazid anti-tuberculosis drug, computational chemistry approaches. Journal of Biomolecular Structure & Dynamics. 2022;**40**:226-235. DOI: 10.1080/07391102.2020.1814871

[46] Stone FGA. Chemical reactivity of the boron hydrides and related compounds. In: Emeleus HJ, Sharpe AG, editors. Academic Press, Advances in Inorganic Chemistry and Radiochemistry. 1960. pp. 279-313

[47] Tao Y, Wu X, Zhang D. Synthesis and solar blind photosensitivity of crystalline boron nanowires. Nanotechnology. 2022;**33**:235601. DOI: 10.1088/1361-6528/ac56f5

[48] Wu Y, Li Y, Chen H, Sun Z, Wang N, Qin J, et al. Growth of single crystalline boron nanotubes in a Cu alloy. Crystal Engineering Communication. 2017;**19**:4510-4518. DOI: 10.1039/c7ce00818j

[49] Xu J, Chang Y, Gan L, Ma Y, Zhai T. Ultrathin single-crystalline boron nanosheets for enhanced electro-optical performances. Advancement of Science. 2015;**2**:1500023. DOI: 10.1002/advs.201500023

[50] Mannix AJ, Zhou XF, Kiraly B, Wood JD, Alducin D, Myers BD, et al. Synthesis of borophenes: Anisotropic, two-dimensional boron polymorphs. Science. 2015;**350**:1513-1516. DOI: 10.1126/science.aad1080

[51] Feng B, Zhang J, Zhong Q, Li W, Li S, Li H, et al. Experimental realization of two-dimensional boron sheets. Nature Chemistry. 2016;**8**:563-568. DOI: 10.1038/nchem.2491

[52] Otten CJ, Lourie OR, Yu MF, Cowley JM, Dyer MJ, Ruoff RS, et al. Crystalline boron nanowires. Journal of the American Chemical Society. 2002;**124**:4564-4565. DOI: 10.1021/ja017817s

[53] Xu TT, Zheng JG, Wu N, Nicholls AW, Roth JR, Dikin DA, et al. Crystalline boron nanoribbons: Synthesis and characterization. Nano Letters. 2004;**4**:963-968. DOI: 10.1021/nl0498785

[54] Tai G, Hu T, Zhou Y, Wang X, Kong J, Zeng T, et al. Synthesis of atomically thin boron films on copper foils. Angewandte Chemie, International Edition. 2015;**54**:15473-15477. DOI: 10.1002/anie.201509285

[55] Wu R, Drozdov IK, Eltinge S, Zahl P, Ismail-Beigi S, Božović I, et al. Large-area single-crystal sheets of borophene on Cu (111) surfaces. Nature Nanotechnology. 2019;**14**:44-49. DOI: 10.1038/s41565-018-0317-6

[56] Liu F, Shen C, Su Z, Ding X, Deng S, Chen J, et al. Metal-like single crystalline boron nanotubes: Synthesis and in situ study on electric transport and field emission properties. Journal of Materials Chemistry. 2010;**20**:2197-2205. DOI: 10.1039/b919260c

[57] Meng XM, Hu JQ, Jiang Y, Lee CS, Lee ST. Boron nanowires synthesized by laser ablation at high temperature. Chemical Physics Letters. 2003;**370**:825-828. DOI: 10.1016/S0009-2614(03)00202-1

[58] Van DL, Mai DTT, Pham-Ho MP, Nguyen MT. A theoretical approach to the role of different types of electrons in planar elongated boron clusters. Physical Chemistry Chemical Physics. 2019;**21**:13030-13039. DOI: 10.1039/C9CP00737G

[59] Tam NM, Duong LV, Pham HT, Nguyen MT, Pham-Ho MP. Effects of single and double nickel doping on boron clusters: Stabilization of tubular structures in B_nNi_m, (n = 2-22, m = 1, 2). Physical Chemistry Chemical Physics. 2019;**21**:8365-8375. DOI: 10.1039/C9CP00762H

[60] Dong X, Jalife S, Vásquez-Espinal A, Barroso J, Orozco-Ic M, Ravell E, et al. Li_2B_{24}: The simplest combination for a three-ring boron tube. Nanoscale. 2019;**55**:7490-7493. DOI: 10.1039/C9CC03732B

[61] Liang WY, Barroso J, Jalife S, Orozco-Ic M, Zarate X, Dong X, et al. $B_{10}M_2$ (M = Rh, Ir): Finally a stable boron-based icosahedral cluster. Chemical Communications. 2019;**55**:7490-7493. DOI: 10.1039/C9CC03732B

[62] Pan S, Barroso J, Jalife S, Heine T, Asmis KR, Merino G. Fluxional boron clusters: From theory to reality. Accounts of Chemical Research. 2019;**52**:2732-2744. DOI: 10.1021/acs.accounts.9b00336

[63] Li Q, Zhao Y, Xu W, Li N. Structure and stability of B_8 clusters. International Journal of Quantum Chemistry. 2005;**101**:219-229. DOI: 10.1002/qua.20290

[64] Dong X, Jalife S, Vásquez-Espinal A, Ravell E, Pan S, Cabellos JL, et al. Li_2B_{12} and Li_3B_{12} : Prediction of the smallest tubular and cage-like boron structures. Angewandte Chemie, International Edition. 2018;**57**:4627-4631. DOI: 10.1002/anie.201800976

[65] Pham HT, Van DL, Pham BQ, Nguyen MT. The 2D-to-3D geometry hopping in small boron clusters: The charge effect. Chemical Physics Letters. 2013;**577**:32-37. DOI: 10.1016/j.cplett.2013.05.041

[66] Tam NM, Pham HT, Duong LV, Pham-Ho MP, Nguyen MT. Fullerene-like boron clusters stabilized by an endohedrally doped iron atom: B_nFe with n = 14, 16, 18 and 20. Physical Chemistry

Chemical Physics. 2015;**17**:3000-3003. DOI: 10.1039/C4CP04279D

[67] Mai DTT, Duong LV, Tai TB, Nguyen MT. Electronic structure and thermochemical parameters of the silicon-doped boron clusters B_nSi, with n=8-14, and their anions. The Journal of Physical Chemistry. A. 2016;**120**: 3623-3633. DOI: 10.1021/acs.jpca.6b00847

[68] Duong LV, Si NT, Hung NP, Nguyen MT. The binary boron lithium clusters B_{12}Lin with n 1-14: In search for hydrogen storage materials Phys. Chemistry and Chemical Physics. 2021;**23**:24866. DOI: 10.1039/ D1CP03682C

[69] Pham-Ho MP, Tan Pham H, Nguyen MT. Boron teetotum: Metallic $[Ti\ (B_6C_xN_y)]^q$ and bimetallic $[Ti_2\ (B_6C_xN_y)]^q$ nine-membered heterocycles with x + y = 3 and $-1 \leq q \leq 3$. The Journal of Physical Chemistry. A. 2018;**122**: 6196-6205. DOI: 10.1021/acs.jpca.8b02713

[70] Pham HT, Muya JT, Buendía F, Ceulemans A, Nguyen MT. Formation of the quasi-planar B_{50} boron cluster: Topological path from B_{10} and disk aromaticity. Physical Chemistry Chemical Physics. 2019;**21**:7039-7044. DOI: 10.1039/C9CP00735K

[71] Pham HT, Nguyen MT. Formation of a bi-rhodium boron tube Rh_2B_{18} and its great CO_2 capture ability. Physical Chemistry Chemical Physics. 2018;**20**:26072-26082. DOI: 10.1039/C8CP03584A

[72] Duong LV, Pham HT, Tam NM, Nguyen MT. A particle on a hollow cylinder: The triple ring tubular cluster B_{27}^+. Physical Chemistry Chemical Physics. 2014;**16**:19470-19478. DOI: 10.1039/C4CP01996B

[73] Tai TB, Nguyen MT. Electronic structure and photoelectron spectra of B_n with n = 26-29: An overview of structural characteristics and growth mechanism of boron clusters. Physical Chemistry Chemical Physics. 2015;**17**:13672-13679. DOI: 10.1039/c5cp01851j

[74] Tai TB, Van DL, Pham HT, Mai DTT, Nguyen MT. A disk-aromatic bowl cluster B_{30}: Toward formation of boron buckyballs. Chemical Communications. 2014;**50**:1558-1560. DOI: 10.1039/ C3CC48392D-

[75] Pham HT, Duong LV, Tam NM, Pham-Ho MP, Nguyen MT. The boron conundrum: Bonding in the bowl B_{30} and B_{36}, fullerene B_{40} and triple ring B_{42} clusters. Chemical Physics Letters. 2014;**608**:295-302. DOI: 10.1016/j. cplett.2014.05.069

[76] Piazza ZA, Hu HS, Li WL, Zhao YF, Li J, Wang LS. Planar hexagonal B_{36} as a potential basis for extended single-atom layer boron sheets. Nature Communications. 2014;**5**:3113. DOI: 10.1038/ncomms4113

[77] Bai H, Ma M, Zuo J, Zhang QF, Bai B, Cao H, et al. Recyclable and superior selective CO_2 adsorption of C_4B_{32} and $Ca@C_4B_{32}$: A new category of perfect cubic heteroborospherenes. Physical Chemistry Chemical Physics. 2019;**21**:15541-15550. DOI: 10.1039/ c9cp02380a

[78] Liu P, Zhang Y, Liu F, Zhou D. Li decorated heteroborospherene C_4B_{32} as high capacity and reversible hydrogen storage media: A DFT study. International Journal of Hydrogen Energy. 2022;**47**:11948-11954. DOI: 10.1016/j.ijhydene.2022.01.208

[79] Zhai HJ, Zhao YF, Li WL, Chen Q, Bai H, Hu HS, et al. Observation of an all-boron fullerene. Nature Chemistry. 2014;**6**:727-731. DOI: 10.1038/ nchem.1999

[80] Shakerzadeh E, Biglari Z, Tahmasebi E. M@B40 (M=Li, Na, K) serving as a potential promising novel NLO nanomaterial. Chemical Physics Letters. 2016;**654**:76-80. DOI: 10.1016/j.cplett.2016.05.016

[81] Shakerzadeh E, Kazemimoghadam F. Magnesiation of bare and halides encapsulated B_{40} fullerenes for their potential application as promising anode materials for Mg-ion batteries. Applied Surface Science. 2021;**538**:148060. DOI: 10.1016/j.apsusc.2020.148060

[82] Kaur R, Kaur J. The electronic transport properties of B_{40} fullerenes with chalcogens as anchor atoms. Journal of Molecular Modeling. 2017;**23**:351. DOI: 10.1007/s00894-017-3520-8

[83] Shakerzadeh E, Yousefizadeh M, Bamdad M. Electronic and nonlinear optical features of first row transition metals-decorated all-boron B_{40} fullerene: A promising route to remarkable electro-optical response. Inorganic Chemistry Communications. 2020;**112**:107692. DOI: 10.1016/j.inoche.2019.107692

[84] Yang Y, Zhang Z, Penev ES, Yakobson BI. B_{40} cluster stability, reactivity, and its planar structural precursor. Nanoscale. 2017;**9**:1805-1810. DOI: 10.1039/c6nr09385j

[85] Kaur H, Kaur J, Kumar R. Materials today: Proceedings a review on all boron fullerene (B_{40}): A promising material for sensing and device applications. Mater Today Proc. 2022;**48**:1095-1102. DOI: 10.1016/j.matpr.2021.07.465

[86] Leśnikowski ZJ. Recent developments with boron as a platform for novel drug design. Expert Opinion on Drug Discovery. 2016;**11**:569-578. DOI: 10.1080/17460441.2016.1174687

[87] Ali F, Hosmane NS, Zhu Y. Boron chemistry for medical applications. Molecules. 2020;**25**(828):1-24. DOI: 10.3390/molecules25040828

[88] Fink K, Uchman M. Boron cluster compounds as new chemical leads for antimicrobial therapy. Coordination Chemistry Reviews. 2021;**431**:213684. DOI: 10.1016/j.ccr.2020.213684

[89] Kwiatkowska A, Sobczak M, Mikolajczyk B, Janczak S, Olejniczak AB, Sochacki M, et al. SiRNAs modified with boron cluster and their physicochemical and biological characterization. Bioconjugation Chemistry. 2020;*24*:1017-1026

[90] Han Y, Wen P, Li J, Kataoka K. Targeted nanomedicine in cisplatin-based cancer therapeutics. Journal of Controlled Release. 2022;**345**:709-720. DOI: 10.1016/j.jconrel.2022.03.049

[91] Tsvetkova D, Ivanova S. Application of approved cisplatin derivatives in combination therapy against different Cancer diseases. Molecules. 2022;**27**:2466. DOI: 10.3390/molecules27082466

[92] Shimada M, Itamochi H, Kigawa J. Nedaplatin: A cisplatin derivative in cancer chemotherapy. Cancer Management and Research. 2013;**5**:67-76. DOI: 10.2147/CMAR.S35785

[93] Musiałek MW, Rybaczek D. Hydroxyurea—The good, the bad and the ugly. Genes (Basel). 2021;**12**:1096. DOI: 10.3390/genes12071096

[94] Dawson W, Degomme A, Stella M, Nakajima T, Ratcliff LE, Genovese L. Density functional theory calculations of large systems: Interplay between fragments, observables, and computational complexity. Wiley Interdisciplinary Reviews: Computational Molecular Science. 2022;**12**:1574. DOI: 10.1002/wcms.1574

[95] Maitra NT. Double and charge-transfer excitations in time-dependent density functional theory. Annual Review of Physical Chemistry. 2022;**73**:117-140. DOI: 10.1146/annurev-physchem-082720-124933

[96] Sim E, Song S, Vuckovic S, Burke K. Improving results by improving densities: Density-corrected density functional theory. Journal of the American Chemical Society. 2022;**144**:6625-6639. DOI: 10.1021/jacs.1c11506

[97] Urrego-Ortiz R, Builes S, Calle-Vallejo F. Impact of intrinsic density functional theory errors on the predictive power of nitrogen cycle electrocatalysis models. ACS Catalysis. 2022;**12**:4784-4791. DOI: 10.1021/acscatal.1c05333

[98] Nagy Á. Fisher information and density functional theory. International Journal of Quantum Chemistry. 2022;**122**:e26679. DOI: 10.1002/qua.26679

[99] Jang T, Paik D, Shin SJ, Kim H. Density functional theory in classical explicit solvents: Mean-field QM/MM method for simulating solid–liquid interfaces. Bulletin of the Korean Chemical Society. 2022;**43**:476-483. DOI: 10.1002/bkcs.12485

[100] Maihom T, Sittiwong J, Probst M, Limtrakul J. Understanding the interactions between lithium polysulfides and anchoring materials in advanced lithium-sulfur batteries using density functional theory. Physical Chemistry Chemical Physics. 2022;**24**:8604-8623. DOI: 10.1039/d1cp05715d

[101] Yang Y, Wang J, Shu Y, Ji Y, Dong H, Li Y. Significance of density functional theory (DFT) calculations for electrocatalysis of N_2 and CO_2 reduction reactions. Physical Chemistry Chemical Physics. 2022;**24**:8591-8603. DOI: 10.1039/d1cp05442b

[102] Pernal K, Hapka M. Range-separated multiconfigurational density functional theory methods. Wiley Interdisciplinary Reviews: Computational Molecular Science. 2022;**12**:e1566. DOI: 10.1002/wcms.1566

[103] Tölle J, Neugebauer J. The seamless connection of local and collective excited states in subsystem time-dependent density functional theory. Journal of Physical Chemistry Letters. 2022;**13**:1003-1018. DOI: 10.1021/acs.jpclett.1c04023

[104] Sook N, Kee Y. Assessment of CCSD (T), MP2, and DFT methods for the calculations of structures and interaction energies of the peptide backbone with water molecules. Chemical Physics Letters. 2017;**687**:23-30. DOI: 10.1016/j.cplett.2017.08.063

[105] Si NT, Nhung NTA, Bui TQ, Nguyen MT, Nhat PV. Gold nanoclusters as prospective carriers and detectors of pramipexole. RSC Advances. 2021;**11**:16619-16632. DOI: 10.1039/D1RA02172A

[106] Hazrati MK, Javanshir Z, Bagheri Z. $B_{24}N_{24}$ fullerene as a carrier for 5-fluorouracil anti-cancer drug delivery: DFT studies. Journal of Molecular Graphics & Modelling. 2017;**77**:17-24. DOI: 10.1016/j.jmgm.2017.08.003

[107] Tao Y, Chan HF, Shi B, Li M, Leong KW. Light: A magical tool for controlled. Drug Delivery. 2021;**30** (1-52). DOI: 10.1002/adfm.202005029

[108] Tang Y, Wang G. NIR light-responsive nanocarriers for controlled release. Journal of Photochemistry and Photobiology, C: Photochemistry

Reviews. 2021;**47**:100420. DOI: 10.1016/j.jphotochemrev.2021.100420

[109] Kaur J, Kumar R. Borospherene - based biomarker for DNA sequencing : A DFT study. Journal of Computational Electronics. 2021;**20**:1916-1929. DOI: 10.1007/s10825-021-01731-6

[110] Kaur J, Kumar R, Vohra R, Sawhney RS. Density functional theory investigations on the interaction of uracil with borospherene. Bulletin of Materials Science. 2022;**45**:22. DOI: 10.1007/s12034-021-02595-z

[111] Rastgou A, Soleymanabadi H, Bodaghi A. DNA sequencing by borophene nanosheet via an electronic response: A theoretical study. Microelectronic Engineering. 2017;**169**:9-15. DOI: 10.1016/j.mee.2016.11.012

[112] Yong Y, Su X, Kuang Y, Li X, Lu Z. B_{40} and M@B_{40} (M=Li and Ba) fullerenes as potential molecular sensors for acetone detection: A first-principles study. Journal of Molecular Liquids. 2018;**264**: 1-8. DOI: 10.1016/j.molliq.2018.05.046

[113] Xiao C, Ma K, Cai G, Zhang X, Vessally E. Borophene as an electronic sensor for metronidazole drug: A computational study. Journal of Molecular Graphics & Modelling. 2020;**96**:107539. DOI: 10.1016/j.jmgm.2020.107539

[114] Noroozi Z, Rahimi R, Solimannejad M. A computational study for the B_{30} bowl-like nanostructure as a possible candidate for drug delivery system for amantadine. Computational & Theoretical Chemistry. 2018;**1129**:9-15. DOI: 10.1016/j.comptc.2018.02.016

[115] Shakerzadeh E. Quantum chemical assessment of the adsorption behavior of fluorouracil as an anticancer drug on the B_{36} nanosheet. Journal of Molecular Liquids. 2017;**240**:682-693. DOI: 10.1016/j.molliq.2017.05.128

[116] Kamalinahad S, Soltanabadi A, Gamallo P. B_{36} bowl-like structure as nanocarrier for sulfonamides: A theoretical study. Monatshefte für Chemie. 2020;**151**:1785-1796. DOI: 10.1007/s00706-020-02705-3

[117] Zheng Y, Shan K, Zhang Y, Gu W. Amino acid-functionalized borospherenes as drug delivery systems. Biophysical Chemistry. 2020;**260**:106407. DOI: 10.1016/j.bpc.2020.106407

[118] Yunyu Z, Jameh-Bozorghi S. Endohedral metalloborospherenes as promising drug delivery systems. Journal of Coordination Chemistry. 2020;**73**:1425-1435. DOI: /10.1080/00958972.2020.1788000

[119] Deng S, Jiang Q, Wang Y, Lu X, Zhu Y, Zhang Y, et al. C_4B_{32} nanocluster as a drug delivery system for nitrosourea anticancer drug: A first-principles perception. Molecular Physics. 2021;**119**:3. DOI: 10.1080/00268976.2020.1808906

[120] Luo Q, Gu W. Novel borospherenes as cisplatin anticancer drug delivery systems. Molecular Physics. 2020;**118**:23. DOI: 10.1080/00268976.2020.1774088

[121] Zhang L, Qi ZD, Ye YL, Li XH, Chen JH, Sun WM. DFT study on the adsorption of 5-fluorouracil on B_{40}, B_{39}M, and M@B_{40} (M = Mg, Al, Si, Mn, Cu, Zn). RSC Advances. 2021;**11**: 39508-39517. DOI: 10.1039/D1RA08308B

[122] Zhang L, Ye YL, Li XH, Chen JH, Sun WM. On the potential of all-boron fullerene B_{40} as a carrier for anti-cancer drug nitrosourea. Journal of Molecular Liquids. 2021;**342**:117533. DOI: 10.1016/j.molliq.2021.117533

[123] Shakerzadeh E. Efficient carriers for anticancer 5-fluorouracil drug based on the bare and M-encapsulated (M=Na and Ca) B_{40} fullerenes. In silico investigation. J Mol Liq. 2021;**343**:116970. DOI: 10.1016/j.molliq.2021.116970

[124] Shakerzadeh E. Endohedral M@B_{40} (M= Na and Ca) metalloborospherenes as innovative potential carriers for chemotherapy melphalan drug: A theoretical study. Applied Organometallic Chemistry. 2021;**35**:e 6411. DOI: 10.1002/aoc.6411

[125] Shakerzadeh E. Li@B_{40} and Na@B_{40} fullerenes serving as efficient carriers for anticancer nedaplatin drug: A quantum chemical study. Computational & Theoretical Chemistry. 2021;**1202**:113339. DOI: 10.1016/j.comptc.2021.113339

Chapter 7

Investigation of Boron-Based Ionic Liquids for Energy Applications

Gülbahar Bilgiç

Abstract

As a result of its electron-deficient nature, boron is utilized as the building block of ionic liquids (ILs) in energy applications (EAs) by composing a number of anions, cations, and negatively charged clusters chemically, electrochemically, and thermally. Anionic boron clusters, for example, feature distinctive polycentric bonding and are an important component of low-viscosity room temperature ILs (RTILs) utilized in electrochemical devices. In this context, recent breakthroughs in the synthesis of boron-containing ILs and various materials derived from them provide a powerful opportunity for further investigation in the field of energy research to improve and develop the properties of these boron-based ILs (BBILs). This review has provided a brief summary of boron atom or molecule-based ILs with unique features that make them ideal candidates for particular EAs. This perspective can guide further research and development of the unique properties of green and halogen-free BBILs.

Keywords: boron, ionic liquids, energy applications

1. Introduction

With the increasing utilization of solar and wind energy worldwide, the intermittent nature of these resources restricts the sustainability of the energy supply. Due to environmental concerns regarding air quality, reducing the impact of fossil fuel use has required the development of new energy technologies for alternative energy options. In addition, minimizing the effect of fossil fuel consumption owing to environmental concerns regarding air quality issues mandates the development of new technologies for alternative energy solutions. There is great potential for discovering new energetic materials and employing them in applications in all of these energy technology development initiatives. BBILs are a novel class of energetic materials with a wide variety of characteristics that may be tailored to specific applications in the energy area [1].

Paul Walden first discovered ILs in 1914 and determined that the ethylammonium nitrate salt had a low melting point of 13°C [2–4]. In 1992, Wilkes and Zaworotko created air and water-stable ILs, known as the second generation, utilizing 1ethyl-3-methylimidazolium cation ($[EMIM]^+$) and tetrafluoroborate ($[BF_4]^-$) or hexafluorophosphate anion ($[PF_6]^-$). They discovered that whereas these ILs are normally water insensitive, prolonged exposure to moisture produces alterations in their

IntechOpen

characteristics [4]. The continuous fascination with ILs arises from their excellent features in future research and potential applications [1, 5, 7].

ILs are a novel family of solvents with melting temperatures less than 100°C, consisting of a combination of a bulky cation and an inorganic or organic anion [1, 2, 4–10]. An advantage of low melting point solvents is that they have negligible vapor pressures.

As a result, unlike liquids, ILs do not evaporate under standard conditions. ILs are salts with high thermal stability (most ILs are stable up to about 300–400°C) [9], chemical stability, nonflammability, a broad electrochemical window (2–6 V) [10, 11], and high electrical conductivity (1.3–8.5 mS cm^{-1}) [12]. ILs are also less hazardous and effective solvents for both organic and inorganic compounds.

ILs are prominent solvents in electrochemistry due to the constituent anions being oxidized at sufficiently large potentials and organic cations being reduced at low potentials. ILs are generally aprotic, so problems with hydrogen ions that occur in protic solvents can be avoided. In recent years, ILs have great attention as an environmentally friendly ("greenness") liquid that is a candidate to replace commonly used, solvent-based, volatile, and flammable electrolytes that tend to be corrosive [6]. When creating ILs for applications, commonly at least one ion is weakly coordinated; either the cation or the anion is weakly coordinated; and in some ILs, both the ions are weakly coordinated. Solvent properties of ILs vary depending on the nature of the ions in their structure; anions with high charge density and organic cations with short alkyl chains stabilize more polar molecules [13, 14]. ILs have advantages as well as disadvantages. **Table 1** shows the summary of the physicochemical properties of ILs. The main disadvantages are high density, high viscosity, and low conductivity.

ILs are liquids showing high ion density. The density of ILs typically ranges between 1.2 and 1.5 g cm^{-3}, while certain ILs, such as those based on the dicyanamide anion, has a density of less than 1 g cm^{-3}. Because viscosity is important in conductivity and diffusion, conductivity decreases as viscosity increases. ILs have a significantly higher viscosity (30–50 cP) than water (H_2O = 0.89 cP at 25°C) [15].

(30–50 cP) than water (H_2O = 0.89 cP at 25°C) [15]. The conductivities at room temperature for ILs are in the wide range of from 0.1 to 18 mS cm^{-1}, but even at the highest conductivity, they are much lower compared to conventional aqueous

Properties	Values	Advantages	Disadvantages
Low melting point	< 100°C	Liquid at ambient temperature Wide temperature interval for applications	High viscosity
Nonflammability		Non-volatiliy Thermal stability Flame retardancy	Less toxic
Composed by ions	Cation and an inorganic or organic anion	Unlimited combinations possible A wide electrochemical window High ion conductivity Designable/Tunable	High ion density Low conductivity (compared to conventional aqueous electrolyte solutions)

Table 1.
Summary of the physicochemical properties of ILs (adapted from Refs [13–16]).

electrolyte solutions. Dilution of pure ILs with molecular fluids provides an increase in the conductivity of the medium. For example, pure [EMIM][BF_4^-] has a specific conductivity of 14 mS cm^{-1}, while 2 moles of dm^{-3} solution in acetonitrile shows a conductivity of 47 mS cm^{-1}. In this context, it can be said that diluting pure ILs with a molecular diluent also reduces the viscosity of the mixture [16].

At room temperature or below, RTILs are typically composed of organic or inorganic anions with weak basic properties and organic cations with low molecular symmetry [11, 13, 16]. It is a liquid group that offers more advantages over organic electrolyte solutions, including low vapor pressure, nonflammability, electrochemical and chemical stability, and high ionic conductivity. RTILs are currently of great interest in both academia and industry. To date, thousands of RTILs with unique physical features and functionalities have been created for use in energy applications. Some special boron-based RTILs have been investigated in EAs such as electrolyte materials in lithium batteries, fuel cells, and solar cells, as they exhibit ionic conductivity values of more than 10–2 S cm^{-1} at room temperature [11, 13, 16–22].

This review emphasizes on the highlights of boron's ability to facilitate the development of BBILs in EAs such as an atom and small molecule activation toward hyperbolic fuel additive, dye synthesized solar cells (DSSCs), advanced secondary batteries, hydrogen production and storage, electrolyte materials for electroreduction, and CO_2 capture.

The first chapter emphasizes the role of boron cations and anions in the production of BBILs designed for usage in various EAs. Following that, examples of boron cations and anions found in the literature are given. Then, BBILs utilized in EA were thoroughly explored. Before detailing the selected examples of each area, a brief introduction was made and then closed with a brief perspective. Ultimately, this review highlights BBILs with less toxic and less expensive starting materials for future energy demands, as well as the possibility for ILs to play an essential part in meeting some of the future difficulties.

1.1 Boron's role in ILs

With the increasing demand for renewable energy and green chemistry, boron has been playing a key role in energy-related research, from synthesizing energy-rich molecules to energy storage to converting electrical energy to light. In this regard, specifically constructed ILs have attracted the attention of researchers by including a variety of tribologically active elements such as sulfur, nitrogen, phosphorus, nitrogen, zinc, molybdenum, halogens, boron, and so on. Among these elements, boron's versatile chemistry makes it prominent in its use in EAs. Given the growing interest in boron chemistry, it is critical to understand why and how boron may become a favored element in IL functionalization [22–31].

Boron is one of the few elements known to show excellent harmony with ILs energy applications [22, 31]. **Table 2** lists the physical, atomic, and other characteristics of boron. Boron, symbol B, atomic number 5, and group 3A of the periodic table, is not found in elemental form in nature but may be produced in pure form by several processes [24, 25]. This position of boron between metals and nonmetals allows it to be employed in a wide range of research areas. Boron has an empty p orbital, so it has an electron deficiency. It has a sensitivity to undergo chemical reactions to saturate the coordination sphere and valence shell. Owing to the chemical properties and orbital nature of boron allow the formation of many useful ILs, including neutral, anionic, and cationic species.

Boron ($_5B$)	
Phase	Solid
Atomic number	5
Electron configuration	$1s^2 2s^2 2p^1$
Allotropes	α-rhombohedral, α-tetragonal boron, β-rhombohedral boron, β-tetragonal boron,γ-orthorhombic
Isotope	^{10}B, ^{11}B
Density (liquid)	2.08 g/cm^3
Melting point	2076°C
Molar heat capacity	11.087 J/(mol·K)
Atomic weight	10.81
Electronegativity	2.04
Thermal conductivity	27.4 W/(m·K)

Table 2.
Physical, atomic and other properties of boron.

While forming a compound, the empty p-orbital of the boron can lead to significant delocalization and is attacked readily by nucleophiles such as water or halides [22–38]. As a result, by bond cleavage of the neutral tri-coordinated borate, boron readily forms trivalent compounds with electrophile molecules such as oxides, sulfides, nitrides, and halides. Moreover, fluoride (F^-) and boron trifluoride $[BF_3]^-$ may combine to create $[BF_4]^-$ the anionic tetracoordinated borate species [31].

Because boron has an intrinsic electron deficiency, it is primarily defined by its Lewis acidity since it easily forms adducts by seizing electron pairs from Lewis bases [22, 28–30]. However, when certain conditions are met, the boron atom can become negatively charged or polarized and therefore, act as a nucleophile or Lewis base [27–30]. For example, boron and hydrogen combine to form many borane anions. The high hydrogen capacity of these borane anions makes them a suitable material for hydrogen storage. They are also possible candidates for ILs employed in electrochemical devices due to their unsymmetrical borate anions and bulky and anionic boron aggregates.

Many boron-atom/molecule-based ILs can be developed by tailoring cation-anion ion couples for specific purposes. Hitherto, elemental boron, carboranes, and organoboron compounds (cationic borinium (R2B1L2), borenium (R2B1L1), and boronium ions (R2B1), tetrahedral boron anions ($[BH_4^-]$, $[B_{12}H_{12}^-]$), orthoborate anion ($[BO_3^-]$) were often employed in the synthesis of BBILs [22, 23, 28, 31]. Because of its electron-deficient nature, boron forms a series of highly chemically, electrochemically, and thermally stable anions and negatively charged boron clusters. For instance, organoboron anion compounds from the boron ion family have been frequently used in the formation of stable ILs in recent years, as they can dissolve in solvents with a low dielectric constant with reasonable solubility [31]. It has been indicated that carborane-based ILs with relatively high boron content, such as 1-carba-closo-dodecaborate $[CB_{11}H_{12}]^{-1}$, are exceptionally stable toward oxidation and coordination reactions due to their unique molecular structures [31, 32, 39–46].

As shown in **Figure 1**, designing BBILs for different EAs is often done by tailoring component ions with functional groups. The ion structure (like a positive-negative charge, small-large) greatly affects the physicochemical properties of ILs and, ergo, also affects the application to be used. For example, adding hydrophobic alkyl chains to its component ions causes changes in the chemical properties of ILs, increasing the viscosity and melting temperature that are undesirable in electrochemical devices. Therefore, it is necessary to achieve functionalization without destroying the basic properties of BBILs.

1.2 The choice of boron cations and anions in the formation of ionic liquids

The physicochemical characteristics of all ILs, not just boron-related ILs, depend on the nature of cations and anions and their combination. Generally, when designing ILs, properties such as melting point, conductivity, viscosity, air and moisture stability, hydrophobicity, miscibility with water, and density must be adjusted and optimized for the application to be utilized. According to Plechkova and Seddonna et al. [5], the cation is responsible for the physical properties of an IL, for example, density, viscosity, and melting point, while the anion is accountable for the chemical properties and reactivity. For example, halides as anions with hydrogen in position 2 (CH ⋯ X−) lead to higher melting points. For this reason, it can be said anions play a critical role in higher melting and lower melting points, as well as cations [34]. The anion type has a significant effect on the hydrophobicity and hydrophilic tuning tunable of IL. Fluoridation of the anion leads to promoted hydrophobicity and weaker hydrogen

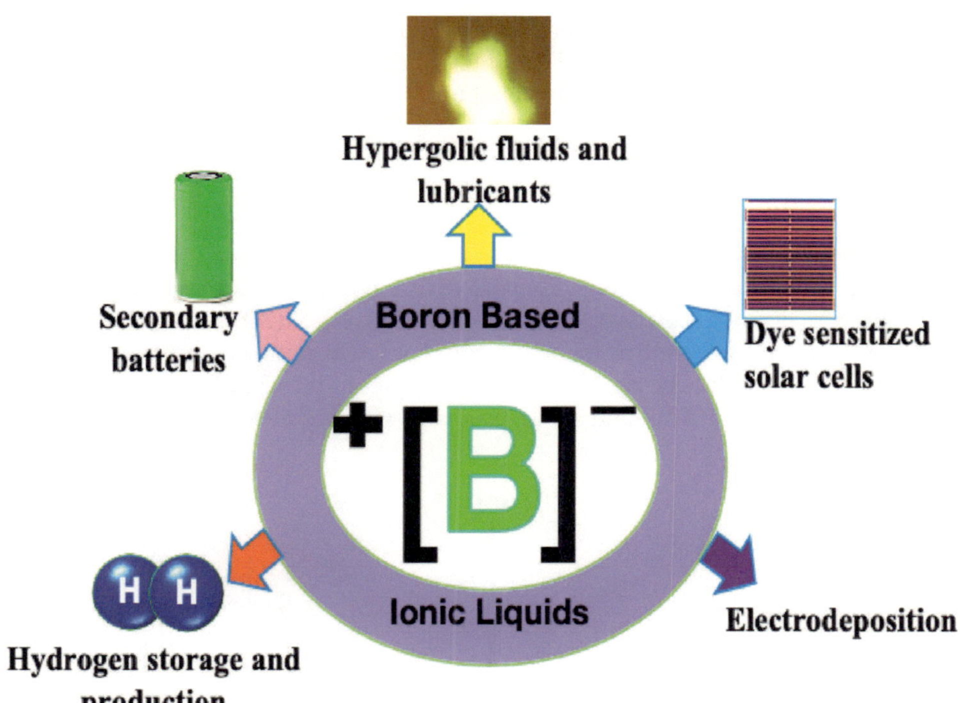

Figure 1.
Modification of BBILs for energy-related applications.

bonding, and hence lower melting temperatures; it also increases thermal and electrochemical stability. Besides halides, the inorganic anions $[BF_4]^-$ and $[PF_6]$– are often utilized in designing ILs. The modification of the cation by N-alkylation is critical in adjusting ILs. Increasing cation size and chain length lowers IL melting values while increasing viscosity. Also, the cation's branched alkyl chains induce a greater melting temperature in IL. In this context, melting points appear to be more connected to cation asymmetry [13] because this asymmetry causes packing inefficiency and hence prevents crystallization. Asymmetric cation salts with just C1 symmetry, for example, have lower melting temperatures than identical salts containing C2v symmetry cations [36]. It has been noted that the absence of strong hydrogen bonding correlates to lower IL melting temperatures [36].

The most commonly used cations in the formation of ILs are N,N′-dialkylated imidazolium ions, due to their ease of synthesis and favorable physical and chemical properties. Cations of imidazolium impart low melting points, high conductivity, and low viscosity properties to ILs [34]. Quaternary ammonium is also used as the cation for some EAs. The use of ILs using 1,3-dialkyl imidazolium cations in high-energy electrochemical devices such as lithium-ion batteries is not possible due to the electrochemical instability of the cation. Therefore, ILs containing quaternary ammonium ions are preferred as they are more resistant to reduction and oxidation [37, 38].

Inorganic anions with which the ILs decompose endothermically, while organic anions lead to exothermic thermal decomposition [39]. The thermal stability of halides is substantially lower (300°C). Many of these imidazolium salts display supercooling and are liquid at room temperature. More symmetric cations lead to higher melting points and vice versa. A longer alkyl chain (propyl as compared to ethyl) lowers the melting points and branching of the chain (i-propyl) raises the melting point [39]. Considering the aforementioned features, ILs synthesized by changing the boron anion and cation are potential candidates for the field of energy.

Boron anion and boron cation-based ILs are described in further depth in the next section.

1.2.1 Boron anions based ILs

Over the past few decades, boron anion families of ILs have been prominent in the field of energy due to their high thermal stability, low flammability, negligible vapor pressure, and wide electrochemical window [8, 40–53]. One of the liquid salts borates is the most commonly used negatively charged boron compound as anions in BBILs [22, 23, 35–37]. **Figure 2** illustrates examples of tetrahedral boron (borate) anions in ILs. The negative charge of ILs produced with tetrahedral boron anions is

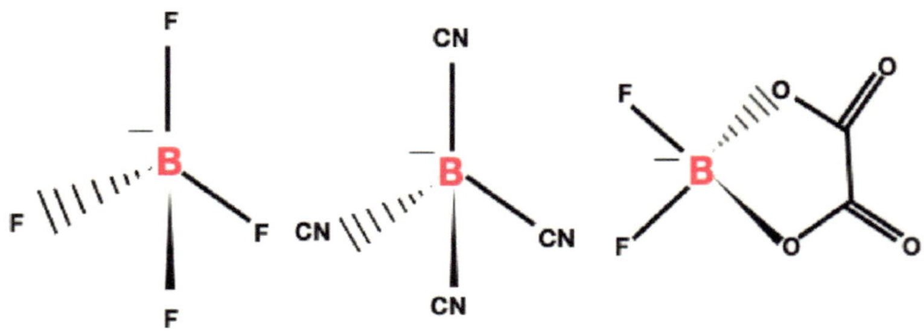

Figure 2.
Tetrahedral boron anions (adapted from Refs [22, 31, 35]).

compensated for by positively charged groups such as ammonium. Mixed borates containing $[BF_4]^-$ as an anion can react with trimethylsilyl ethers of acids to generate IL, as shown in the figure.

The $[BF_4]^-$ anion is uncoordinated in an aqueous solution; therefore, they are called weakly coordinated anions. A modest negative charge and strong charge delocalization are some desirable features for weakly coordinated anions. Nonnucleophilicity and the minimally base surface of the anion also lead to poor coordination [8, 21, 40].

Anions should not be divided into small moieties in ILs, such as $[BF_4]^-$ which loses a fluoride when attacked by a nucleophile. Therefore, kinetic and thermodynamic stability are also significant points. Anions must also resist oxidation because the corresponding electrophilic cation often acts as an oxidizing agent [8].

As an alternative to Lewis acid/base pair anions that function in a combination of a Lewis acidic boron component and a Lewis base, derivatives of polyhedral anions such as 1-carba-closo-dodecaborate ($[CB_{11}H_{12}]^-$)are used (see **Figure 3**) [1, 8, 40].

Polyhedral borane clusters and three-dimensional carboranes have attracted attention in recent years with an increasing interest in ILs as materials and building blocks because of their great chemical and thermal stability [35].

Because of their inertness, dipolarity, and high symmetry, carboranes (most known: closo ($[C_2B_nH\textit{n}_{+2}]$), nido ($[C_2B_nH\textit{n}_{+4)}]$), arachno ($[C_2B_nH\textit{n}_{+6}]$)) as a new class of weakly coordinating anions can be considered three-dimensional analogs of benzene. Carboranes are promising candidates in ILs as anions due to their size, spherical shape, remarkable chemical stability, and only weakly coordinated B-H groups [41]. Carboranes are electron-delocalized organometallic clusters composed of carbon (C), boron (B), and hydrogen (H).

Carboranes are promising candidates in ILs as anions due to their size, spherical shape, remarkable chemical stability, and only weakly coordinated B-H groups [41]. Carboranes are electron-delocalized organometallic clusters composed of carbon (C), boron (B), and hydrogen (H). The general formula of carboranes is represented by $C_2B_nH\textit{n}_{+}\textit{m}$, in which n is an integer; *n* range from 3 to 10. Carboranes are synthesized by adding one-carbon reagents (i.e., cyanide, isocyanides, and formaldehyde) to boron hydride clusters. For example, monocardiodecaborate ($[CB_{11}H_{12}]^-$) is produced from decaborane and formaldehyde, followed by the addition of borane

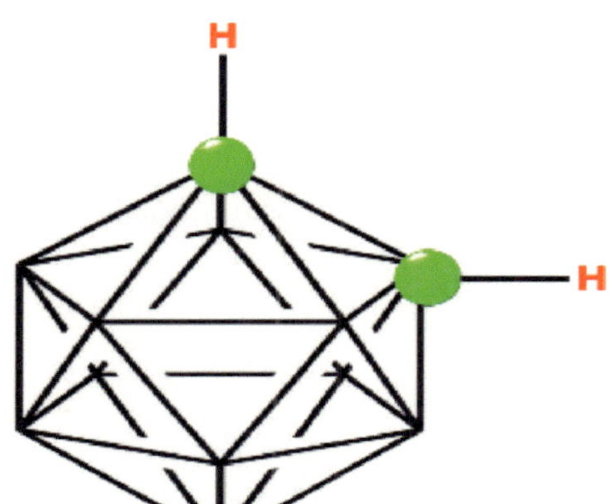

Figure 3.
Closo carborane (a green circle represents a C–H unit or a C in the cases where a charge is specified. And another corner one B-H unit closo-Carborane. Adapted from Refs [41, 42, 45]).

dimethylsulfide [42–45]. Monocarboranes are precursors of poorly coordinated anions [41, 42]. The resulting poor coordination is established in a charge distribution over the entire 12-cornered carborane anion. The 10-sided anion $[CB_9H_{10}]^-$ and its halogenated and methylated derivatives coordinate more strongly than the 12-sided carboranes. Due to their specific properties, these boron molecules are some of the most inert and least nucleophilic anions currently known.

Cyanoborate anions ($[B(CN)]^-$) (see **Figure 4**) have become an important class of building blocks in materials science, especially for ILs that are used as components of electrolytes for electrochemical devices. Also, in recent years, low viscosity room temperature cyanoborate based ILs have been studied in EAs, especially dye-sensitized solar cells (DSSCs) and as a fuel additive for hyperbolic liquids. Cyanoborate chemistry has been popularized as an important topic in the last years, beginning with two independent research on the successful synthesis of the tetracyanoborate anion $[B(CN)_4]^-$ in 2000 [53–56].

A wealth of $[B(CN)_4]^-$ with different substituents in addition to CN group (s) bonded to boron, such as hydrogen, halogen, alkyl, and alkoxy, have been developed, which makes possible the tuning of features of compounds. The easily accessible alkali metals are convenient starting materials for the preparation of cyanoborates with various organic, and inorganic compounds.

Furthermore, cyanoborates are promising starting compounds for the synthesis of other boron species, for example, the weakly coordinating tetrakis(trifluoromethyl) borate anion $[B(CF_3)^4]^-$ and the boron-centered nucleophile $B(CN)_3^{2-}$. Detailed studies on this subject will be described in more detail in the next section.

Moreover, recent studies have shown that orthoborate anion-based ionic liquid combinations possess outstanding electrolyte characteristics and are attractive solvents for lithium-ion battery solvents. **Figure 5** depicts the simplest orthoborate ion, $[BO_3]^{3-}$. Orthoborate is an anion derived from orthoboric acid ($B(OH)_3$). It is a very weak monobasic that functions solely by hydroxyl-ion acceptance rather than proton donation. Many organic molecules have boron-oxygen bonds (B-O), europium borate ($Eu(BO_2)_3$), chromium borate ($CrBO_3$), beta barium borate (β-BaB_2O_4), gadolinium

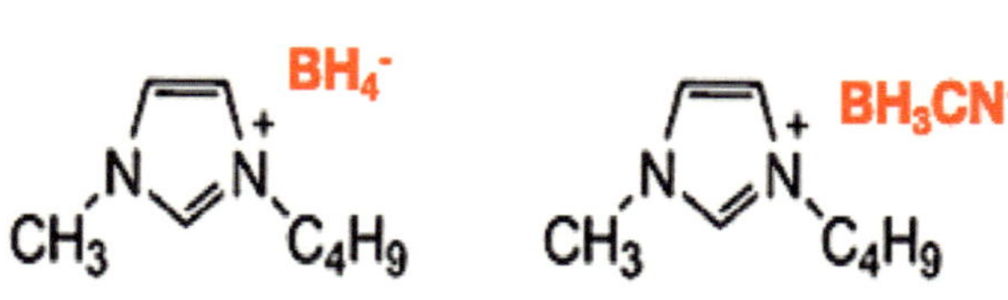

Figure 4.
Borohydride and cyanoborate anion-based ionic liquid (adapted from Refs [53, 55]).

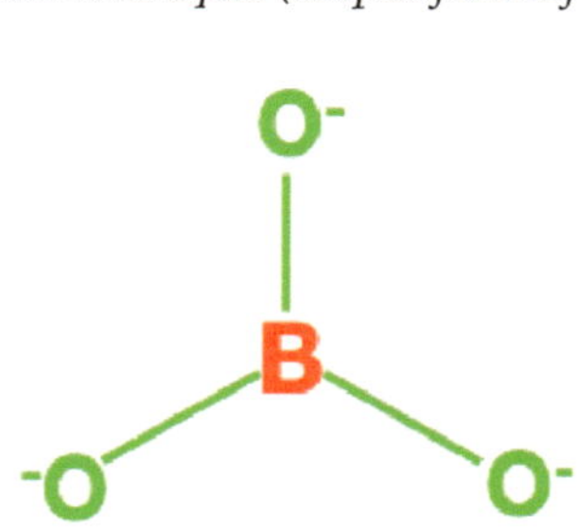

Figure 5.
Simplest orthoborate anion.

orthoborate ($GdBO_3$), and polyborate $[B_3O_9]_9^-$ ion as anions are examples of such compounds [54, 55].

1.2.2 Boron cation based ILs

Studies indicate boron has a significantly higher electropositive charge than most of the donor elements in the ligand (L). After the discovery of cationic boron species, the developed ILs are hard-to-find, highly electrophilic species with high Lewis acidity and reactivity, which are essential in boron chemistry [28, 47]. Boron cation-based ILs, which have high melting points, are used as electrolytes that must be liquid at sub-zero temperatures for electric vehicle applications that must be liquid at ambient temperature battery applications and electrochemical devices due to their relatively low viscosity. Furthermore, the integration of cationic boron centers in organic heterocycles or transition-metal metallocenes is also providing opportunities for the discovery of novel redox-active and optical materials [28].

In recent years, tetracoordinate borocations have had relative stability, which arises from a filled octet and a complete coordination sphere. Because of their larger electronic deficiency and coordinative unsaturation, they are also known to be more reactive than neutral borates [23, 28, 41]. In literature, commonly boron cations are divided into three main groups [28]: Borinium (two-coordinate), borenium (three-coordinate), and boronium (four-coordinate) cation ions (see figure, where: L is a Lewis base; R is substituents, based on the coordination number at boron) [28, 34–47] (see **Figure 6**).

Borinium cations are typically bi-coordinate species bonded by two R that can compensate for the electron deficiency in boron via p donation. Since these species can only have two valence electrons, an additional electron pair must be ensured by an electron donor L (i. e., N or O). That is, at least one L capable of p-bonding must be present to stabilize these cations [46]. In addition, borinium compounds are quite reactive compared to borenium and boronium.

On the other hand, borenium cations are tri-coordinate species containing two bonded R, one L and a third coordination site [28].

The boroniums are the third and most prevalent class of boron cations. The simplicity with which the boronium ions are manufactured and the structural diversity embodied in them provide an essential pragmatic basis for formulating IL. In 2010, Rüther et al. [48] brought out boronium-cation-based RTILs as novel electrolytes for rechargeable lithium batteries. These boronium-containing RTILs exhibited good conductivities and electrochemical windows (4.3–5.8 V). They are stable up to 238

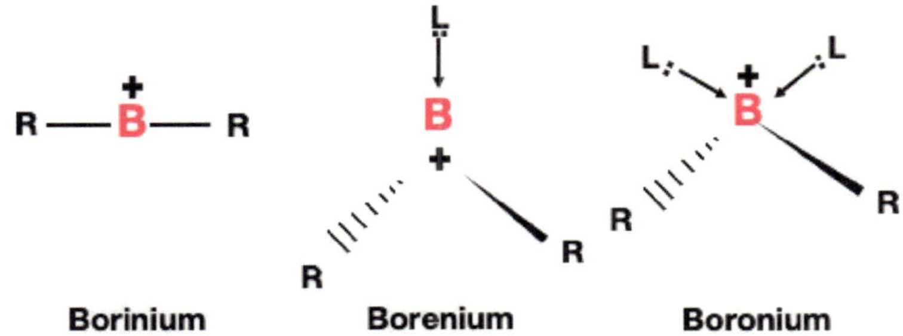

Figure 6.
Boron cation species (adapted from Ref. [44].)

and 335°C, respectively, and have enabled reversible charge-–discharge cycles in batteries with high-capacity retention [48]. Among the boron cations, borinium ions have the most electron deficiencies and are the most chemically unstable species. Thus, while extensive research has been conducted on three-and four-coordinate boron cations, there are only limited examples of borinium ions. Borinium ions generated by electron ionization of borane or borate precursors have been reported to react aggressively and selectively with organic substrates in several cases. However, only a few ionic products are produced, which give valuable structural information on the substrates [46]. Donor ILs in boronium and borinium often serve to extinguish the boron's positive charge. Borenium cations, on the other hand, are better ideal for energy investigations because of the additional stability and electron density offered by donor L. This is evident in the number of studies that describe the production of species or use them as intermediates in various chemical processes [46–48]. Finally, cationic boron compounds are an uncommon but important species in boron chemistry. Emerging research shows that cationic boron compound chemistry is on the verge of a quantum leap in activity; we hope that this review will motivate additional effort in this interesting area of EA.

2. Boron-based ionic liquids for energy-related applications

2.1 Dye sensitized solar cells

Dye-sensitive solar cell technologies continue to be the focus of scientific and industrial research as one of the photovoltaic devices that provide the opportunity to benefit from the sun, which is one of the renewable energy sources. The DSSC projects are widely oriented toward increasing power generation and low-cost generation. The numerous critical processes for DSSC power production mainly occur at the nanocrystal/dye/electrolyte interface (see **Figure** 7), which is now a significant research focus in this field. The maximum efficiency (14.3%) [51] has been obtained using a ruthenium complex in combination with an electrolyte [52, 53]. However, the limited supply of ruthenium and volatile organic solvents are a significant concern for the long-term stable operation of DSSCs. Due to their ideal properties of minimal

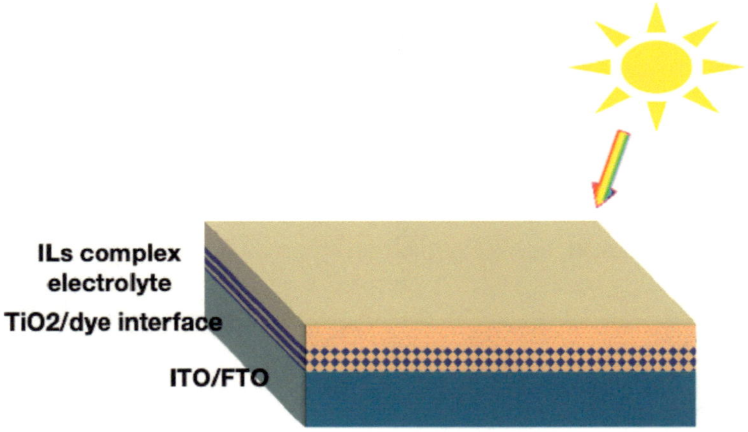

Figure 7.
A systematic view of functionalization ILs in DSSCs (adapted from Ref. [49–53]).

vapor pressures, high conductivity, and thermal stability, RTILs have emerged as a possible choice for enhancing the efficiency of DSSCs. ILs-based DSSCs with ruthenium complexes as sensitizers have already demonstrated impressive photovoltaic performance and stability.

The properties of ILs make them appropriate for use in perovskite solar cells to address significant device efficiency and stability issues, as in DSSC. It was proposed that ILs form a protective layer on the perovskite film to provide the device with moisture and thermal stability. However, no extensive research on boron-containing ILs used in perovskite solar cell technology has been conducted [54, 57]. The DSSC conversion efficiency still lags behind that of organic solvent-containing DSSCs. The fundamental cause of this poor performance is the ILs' high viscosity, which results in mass transfer constraints on the photocurrent under sunlight. To overcome this problem, it uses a combination of a low viscosity IL and a redox-active salt as the electrolyte to optimize the diffusion rate of the redox couple. These additives optimize the kinetics of the processes of electron injection that occur at the photoanode. The nature of IL, especially basic anions such as dicyanamide, can significantly affect the position of the conduction band edge and thus the open-circuit voltage (Voc) of the device. The most efficient IL-based system reported to date employs the I^-/I_3^- redox couple in conjunction with one of a eutectic solution of imidazolium iodide salts [C2mim], yielding good stability $[B(CN)_4]^-$ [53–56, 58]. According to research, when combined with titania/electrolyte in DSSCs, $[B(CN)_4]$ exhibits a higher photocurrent response. $[B(CN)_4]^-$ anions cause a downward displacement of an electrolyte-immersed film's conduction-band edge, resulting in a more favorable energy balance at the titania/dye interface and thus a better exciton dissociation efficiency [53].

Organic dye-sensitized $[B(CN)_4]^-$ IL-based solar cells were reported by Daibin Kuang et al [55]. It has been claimed that the first rapor and organic DSSC combined with these newly developed ILs (without solvent) obtained a conversion efficiency of 7.2% to electrical power. A molecularly tailored indoline sensitizer with 1-ethyl-3-methyl-imidazolium tetracyanoborate ([EMIB $(CN)_4$]) was used in the electrolyte. This is the first time an ionic-liquid electrolyte has been used to achieve such high efficiency for organic dye-based ILs.

Magdalena Marszalek et al. [56] prepared and characterized the $[B(CN)_4]^-$ anion-based series of ILs for use in DSSC applications. These new fluorous-free ILs were composed of $[B(CN)_4]^-$ anions and cations such as imidazolium and ammonium. Synthesized and characterized novel BBILs were evaluated as electrolyte additives in DSSC, with obtained efficiencies of 7.35 and 7.85% under 100 and 10% sun, respectively, in combination with the standard Z907 dye [56].

Promising alternative electrode materials with BBILs may include nanostructures that may have lower charge transfer resistance for I^-/I_3^- relative to a molecular solvent in an IL electrolyte. However, reducing the charge transfer resistance to obtain acceptable performance under full solar radiation is extremely important for device development, and this field of study remains open to considerable advancement.

2.2 Boron-based ionic liquids in advanced battery technologies (Lithium, sodium and magnesium ion batteries)

The properties of ILs, such as inflammability, extremely wide liquid range, and especially the stability of some types of cations and anions at high anodic potentials, make them suitable for lithium batteries. ILs are mixed with appropriate lithium salts to make electrolytes for lithium-ion batteries. The choice of electrolyte used for lithium battery production is critical and determined by a variety of factors, including safety and "greenness." Because RTILs are non-volatile and non-flammable, they are more appealing as lithium battery electrolytes than conventional organic liquid solvents. The structural chemistry of boron and oxygen compounds, the building blocks of orthoborates, is characterized by extraordinary complexity and diversity. Lithium borate (see **Figure 8**), one of the metal borates, is one of the building blocks of IL, which is used as a lithium-ion battery electrolyte. Orthoborate-based ILs have been reported to be more efficient than conventional salts such as $LiPF_6$ as they offer several advantages such as halogen-free, non-toxic, good thermal stability, and high compatibility with cathode materials [59–62].

Faiz Ullah Shah et al. [61] described the ion transport mechanism of a ternary combination of phosphonium bis(salicylato)borate IL, diethylene glycol dibutyl ether, and a lithium bis(salicylato)borate (Li[BScB]) salt for lithium-ion batteries. The ion transport properties and viscosity of the orthoborate-based halogen-free ionic liquid hybrid electrolytes were investigated. In the investigation, the lithium bis(salicylato)borate salt was dissolved in a combination of IL and diethylene glycol dibutyl ether. Diethylene glycol dibutyl ether has a flashpoint at 118°C and is miscible with ILs. This research is the first to look at the ion transport processes of orthoborate-based ionic liquid hybrid electrolytes. As a consequence, the maximal solubility of Li [BScB] salt in a combination of ILs and diethylene glycol dibutyl ether at room temperature was determined to be 1.0 mol kg^{-1}. The viscosity of the combination was 1000 times lower than that of the neat phosphonium bis(salicylato)borate ionic liquid. However, no differences in ionic conductivity were observed between the combination and the neat phosphonium bis(salicylato)borate ionic liquid.

Liang, Fuxiao et al. [62] have investigated a novel BBIL electrolyte for high voltage lithium-ion batteries with outstanding cyclic stability. According to the study, adding an appropriate amount of N-propyl-N methylpiperidinium difluoro(oxalate)borate ($PP_{13}DFOB$) to an electrolyte containing lithium bis(trifluoromethanesulfonyl)-imide (LiTFSI) results in a high discharge capacity. In addition, a promising strategy is presented to mitigate aluminum corrosion of boron-based ionic liquid electrolytes and further improve the cycling performance of lithium-ion batteries at high cut-off voltages in the paper.

The construction of electrochemical cells requires the use of target carrier ions such as lithium cations, protons, or iodides. That is, a matrix that transports these target ions is essential for such applications. Several solutions have been suggested

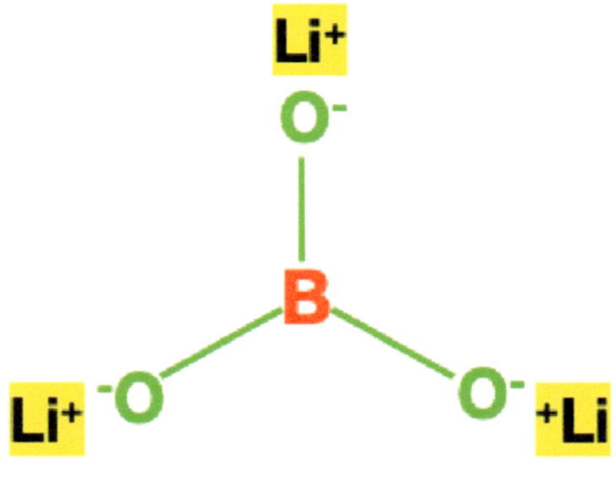

Figure 8.
Lithium orthoborate.

to facilitate selective lithium-ion transport in ILs. Several techniques have been described to accomplish selective lithium-ion transport in ILs. For instance, zwitter ions, which have a cation and an anion in the same molecule, have been proposed to inhibit IL component ion movement under an electrical potential. An IL with alkyl borane units has been synthesized to trap anions by the interaction between boron and the anion [63].

A few publications have subsequently explored variations of IL cation (dominated by small aliphatic and cyclic ammonium cations) and anion (dominated by bis-(trifluoromethanesulfonyl)-imide [$TFSI^-$] or $[BF_4]^-$ combinations with a range of lithium battery anodes and cathodes.

Clarke-Hannaford et al. [64] have conducted enhanced cycling performance in lithium metal batteries of boronium-cation-based ILs. They investigated using a combination of density functional theory calculations and ab initio molecular dynamics simulations, the chemical stability and reaction mechanisms between the Li surface and $[NNBH_2]$[TFSI] and dihydroborate ($[NNBH_2]^+$) to understand the existence of the solid electrolyte interphase layer formed using boronium-cation-based ILs. Simulations showed that the surface interaction with the $[NNBH_2]^+$ cation is weak, and its anions easily dissociated to create numerous chemical species (LiF, Li_2O, and Li_2S). The results provided evidence that the $[NNBH_2]^+$ cation is stable against a lithium metal surface and that self-dissociation of the cation is unlikely to occur. Enhanced cycling performance in lithium metal batteries, a stable solid electrolyte interphase $[NNBH_2]^+$ cation, usually formed on the Li surface, has been shown to have similar properties compared to commonly used cations, helping to explain the positive performance of boronium cation-based ILs.

Research has been done on Na metal to make second-generation batteries a safer, lower-cost option for energy storage. Regarding sodium batteries, Nikitina et al. [49] [bmim] showed that sodium salt, unlike lithium, has only a minor effect on the conductivity, dielectric properties, and viscosity of sodium tetrafluoroborate ($NaBF_4$) in ILs. Electrical conductivities, densities, viscosities, and molar conductivities of [bmim]$[NaBF_4]$ ionic liquid were measured in the wide temperature range of (278.15–358.15) K and (238.15–458.15) K, respectively. The values for viscosities and conductivities were described by the Vogel-Fulcher-Tammann equation. The Walden plots, $\log(\Lambda)$ vs. $\log(\eta{-}1)$, for the $NaBF_4$ solutions, coincide with the straight line found for neat [bmim]$[BF_4]$, indicating that the solute had only a limited impact on the structure of the ionic liquid. Also, they found that the dielectric properties of the most concentrated $NaBF_4$ solution (0.1739 $mol \cdot kg^{-1}$) are identical to those of pure [bmim]$[BF_4]$.

Basile et al. [21] reported on room temperature ionic liquid comprising the dicyanamide anion as a successful electrolyte system for sodium metal batteries that do not contain expensive fluorinated species. At a current density of 10 $\mu A\ cm^{-2}$, the effects of sodium plating and stripping from Na metal electrodes were examined in a symmetrical Na|electrolyte|Naconfiguration. The presence of residual water molecules in the ionic liquid electrolyte was seen to have a significant impact on the surface film and plating/stripping behavior. The increase in moisture content from 90 to 400 ppm has hampered both electrodeposition and electrodissolution of the Na^+/Na. They also used cyclic voltammetry on Ni electrodes at various Na salt concentrations to further understand the mechanism. As a result, the water concentration in this pyrrolidinium ionic liquid alters the Na electrochemistry.

For rechargeable magnesium batteries, Guo and colleagues [18] have developed a boron-based electrolyte system with outstanding electrochemical performance, formed through the reaction of tri (3,5-dimethylphenyl)borane (Mes_3B) and PhMgCl in tetrahydrofuran. In the study, the structure-function correlations of the novel electrolyte were investigated, as well as the identification of the equilibrium types in the solution using NMR, single-crystal XRD, fluorescence spectra, and Raman spectroscopy. Moreover, the electrochemical stability of various current collectors, air sensitivity, and charge-discharge performance of a Mg-Mo_6S_8 battery in the electrolyte are analyzed. As a consequence, fluorescence and Raman spectroscopy investigations revealed that the Mes_3B–$(PhMgCl)_2$ electrolyte's strong anodic stability (about 3.5 V vs. Mg reference electrode) is due to non-covalent interactions between the anion $[Mes_3BPh]^-$ and Ph_2Mg. Motivated by this finding, the researchers proposed a reversible electrochemical technique of Mg intercalation into a Mo6S8 cathode, indicating that the novel boron electrolyte may be used in rechargeable Mg battery systems.

Carter TJ et al. [19] studied a combined carboranyl magnesium halide and closo-borane electrolyte for unconventional electrolyte system optimization in Mg batteries. In the results, they found that closo-borane compounds can function as high-oxidative-stability magnesium-battery electrolytes while maintaining compatibility with magnesium-metal anodes. Also, the carboranyl magnesium halide demonstrated compatibility with magnesium-metal anodes and outstanding oxidative stability (3.2 V vs. Mg) on non-noble-metal electrodes in the study.

2.3 Hypergolic fuels

The concept of hypergolicity is that one chemical (fuel) reacts spontaneously when it comes into contact with another (oxidizer). Hypergolic ILs tend to have low volatility and high thermal and chemical stability that could allow the utilization of these substances as bipropellant fuels under different conditions. Meanwhile, the adjustment of the oxidizer/fuel ratio, the order of adding the fuel and oxidizer, and the ignition temperature should be considered when evaluating new hypergolic fuels.

In the formation of hypergolic boron-based ILs, boron compounds act as triggers, while cations (such as imidazolium) promote hypergolic firing. Recently BBILs frequently have been investigated as components in hypergolic fuel for rocket applications. In order to improve ignition performance and hydrolytic stability, electron-withdrawing moieties such as -CN group and nitrogen heterocyclic ring are often used to tune anions. A strategy of bridging another BH3 moiety or adding phosphorus atoms to anions can also improve the physico-chemical properties of hypergolic ILs.

According to studies in the literature, $[BH_4^-]$ and $[B(CN)H^{-3}]$ were commonly used as anions.

In the design of hypergolic [65–70], hypergolic fuels with $[BH_4]^-$ and $[B(CN)H]^{-3}$ (ultra-fast spontaneous combustion with HNO_3 oxidizers) and amines with borane-based ILs (low ignition delays) are promising.

Zhang et al. [67] reported the synthesis of water-stable hypergolic ILs ($(B(CN)^-$ based) in aqueous media. The B-H bond is unquestionably accountable for the hypergolic nature of compounds based on borohydride and cyanoborate. They stated that dicianoborate-based ILs exhibited similar phase transition temperatures, analogous thermal breakdown temperatures, and lower densities when compared to other hypergolic ILs containing amine anions. As a result, the novel boron-containing anionic ILs had substantially lower viscosities (12.4 mPa s) and ignition delays (4 ms) than nitrocyanamide and dicyanamide hypergolic ionic liquids.

Bhosale et al. [69] synthesized boron-based B-H single bond-rich ILs and investigated hypergolic reactivity. Notably, 1-ethyl-3-methyl imidazolium borohydride ([EMIM]$[BH_4]^-$ IL) distinguished itself by having a short ignition delay time of 18.5 ms. The ignition delay time of [EMIM] $[BH_4]^-$ and methyl imidazolium borane (1:1, w/w) as a combination of IL and fuel was determined to be around 35 ms. In addition, two algorithms were utilized to calculate the gas-phase heat of formation and the specific impact. One of the most promising possibilities for next-generation green hypergolic fluid and hybrid rocket propulsion has been identified as ILs rich in B-H single bonds.

Another hypergolic fuel study belongs to Li et al. [70]. They employed borane derivatives as additives to enhance the ignition delay of borohydride-based hypergolic ILs. First, they synthesized borohydride-based ILs (Amim-BH_4 and Bmim-BH_4) and then investigated the ignition delay time of these ILs with borane additives. The synthesized borohydride-based hypergolic ILs exhibited the shortest ignition delay time compared to any known hypergolic IL in the study. Also, they discovered that the triethylamine-borane combination is the most effective hypergolic additive studied to date. While the ID times of borohydride-based ILs are as short as 2 ms, IL solutions of borane-based additives have the lowest ID of 3 ms.

Based on the above explanations, borohydrides and boranes have great potential not only for ignition acceleration but also for alternative fuels due to their higher densities, lower vapor pressures, and adjustable heats of formation and viscosities. Although hydrazine and its derivatives are poisonous and carcinogenic, this new IL family offers a more environmentally friendly option. The increasing use of boron and its analogs to treat hypergolic ILs has resulted in more cost-effective solutions. However, in propellant systems, hydrazine and its derivatives remain the fuel of choice, despite containing a class of acute carcinogens and toxicants that exhibit extremely high vapor pressures and require expensive handling procedures and costly safety precautions.

2.4 BBILs for hydrogen storage and production

Having excellent hydrogen densities, safety, and release rate characteristics, many boron molecules can retent substantially more releasable hydrogen by weight and volume than pure liquid hydrogen [22]. Boron compounds are used as anions in ILs for hydrogen-containing applications, along with cations such as N,N′-dialkylimidazolium, N-alkylpyridinium, tetraalkylammonium, and tetraalkylphosphonium [35]. Particularly for commercially accessible sodium borohydride ($NaBH_4$), hydrogen technologies provide easy and compact power for portable devices and backup power systems.

Amir Doroodian et al. [71] conducted one of these investigations. They presented the first ionic liquid based on methylguanidinium borohydride ((N_3H_8C)C + BH_4) with effective hydrogen storage capacity. It is stated in the study that a liquid electrolyte is an ionic liquid that releases 9.0% by weight H_2 under both thermal and catalytic conditions.

Developing technologies based on metal borohydrides ($M(BH_4)n$) and amine boranes (AB, NH_3BH_3) on the other hand, have the potential to produce greater power densities than established sodium borohydride systems [72]. AB (see **Figure 9**) is a class of borohydride and one of the leading candidates for chemical hydrogen storage, which can release 19.6 wt% H_2 when heated (at 85°C) due to its high hydrogen

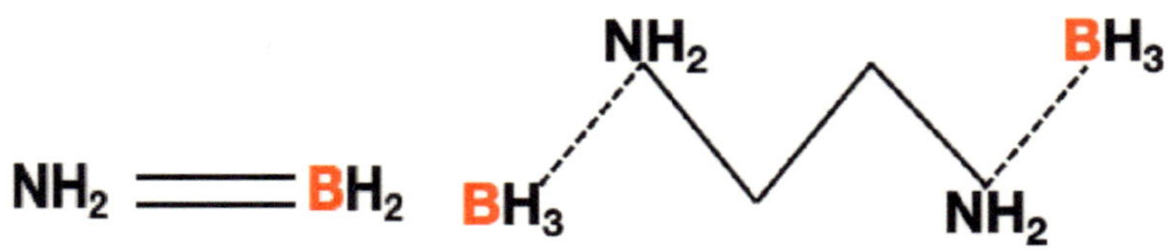

Figure 9.
Structure of aminoborane and diamine bisborane.

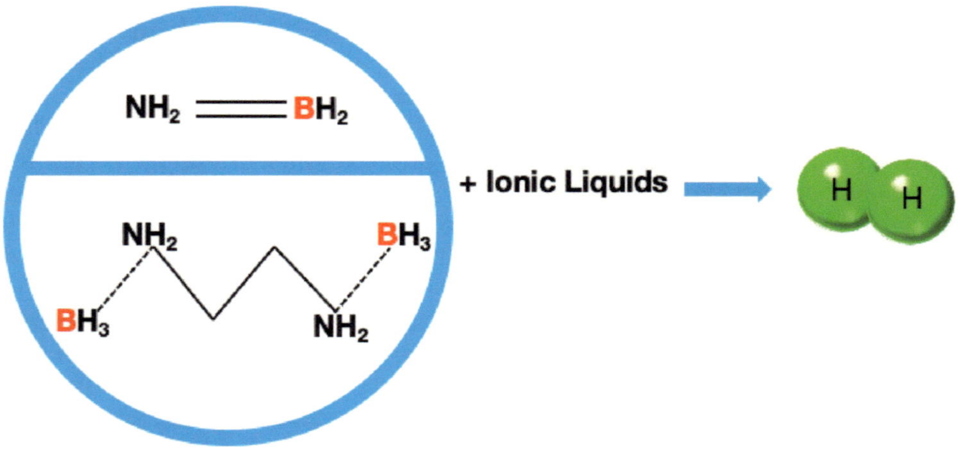

Figure 10.
Aminoborane and diamine bisborane mix ILs for degeneration of hydrogen.

content [73]. A feasible AB regeneration technique, as well as a quick and regulated H_2 release rate are required for AB mix ILs to be suitable for hydrogen storage. Several approaches can accelerate the release of AB H_2 mechanically, including activation with transition metal catalysts. **Figure 10** depicts the systematic figure of AB and diamine bisborane mix ILs for the degeneration of hydrogen.

Daniel et al. [74, 75] showed that base-supported AB increases the rate of H2 release in ILs by promoting the anionic dehydropolymerization mechanism. They reported AB reactions in 1-butyl-3-methylimidazolium chloride that took 171 min at 85°C and only 9 min at 110°C to produce equivalents of H_2. Moreover, ionic-liquid solvents were shown to be more beneficial than other solvents because they minimized the development of unwanted compounds like borazine.

The effect of ethylenediamine bisborane (EDAB), one of the AB derivatives, on the hydrogen release rate in combination with ILs has been studied in the literature. Many research have been conducted to improve dehydrogenation, reduce dehydrogenation temperatures, and enhance equivalent H2 production [75–77]. In one of these studies, Debashis Kundu et al. [77] reported thermal dehydrogenation of EDAB in [BMIM] sulfate-based ILs. The time-resolved and temperature-resolved dehydrogenation of EDAB/IL systems have been carried out and characterized by NMR characterization in the study. Also, the equal quantity of equivalent hydrogen created per mole of EDAB injected into the system was computed. The results showed that IL-facilitated dehydrogenation released a higher amount of equivalent hydrogen than dehydrogenation with pure EDAB at 120°C. At 100°C, the EDAB/[BMIM] [HSO_4] system released 3.92 cumulative equivalent hydrogen.

2.4.1 Hydrogen generation by water splitting

A significant trend in the creation of hydrogen by electrolysis from a mixture of ILs and water, a renewable resource, has lately emerged. Water splitting includes the simultaneous oxidation and reduction of water [78–80]. Water ILs serve as an electrolyte and a solvent in the water-splitting process reactant. Water is merely one of several solutes present in the reaction mixture. When used in water separation, IL electrolytes have produced unique and surprising outcomes.

Roberto F. de Souza et al. [80] tested different electrocatalysts (molybdenum, nickel, and chromium) using aqueous ILs such as [$BMIMBF_4$] for hydrogen production by water electrolysis. The hydrogen evolution reaction (HER) was carried out at ambient temperature with a potential of −1.7 V. For the Mo electrode, a Hoffman cell apparatus with a current density value of 77.5 mA/cm^{-2} in water electrolysis was used. The system efficiency for all tested electrocatalysts was found to be very high, ranging between 97.0 and 99.2%. The findings indicate that hydrogen production in BMIM[BF_4]$^-$ aqueous solution can be performed with inexpensive materials at room temperature, making this process economically viable.

Hydrogen bonds are more readily broken when water molecules are dissolved in a suitable IL, thus increasing the free energy and resulting in a lower energy input required in the water-splitting reaction. The free energy can approach that of gaseous water, which requires a lower free energy input of around 10 kJ mol^{-1}. Such a process must include a thermodynamically endothermic water dissolution process. As an example, [C_2mim] [$B(CN)_4$] based ILs have been proposed. However, no data on the thermodynamic process in BBILs is yet available. Also, the four-electron water oxidation reaction (see **Figure 11**) is much heavier than the water reduction reaction. That is why there has recently been a significant effort focused on developing and understanding water oxidation electrocatalysts.

2.5 Non-aqueous boron-based electrolyte for electrodeposition

This section discusses the electrodeposition of metals from non-aqueous BBILs. The application of non-aqueous BBILs makes it feasible to obtain cathodic residues of metals or metallic alloys that cannot be deposited by conventional electrolysis of aqueous solutions. Since the hydrogen overpotential is relatively low, it does not allow some electrodeposition reactions to occur in aqueous solutions. Due to the intrinsic conductivity of ILs, many metals and semiconductors may be electrodeposited directly from an IL solution of metal ions or metal-ILs ion complexes using standard electrowinning techniques. Because of the low volatility of IL, the procedure may be carried out at high temperatures exceeding 100°C. The number of studies published on non-aqueous solution electrodeposition has grown in recent years [33].

Yanna NuLi and colleagues [81] investigated highly reversible magnesium precipitation and dissolution processes in the ionic liquid of 1 Mg $(CF_3SO_3)_2$ and 1-butyl-3-methylimidazolium tetrafluoroborate [$BMIMBF_4$]. According to the scanning electron microscopy (SEM) data, micrometric-sized, virtually pyramidal-shaped magnesium deposits appeared, and when the magnesium dissolved, the electrode became clean and film-free. Cyclic voltammograms of Mg accumulation-dissolution also show that these reactions are reversible. [$BMIMBF_4$] demonstrated especially promising properties in terms of electrochemical window (Pt vs. 4.2 V) and

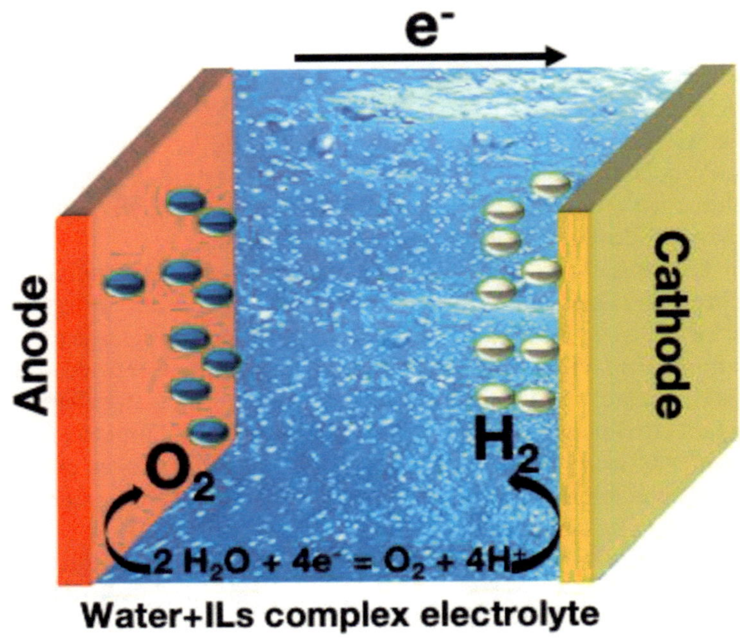

Figure 11.
Water electrolysis in ILs electrolyte mixture.

magnesium deposition-dissolution efficiency. As a result, this method may be acceptable for application in rechargeable Mg batteries.

Survilien et al. [82] studied the electrodeposition of the chromium from the ionic liquid [$BMIMBF_4$] and chromium chloride. The cathodic process of chromium electrodeposition procedure from the ionic liquid used was as follows: electrochemical reduction of water molecules, followed by chemical degradation of [BF4] ions. According to the findings of this investigation, [$BMIMBF_4$] can be deemed promising for hazardous Cr (VI) baths for black chrome plating.

Menzel et al. [83] examined the development of the mechanism of the ZnO nanowire growth model by studying [$BMIMBF_4$] as an ionic liquid. They concentrated on the effects of IL sources on nanowire development in this work. They discovered that because B is an n-type source of ZnO material, IL-promoted growth may be employed to recruit donors. Electrical measurements of XPS and ZnO nanowires validated these hypotheses. Electrical measurements revealed that ionic liquid-assisted growth improved electrical conductivity (=0.09 cm). [$BMIMBF_4$] IL-assisted nanowire growth showed that boron was significantly involved in the alteration of nanowire growth properties compared to pure ZnO nanowire growth.

For electrodeposition, another IL, [C_2mim] [BF4], was utilized. Steichen M and Dale P [84] investigated the electrodeposition of trigonal selenium (t-Se) nanorods from [C_2mim][BF_4]/[C_2mim]Cl at high temperatures ($T > 100$°C).phase, morphology, and crystallinity of Se residues vary, the choice of precursor salts also controls the electrodeposition of selenium. They demonstrated for the first time that t-Se nanorods may be made at high temperatures using a template electrodeposition route from [C_2mim][BF_4]/[C_2mim]Cl. The crystal quality of t-Se nanorods improves when the temperature rises above 100°C.

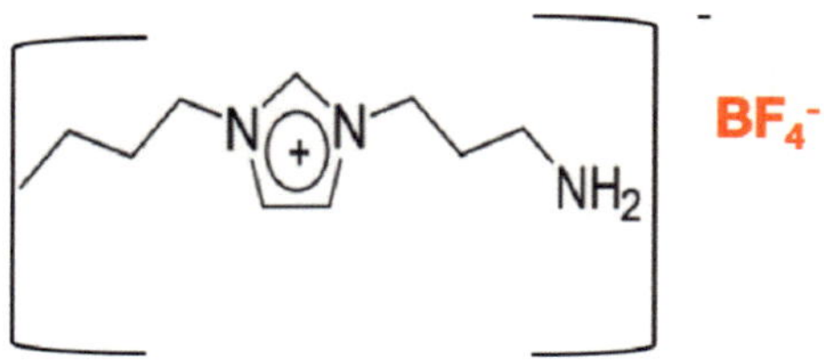

Figure 12.
Structure of ILs designed capture of CO_2 (adapted from Ref. [85, 86]).

According to the researchers, photoactive t-Se nanorods with p-type conductivity have been detected in an ionic liquid for the first time. Photoelectrochemical measurements performed in ionic liquid confirmed the p-type conductivity of t-Se nanorods.

2.6 Carbon dioxide capture using BBILs

The use of ILs to help separate CO_2 from other gases has recently become a hot topic of activity both in academia and industry. ILs' lack of volatility is a particular advantage in CO_2 absorption over molecular liquid absorbers. Carbon capture has become legally mandatory for electric power plants to operate. However, there is a net cost of carbon capture, and these costs are passed on to consumers. For this reason, it is critical to design and synthesize efficient CO_2 capture ILs made from the most basic and inexpensive building blocks, especially since the volumes required to achieve CO_2 capture using ILs would be prohibitively expensive at envisioned scales [85, 86]. In recent years, many publications have appeared on IL-CO_2 interactions of one kind or another (**Figure 12**).

Certain functional groups in the ionic liquid, such as anions of the amide family, absorb CO_2 up to 0.5 mol/mol IL through the traditional carbamate reaction. CO_2 absorption is enabled via similar functionalization through direct contact with amine or anionic functional groups. In addition, several ionic liquids also show potential in other environmentally conscious applications, such as CO_2 capture. Some of these novel "task-specific" ionic liquids have shown promise in CO_2 capture.

Using the ionic liquids 1-*n*-butyl-3-methylimidazolium hexafluorophosphate [BMIM] [PF_6^-] and [BMIM] [BF_4^-], Anthony et al. [87] investigated CO_2 capture. They also compare CO_2 collection utilizing these ILs to conventional monoethanolamine-based technologies. According to the findings, [BMIM] [PF_6^-] is particularly effective in capturing CO_2 from a mixture of N_2 or CH_4. As a result, these ILs are a viable choice for CO_2 capture in order to create an ionic liquid with a carrying capacity comparable to monoethanolamine.

3. Conclusion

Because as the fields of energy transmission, storage, and conversion continue to grow technologically, IL-based systems can provide a platform for safe, clean, durable, and high-energy-density materials, especially halogen-free and fluorine-free. Currently, there are limited studies based on boron atoms or particles, although millions of IL combinations with existing cations and anions are possible. In the field of energy storage, this focus has been mainly on fluorinated anions because of their stability.

There is a need to design halogen-free anions and cations with high electrochemical stability. However, in the context of energy science and technology, it is also necessary to quantitatively assess toxicity, recyclability, and biodegradability, for which there is insufficient information in most cases. For most energy applications of ILs, the cost is another issue for those with high electrochemical stability.

In this context, this review, recent advances in the synthesis of boron-containing ILs and various materials derived from them provide a powerful opportunity for further investigation in the field of energy research to improve and develop the properties of these BBILs. With this perspective, the article can guide further research and development of the unique properties of green and halogen-free BBILs.

Acknowledgements

I am grateful to the anonymous referees for their useful and constructive comments that contributed to the review of this chapter and to the editors of the book for their support in its publication.

Author details

Gülbahar Bilgiç
Faculty of Engineering and Architecture, Department of Energy Systems Engineering, Nevşehir Hacı Bektaş Veli University, Nevşehir, Turkey

*Address all correspondence to: glbhrblg@gmail.com

References

[1] MacFarlane DR, Tachikawa NM, Pringle JM, Howlett PC, Elliott GD, et al. Energy applications of ionic liquids. Energy & Environmental Science. 2014;7:232-250. DOI: 10.1039/c3ee42099j

[2] Walden P. Molecular weights and electrical conductivity of several fused salts. Bulletin of the Imperial Academy of Sciences in St. Petersburg. 1914;**1800**:405-422

[3] Wilkes JS, Zaworotko MJ. Air and water stable 1-ethyl-3-methylimidazolium based ionic liquids. Journal of the Chemical Society, Chemical Communications. 1992:965-967. DOI: 10.1039/c39920000965

[4] Wilkes JS. A short history of ionic liquids-from molten salts to neoteric solvents. Green Chemistry. 2002;**4**:73-80. DOI: 10.1039/b110838g

[5] Plechkova NV, Seddon KR. Applications of ionic liquids in the chemical industry. Chemical Society Reviews. 2008;**37**:123-150. DOI: 10.1039/b006677j

[6] Rogers RD, Seddon KR. Ionic liquids-solvents of the future? Science. 2003;**302**:792-793. DOI: 10.1126/science.1090313

[7] Ranke J, Stolte S, Störmann R, Arning J, Jastorff B. Design of sustainable chemical products the example of ionic liquids. Chemical Reviews. 2007;**107**:2183-2206. DOI: 10.1021/cr050942s

[8] Forsyth SA, Pringle JM, MacFarlane DR. Ionic liquids-an overview. Australian Journal of Chemistry. 2004;**57**:113. DOI: 10.1071/ch03231

[9] Rabideau B, West K, Davis J. Making good on a promise: Ionic liquids with genuinely high degrees of thermal stability. Chemical Communications. 2018;**54**(40):5019-5031. DOI: 10.1039/c8cc01716f

[10] Ue M, Takeda M, Toriumi A, Kominato A, Hagiwara R, Ito Y. Application of low-viscosity ionic liquid to the electrolyte of double-layer capacitors. Journal of the Electrochemical Society. 2003;**150**(4):A499. DOI: 10.1149/1.1559069

[11] Matsumoto H, Yanagida M, Tanimoto K, Nomura M, Kitagawa Y, Miyazaki Y. Highly conductive room temperature molten salts based on small trimethylalkylammonium cations and bis(trifluoromethylsulfonyl)imide. Chemistry Letters. 2000;**29**(8):922-923. DOI: 10.1246/cl.2000.922

[12] Pérez P, Uhl S, Laux E, Sanglard P, Marti R, Keppner H, et al. Synthesis and structure modification of ionic liquids to optimize their thermoelectric properties. Materials Today: Proceedings. 2018;**5**(4):10298-10305

[13] Seddon K. ChemInform abstract: Room-temperature ionic liquids: Neoteric solvents for clean catalysis. ChemInform. 2010;**28**(9):37, 743

[14] Hayes R, Warr G, Atkin R. Structure and nanostructure in ionic liquids. Chemical Reviews. 2015;**115**(13):6357-6426. DOI: 10.1021/cr500411q

[15] Galiński M, Lewandowski A, Stępniak I. Ionic liquids as electrolytes. Electrochimica Acta. 2006;**51**(26):5567-5580. DOI: 10.1016/j.electacta.2006.03.016

[16] Kalugin ON, Voroshylova IV, Riabchunova AV, Lukinova EV,

Chaban VV. Conductometric study of binary systems based on ionic liquids and acetonitrile in a wide concentration range. Electrochimica Acta. 2013;**105**:188-199. DOI: 10.1016/j.electacta.2013.04.140

[17] Garcia B, Lavallée S, Perron G, Michot C, Armand M. Room temperature molten salts as lithium battery electrolyte. Electrochimica Acta. 2004;**49**:4583-4588. DOI: 10.1016/j.electacta.2004.04.041

[18] Guo Y-S, Zhang F, Yang J, Wang F-F, NuLi Y, Hirano S-I. Boron-based electrolyte solutions with wide electrochemical windows for rechargeable magnesium batteries. Energy and Environmental Science. 2012;**5**:9100

[19] Carter TJ, Mohtadi R, Arthur TS, Mizuno F, Zhang R, Shirai S, et al. Boron clusters as highly stable magnesium-battery electrolytes. Angewandte Chemie International Edition. 2014;**53**:3173-3177

[20] Vardar G, Sleightholme AE, Naruse J, Hiramatsu H, Siegel DJ, Monroe CW. Electrochemistry of magnesium electrolytes in ionic liquids for secondary batteries. ACS Applied Materials and Interfaces. 2014;**6**:18033-18039

[21] Basile A, Yoon H, MacFarlane DR, Forsyth M, Howlett PC. Investigating non-fluorinated anions for sodium battery electrolytes based on ionic liquids. Electrochemistry Communications. 2016;**71**:48-51. DOI: 10.1016/j.elecom.2016.07.014

[22] Huang Z, Wang S, Dewhurst RD, Ignat'ev NV, Finze M, Braunschweig H. Boron: Its role in energy-related processes and applications. Angewandte Chemie International Edition. 2020;**59**:8800-8816. DOI: 10.1002/anie.201911108

[23] Shah FU, Glavatskih S, Antzutkin ON. Boron in tribology: From borates to ionic liquids. Tribology Letters. 2013;**51**:281-301. DOI: 10.1007/s11249-013-0181-3

[24] Oganov AR, Solozhenko VL. Boron: A hunt for Superhard polymorphs. Journal of Superhard Materials. 2009;**31**:285-291. DOI: 10.3103/s1063457609050013

[25] Bilgiç G, Şahin M, Kaplan H. A system design for large scale production of elemental boron by electrochemical deposition. Journal of The Electrochemical Society. 2020;**167**:162513. DOI: 10.1149/1945-7111/abd2d7

[26] Oganov AR, Chen J, Gatti C, Ma Y, Ma Y, Glass CW, et al. Ionic high-pressure form of elemental boron. Nature. 2009;**460**:292. DOI: 10.1038/nature08164

[27] Braunschweig H, Whittell GR. Boron as a bridging ligand. Chemistry—A European Journal. 2005;**11**:6128-6133. 11, 6128. DOI: 10.1002/chem.200500368

[28] Piers WE, Bourke SC, Conroy KD. Borinium, borenium, and boronium ions: Synthesis, reactivity, and applications. Angewandte Chemie International Edition. 2005;**44**:5016-5036. DOI: 10.1002/anie.200500402

[29] Rodrigues Silva D, Azevedo Santos L, Freitas MP, Guerra CF, Hamlin TA. Nature and strength of Lewis acid/base interaction in boron and nitrogen trihalides. Chemistry—An Asian Journal. 2020;**15**:4043-4054. DOI: 10.1002/asia.202001127

[30] Erker G. Tris(pentafluorophenyl) Borane: A special boron Lewis acid for special reactions. Dalton Transactions. 2005:1883-1890. DOI: 10.1039/b503688g

[31] Zhu Y, Hosmane NS. Ionic liquids: Recent advances and applications in boron chemistry. European Journal of Inorganic Chemistry. 2017;**2017**:4369-4377. DOI: 10.1002/ejic.201700553

[32] Axtell J, Saleh L, Qian E, Wixtrom A, Spokoyny A. Synthesis and applications of perfunctionalized boron clusters. Inorganic Chemistry. 2018;**57**(5):2333-2350. DOI: 10.1021/acs.inorgchem.7b02912

[33] Bilgiç G, Korkmaz N, Şahin M, Karadağ A. Synthesis, structural, and electrochemical properties of boron-based ionic liquid. Ionics. 2022;**28**:3289-3300. DOI: 10.1007/s11581-022-04575-7

[34] Davis JH, Ruether T, Dorman SC. (Keynote) boronium based ionic liquids: Salts of boron centered cations as promising salts for electrochemical applications. ECS Transactions. 2013;**50**:293-299. DOI: 10.1149/05011.0293ecst

[35] Nieuwenhuyzen M, Seddon KR, Teixidor F, Puga AV, Viñas C. Ionic liquids containing boron cluster anions. Inorganic Chemistry. 2009;**48**:889-901. DOI: 10.1021/ic801448w

[36] Holbrey JD, Seddon KR. The phase behaviour of 1-alkyl-3-methylimidazolium tetrafluoroborates; ionic liquids and ionic liquid crystals. Journal of the Chemical Society, Dalton Transactions. 1999:2133-2140

[37] Zhou Z-B, Matsumoto H, Tatsumi K. Low-melting, low-viscous, hydrophobic ionic liquids: Aliphatic quaternary ammonium salts with perfluoroalkyltrifluoroborates. Chemistry—A European Journal. 2005;**11**:752-766. DOI: 10.1002/chem.200400817

[38] Tsurumaki E, Hayashi S-Y, Tham FS, Reed CA, Osuka A. Planar subporphyrin borenium cations. Journal of the American Chemical Society. 2011;**133**:11956-11959. DOI: 10.1021/ja2056566

[39] Sowmiah S, Srinivasadesikan V, Tseng M-C, Chu Y-H. On the chemical stabilities of ionic liquids. Molecules. 2009;**14**:3780-3813. DOI: 10.3390/molecules14093780

[40] MacFarlane DR, Pringle JM, Howlett PC. Ionic liquids and reactions at the electrochemical interface. Physical Chemistry Chemical Physics. 2010;**12**:1659. DOI: 10.1039/b923053j

[41] Hosmane NS, Maguire JA. Evolution of C2B4 carborane chemistry: From early years to the present. Organometallics. 2005;**24**:1356-1389. DOI: 10.1021/om0490031

[42] Reed CA. H+, CH3+, and R3si+ carborane reagents: When triflates fail. Accounts of Chemical Research. 2009;**43**:121-128. DOI: 10.1021/ar900159e PMC 2808449. PMID 19736934

[43] Chen Z, King RB. Spherical aromaticity: Recent work on fullerenes, polyhedral boranes, and related structures. Chemical Reviews. 2005;**105**:3613-3642. DOI: 10.1021/cr0300892

[44] Matsumoto H, Kageyama H, Miyazaki Y. Room temperature ionic liquids based on small aliphatic ammonium cations and asymmetric amide anions. Chemical Communications. 2002:1726-1727. DOI: 10.1039/b204046h

[45] Tanaka N, Shoji Y, Fukushima T. Convenient route to monocarba-*closo*-dodecaborate anions. Organometallics. 2016;**35**:2022-2025. DOI: 10.1021/acs.organomet.6b00309

[46] Shoji Y, Tanaka N, Mikami K, Uchiyama M, Fukushima T. A

two-coordinate boron cation featuring C–B+–C bonding. Nature Chemistry. 2014;**6**:498-503. DOI: 10.1038/nchem.1948

[47] Koelle P, Noeth H. The chemistry of borinium and borenium ions. Chemical Reviews. 1985;**85**(5):399-418. DOI: 10.1021/cr00069a004

[48] Rüther T, Huynh TD, Huang J, Hollenkamp AF, Salter EA, Wierzbicki A, et al. Stable cycling of lithium batteries using novel boronium-cation-based ionic liquid electrolytes. Chemistry of Materials. 2009;**22**:1038-1045. DOI: 10.1021/cm9019815

[49] Nikitina VA, Nazet A, Sonnleitner T, Buchner R. Properties of sodium tetrafluoroborate solutions in 1-butyl-3-methylimidazolium tetrafluoroborate ionic liquid. Journal of Chemical & Engineering Data. 2012;**57**:3019-3025. DOI: 10.1021/je300603d

[50] Shea SL, MacKinnon P, Thornton-Pett M, Kennedy JD. Macropolyhedral boron-containing cluster chemistry. Inorganica Chimica Acta. 2005;**358**:1709-1714. DOI: 10.1016/j.ica.2004.07.063

[51] Kakiage K, Aoyama Y, Yano T, et al. Highly-efficient dye-sensitized solar cells with collaborative sensitization by silyl-anchor and carboxy-anchor dyes. Chemical Communications. 2015;**51**:15894-15897. DOI: 10.1039/c5cc06759f

[52] Li F, Jennings JR, Wang X, et al. Influence of ionic liquid on recombination and regeneration kinetics in dye-sensitized solar cells. The Journal of Physical Chemistry C. 2014;**118**:17153-17159. DOI: 10.1021/jp502341a

[53] Zhou D, Bai Y, Zhang J, et al. Anion effects in organic dye-sensitized mesoscopic solar cells with ionic liquid electrolytes: Tetracyanoborate vs dicyanamide. The Journal of Physical Chemistry C. 2010;**115**:816-822. DOI: 10.1021/jp109803n

[54] Ghosh S, Singh T. Role of ionic liquids in organic-inorganic metal halide perovskite solar cells efficiency and stability. Nano Energy. 2019;**63**:103828. DOI: 10.1016/j.nanoen.2019.06.024

[55] Kuang D, Wang P, Ito S, Zakeeruddin SM, Grätzel M. Stable mesoscopic dye-sensitized solar cells based on tetracyanoborate ionic liquid electrolyte. Journal of the American Chemical Society. 2006;**128**:7732-7733. DOI: 10.1021/ja061714y

[56] Marszalek M, Fei Z, Zhu D-R, et al. Application of ionic liquids containing tricyanomethanide [c(cn)3]− or tetracyanoborate [b(cn)4]− anions in dye-sensitized solar cells. Inorganic Chemistry. 2011;**50**:11561-11567. DOI: 10.1021/ic201513m

[57] Shi J, Ran Z, Peng F. Promising four-coordinated organoboron emitters for organic light-emitting diodes. Dyes and Pigments. 2022:110383. DOI: 10.1016/j.dyepig.2022.110383

[58] Kothavale SS, Lee JY. Three- and four-coordinate, boron-based, thermally activated delayed fluorescent emitters. Advanced Optical Materials. 2020;**8**:2000922. DOI: 10.1002/adom.202000922

[59] Chen Z, Lu WQ, Liu J, Amine K. Lipf6/LiBOB blend salt electrolyte for high-power lithium-ion batteries. Electrochimica Acta. 2006;**51**:3322-3326. DOI: 10.1016/j.electacta.2005.09.027

[60] Lee D-J, Hassoun J, Panero S, Sun Y-K, Scrosati B. A tetraethylene glycol

dimethylether-lithium bis(oxalate) borate (TEGDME-LiBOB) electrolyte for advanced lithium ion batteries. Electrochemistry Communications. 2012;**14**:43-46. DOI: 10.1016/j.elecom.2011.10.027

[61] Shah FU, Gnezdilov OI, Gusain R, Filippov A. Transport and association of ions in lithium battery electrolytes based on glycol ether mixed with halogen-free orthoborate ionic liquid. Scientific Reports. 2017;7(1):16340. DOI: 10.1038/s41598-017-16597-7

[62] Liang F, Yu J, Chen J, Wang D, Lin C, Zhu C, et al. A novel boron-based ionic liquid electrolyte for high voltage lithium-ion batteries with outstanding cycling stability. Electrochimica Acta. 2018;**283**:111-120. DOI: 10.1016/j.electacta.2018.06.170

[63] Ohno H, Yoshizawa-Fujita M, Kohno Y. Design and properties of functional zwitterions derived from ionic liquids. Physical Chemistry Chemical Physics. 2018;**20**(16):10978-10991. DOI: doi.org/10.1039/c7cp08592c

[64] Clarke-Hannaford J, Breedon M, Rüther T, Spencer MJ. Stability of boronium cation-based ionic liquid electrolytes on the Li metal anode surface. ACS Applied Energy Materials. 2020;**3**:5497-5509. DOI: 10.1021/acsaem.0c00482

[65] Ignat'ev NV, Finze M. Cyanoborates. European Journal of Inorganic Chemistry. 2019;**2019**:3537. DOI: 10.1002/ejic.201900832

[66] Liu T, Qi X, Huang S, Jiang L, Li J, Tang C, et al. Exploiting hydrophobic borohydride-rich ionic liquids as faster-igniting rocket fuels. Chemical Communications. 2016;**52**:2031-2034. DOI: 10.1039/c5cc09737a

[67] Zhang Y, Shreeve JM. Dicyanoborate-based ionic liquids as hypergolic fluids. Angewandte Chemie. 2010;**123**:965-967. DOI: 10.1002/ange.201005748

[68] Zhang Q, Yin P, Zhang J, Shreeve JM. Cyanoborohydride-based ionic liquids as green aerospace bipropellant fuels. Chemistry—A European Journal. 2014;**20**:6909-6914. DOI: 10.1002/chem.201402704

[69] Bhosale VK, Jeong J, Kwon S. Ignition of boron-based green hypergolic fuels with hydrogen peroxide. Fuel. 2019;**255**:115729. DOI: 10.1016/j.fuel.2019.115729

[70] Li S, Gao H, Shreeve JM. Borohydride ionic liquids and borane/ionic-liquid solutions as hypergolic fuels with superior low ignition-delay times. Angewandte Chemie. 2014;**126**:3013-3016. DOI: 10.1002/ange.201309044

[71] Doroodian A, Dengler JE, Genest A, Rösch N, Rieger B. Methylguanidinium borohydride: An ionic-liquid-based hydrogen-storage material. Angewandte Chemie International Edition. 2010;**49**:1871-1873. DOI: 10.1002/anie.200905359

[72] David M. Schubert Boron Chemistry for Hydrogen Storage. Chapter. Boron Chemistry for Hydrogen Storage In book: Boron Science - New Technologies and ApplicationsChapter: Boron Chemistry for Hydrogen StoragePublisher: CRC PressEditors: N. S. Hosmane, Hydrogen storage. January 2012. DOI: 10.1201/b11199-23

[73] Smythe NC, Gordon JC. Ammonia borane as a hydrogen carrier: Dehydrogenation and regeneration. European Journal of Inorganic Chemistry. 2010;**2010**:509-521. DOI: 10.1002/ejic.200900932

[74] Himmelberger DW, Alden LR, Bluhm ME, Sneddon LG. Ammonia

borane hydrogen release in ionic liquids. Inorganic Chemistry. 2009;**48**:9883-9889. DOI: 10.1021/ic901560h

[75] Himmelberger DW, Yoon CW, Bluhm ME, Carroll PJ, Sneddon LG. Base-promoted ammonia borane hydrogen-release. Journal of the American Chemical Society. 2009;**131**:14101-14110. DOI: 10.1021/ja905015x

[76] Sahler S, Konnerth H, Knoblauch N, Prechtl M. Hydrogen storage in amine boranes: Ionic liquid supported thermal dehydrogenation of ethylene diamine bisborane. International Journal of Hydrogen Energy. 2013;**38**(8):3283-3290. DOI: doi.org/10.1016/j.ijhydene.2012.12.150

[77] Kundu D, Chakma S, Saikrishnan S, Pugazhenthi G, Banerjee T. Molecular modeling and experimental insights for the dehydrogenation of ethylene diamine bisborane using hydrogen sulfate based ionic liquid. Journal of Industrial and Engineering Chemistry. 2019;**70**:472-483. DOI: 10.1016/j.jiec.2018.11.010

[78] Milikić J, Oliveira RC, Tapia A, Santos DM, Zdolšek N, Trtić-Petrović T, et al. Ionic liquid-derived carbon-supported metal electrocatalysts as anodes in direct borohydride-peroxide fuel cells. Catalysts. 2021;**11**:632. DOI: 10.3390/catal11050632

[79] Lima DW, Fiegenbaum F, Trombetta F, de Souza MO, Martini EMA. PtNi and PtMo nanoparticles as efficient catalysts using TEA-PS.BF4 ionic liquid as electrolyte towards HER. Internation al Journal of Hydrogen Energy. 2017;**42**(9):5676-5683. DOI: 10.1016/j.ijhydene.2016.11.166

[80] de Souza RF, Loget G, Padilha JC, Martini EMA, de Souza MO. Molybdenum electrodes for hydrogen production by water electrolysis using ionic liquid electrolytes. 2008:**10**(11);1673-1675. DOI: 10.1016/j.elecom.2008.08.029

[81] NuLi Y, Yang J, Wu R. Reversible deposition and dissolution of magnesium from BMIMBF4 ionic liquid. Electrochemistry Communications. 2005;**7**:1105-1110. DOI: 10.1016/j.elecom.2005.07.013

[82] Survilienė S, Eugénio S, Vilar R. Chromium electrodeposition from [bmim][bf4] ionic liquid. Journal of Applied Electrochemistry. 2010;**41**:107-114. DOI: 10.1007/s10800-010-0218-3

[83] Menzel A, Komin K, Yang Y, Güder F, Trouillet V, Werner P, et al. Ultra-long zinc oxide nanowires and boron doping based on ionic liquid assisted thermal chemical vapor deposition growth. Nanoscale. 2015;**7**:92-97. DOI: 10.1039/C4NR05426A

[84] Steichen M, Dale P. Synthesis of trigonal selenium nanorods by electrodeposition from an ionic liquid at high temperature. Electrochemistry Communications. 2011;**13**:865-868. DOI: 10.1016/j.elecom.2011.05.023

[85] Zeng S, Zhang X, Bai L, et al. Ionic-liquid-based CO_2 capture systems: Structure, interaction and process. Chemical Reviews. 2017;**117**:9625-9673. DOI: 10.1021/acs.chemrev.7b00072

[86] Bates ED, Mayton RD, Ntai I, Davis JH. CO_2 capture by a task-specific ionic liquid. Journal of the American Chemical Society. 2002;**124**(6):926-927. DOI: 10.1021/ja017593d

[87] Anthony JL, Aki SNVK, Maginn EJ, Brennecke JF. Feasibility of using ionic liquids for carbon dioxide capture. International Journal of Environmental Technology and Management. 2004;**4**:105. DOI: 10.1504/ijetm.2004.004624

Chapter 8

Boron Doping in Next-Generation Materials for Semiconductor Device

Linh Chi T. Cao, Luqman Hakim and Shu-Han Hsu

Abstract

The article surveys the most recent achievements starting with the boron doping mechanism, mainly focused on doping in semiconductor materials such as Si, Ge, graphene, carbon nanotube, or other 2D materials. Frequently used doping methodologies are discussed, including ion implantation and solid-phase doping, mainly focused on recent developing techniques of monolayer doping. These doped materials' structural, electronic, and chemical properties are addressed to understand the boron doping effect better. Theoretical and experimental information and data are used to support such atomic-level effects. Therefore, this review can provide valuable suggestions and guidelines for materials' properties manipulation by boron doping for further research exploration.

Keywords: boron doping, monolayer doping, ion implantation, CVD, semiconductors

1. Introduction

In semiconductor technology, doping is a process that introduces delicately controlled amounts of impurities (called dopants) into an intrinsic semiconductor to modify its electrical, optical, and structural properties significantly. The intrinsic semiconductors are pure semiconductors without impurities (typical semiconductors of group IV in the periodic table: Si and Ge), in which the number of excited electrons equals the number of holes. In the doping process, a dopant is added, which could play a role as either a donor to contribute an electron or an acceptor to create a hole with the semiconductor crystal that respectively generates two types of semiconductors: n-type and p-type. The dopants belonging to group III, such as boron (B), aluminum (Al), gallium (Ga), and indium (In), are referred to as acceptors for p-type semiconductors. Moreover, group V elements, including phosphorus (P), arsenic (As), antimony (Sb), bismuth (Bi), and lithium (Li), are donors to contribute free electrons in n-type semiconductors. Boron is a p-type dopant with only three electrons in its valence shell. During the boron incorporation process into the silicon crystal, the one atom of boron can bond with four silicon atoms. Still, since boron only has three free electrons to provide, a hole is created. This hole acts like a positive charge, so boron-doped (B-doped) semiconductors are referred to as p-type semiconductors (**Figure 1a**). In the p-type semiconductors, the holes, like the positive charge, attract

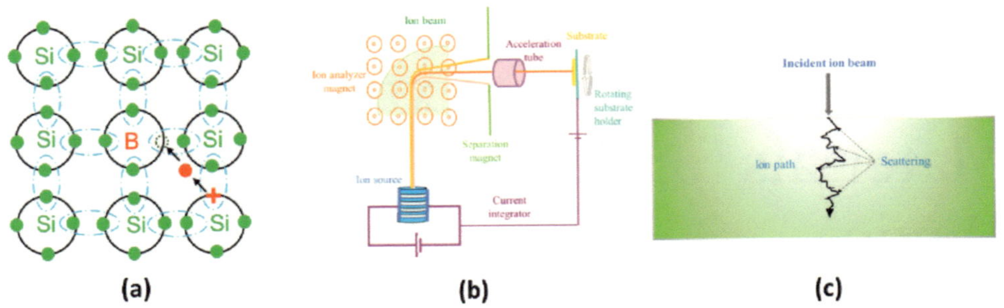

Figure 1.
Schematic of (a) boron-doped silicon, (b) an ion implanter, and (c) ion penetration path into a silicon substrate.

electrons. But when an electron moves into a hole, the electron leaves a new hole in the previous position. Thus, in a boron-doped semiconductor, the holes constantly move around inside the crystal as electrons continuously try to fill them. This appears like the moving of the positive carrier.

The unstoppable development of electronic technology demands the detailed design and effective performance of microelectronics. The formation of shallow and low resistivity junctions is required for contact resistance reduction and leakage current consideration. The precise control of dimension and dopant concentration of source/drain region to achieve a high shallow doping efficiency is crucial for junction fabrication [1–3]. Shallow doping could create doped layers with depths ranging about dozens of nanometers. It required low-energy ions for implantation by considering thermal redistribution [4]. At low energy, the penetrated navigation of ions was mainly directed along crystalline channels rather than moving randomly into semiconductors [5]. Boron is one of the essential dopants for shallow doping in silicon because of its good diffusivity [3].

Boron doping has grabbed attention for several decades. Studies can be classified as ion implantation, solid-phase doping, monolayer doping, and other methods such as sputtering and chemical solution mixing. These techniques are used widely in semiconductor technology. This chapter studied the characteristics of widely used and recently developing methods, such as monolayer doping, by showing the advantages and disadvantages of these doping techniques to give an overall sight of the doping methodology of boron, so it is easier to choose and use suitable doping techniques to meet a specific requirement in further boron-doping application.

2. Boron doping methodology

2.1 Ion implantation

2.1.1 The general principle of boron ion implantation.

Ion implantation is a material surface modification process by introducing a dopant, also called an impurity, into a solid substrate at a low temperature. In boron implantation, boron atoms are ionized into cations, which are accelerated and injected into a solid substrate at a depth from dozens to hundreds of nanometers by an intense electric field, consequently modifying the mechanical, chemical, or electrical properties of the target material [6, 7]. The usage of ion implantation in doping semiconductors was described first by William Shockley in 1954, but it gained noticeability only until the late 1970s and entered mass production [8]. Ion implantation has been used to dope boron in various semiconductors ranging from Si/Ge, 2D materials such as graphene, hBN, carbon nanotube, metal oxide, TiNi, TiAlNi, etc. An ion implant system, a so-called implanter, is very complicated that used to ionize, select, and accelerate ions for implantation, as shown in **Figure 1(b)**.

It allows preciseness to control the penetration depth of boron atoms into the substrate in the ion implantation process. At first, the boron is ionized by electron impact in an ion source that contains a plasma generated by microwave radiation or radiofrequency (RF). The boron ions are extracted from the ion source using electromagnetic fields to form the ion beam, which is directed into a mass analyzer magnet. The beam is centered and bent at a right angle. The radius of the ion bend is determined by analyzing the ions' electromagnetic field characteristics in a high vacuum environment to avoid the ambient gas molecules that could affect the mass-to-charge charge ratio. Therefore, boron ions are selected from different ions in the ion source to exit the mass analyzer using an electromagnetic lens. The ion beam of boron atoms is accelerated to high energies (toping up from sub-keV to MeV values) and steered to inject onto the target substrate using electromagnetic fields. This process must be carried out in a high vacuum environment to avoid the ambient gas molecules that could affect the linear free travel of the ions. When boron ion reaches the crystal surface, the penetration of boron ions into the crystal matrix is proportional to its angle of incidence and energy. The path of ions is not a linear line but follows a "lightning" line through the crystal (**Figure 1c**). The concentration of dopant atoms corresponds to the penetrated depth into the substrate obeying a Gaussian distribution as shown in **Figure 2**.

The average value of total path length is termed the range R that is considered at both horizontal and vertical motions. The average depth of profile is known as the projected range R_p, which featured for ion energy and mass of dopant with a standard deviation ΔR_p. The ion concentration N(x) at depth x can be described by equation in **Figure 2**, where N_p is peak concentration, R_p is the projected range, and ΔR_p is the standard deviation.

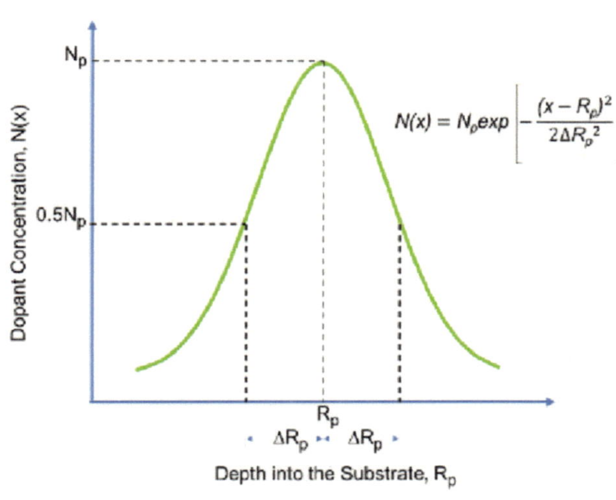

Figure 2.
Representative dopant profile in a substrate undergoing ion implantation.

The implanted dose Q that is required to satisfy N_p and R_p is calculated by the below Eq. (1):

$$Q = \int N(x)\,dx = \sqrt{2\pi}\,N_p\,\Delta R_p \tag{1}$$

The two factors affecting the boron implant process that can be controlled to adjust the implantation conditions are implant energy and boron dosage (fluence). These two parameters can examine the range (depth) distributions of implanted ions. Moreover, the characteristics of dopants are essential for the implant process. Unlike heavy ions formed by Sb, As, and P, light boron ions are transferred easily into the crystal, making boron ions stop at a more profound distance than at the same energy condition.

During the ion penetration, the irradiation energy is enough to break the lattice matrix of the semiconductor to create defects. Still, the atoms sometimes could not substitute lattice sites and could be stranded in interstitial positions. Post-implant annealing is vital to stimulate boron by replacing the boron atoms in the crystal lattice positions. This process will also help repair any damage induced in the crystal matrix by the extreme collisions of the high-energy boron ions and somewhat widen the allocation of boron [9]. Once situated into the lattice, boron will work as an acceptor to improve the electrical properties of original semiconductors.

Ion implantation in crystalline solid created the different types of defects, including (1) a Frenkel defect, a type of point defects in crystalline solid, are interstitials (self-interstitials), and vacancies (substitutional points) created from breaking lattice sites illustrated in **Figure 3a** (2) interstitial and vacancy clusters formed by the combination of interstitials and vacancies, (3) the complexes of defects-dopant resulting from the interaction between defects and dopants, (4) amorphous states, in which regular lattice is destroyed thoroughly after implantation. All these defect forms are determined in boron implanted silicon corresponded with different implantation conditions. Silicon interstitials are typically dominant defects that are created from boron implantation in silicon; besides that, we also found the existence of Si interstitial clusters, boron-Si interstitial clusters, which are products of the interaction of Si interstitials with each other, and Si interstitials with boron implanted atoms [10–12].

Depending on implantation conditions (implantation dose, energy, and annealing), the implantation-annealing damage can also induce the formation of extended defects

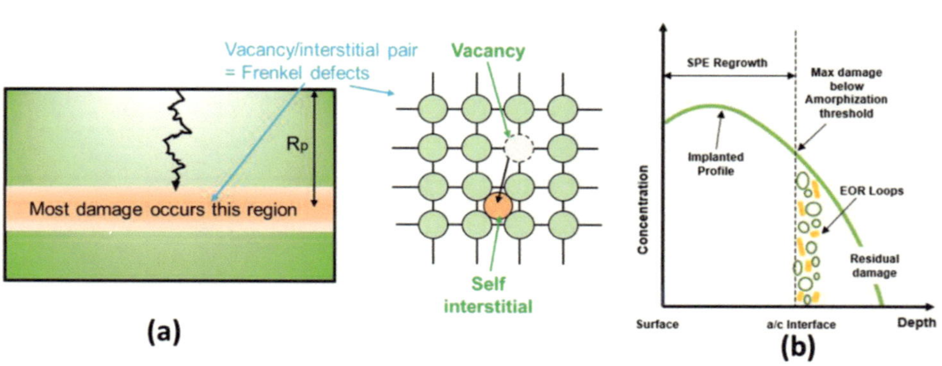

Figure 3.
(a) The formation of a vacancy/interstitial pair of implantation damage in crystalline solid. (b) EOR dislocation loops that create at the a/c interface after solid-phase epitaxial regrowth.

that are divided into three main types, including dislocation loops and rod-like defects {311} and stacking faults [14–17]. The {311} defects are noticed with a long, thin rod-like shape; hence, they are also called rod-like defects. These rod-like defects are collections of silicon atom ribbons that arrange lying on {311} planes and extend in the <110> directions to create planar defects. The dislocation loops, like its name, are the deformed structural defects formed by the precipitation of an extra circular atomic layer of silicon atoms on a {111} plane. The stacking faults are crystallographic defects resulting from the disordering of stacking planes [16, 18]. These extended defects are different forms of Si interstitial clusters formed from the combination of Si interstitials, and they can survive even after thermal annealing [19].

Apart from implantation defects, a dopant diffusion phenomenon was found during annealing, and it is enormously different from normal equilibrium diffusion. They discovered that this phenomenon was more vital than at low temperatures of annealing and slowed down at higher temperatures. By its features, this phenomenon has been named Transient Enhanced Diffusion (TED) since the 1980s [16]. TED is one of the main problems affecting the reduction of boron activation during thermal annealing. The enhanced boron diffusion causes the spreading of the boron profile and the deepening of the junction. It has been found that TED has a profound relationship with the presence of excess Si-self-interstitials in silicon [11].

High-energy implantation is typically favored to obtain superconductors. However, this could cause the lattice disorder after implantation. The structural damages caused by boron implantation at high energies in silicon were investigated and classified. They found three regions of the damaged layer that are situated along with the silicon's depth: the near-surface crystalline region, the severely damaged region, and the tail zone of the damaged layer after boron implantation at 1×10^{15} ions/cm^2 [15]. In other materials, such as diamonds, the accumulation of lattice defects is the main problem of ion implantation. The accumulation of multiple defects generated a lot of vacancies, represented by damage density (vacancies/cm^3). A considerable damage density in diamonds is caused by high-energy boron implantation around MeV and the thermal annealing process rather than restoring the diamond structure, and it causes the graphitization in diamonds [20–22]. These defects resulted from the high-energy collision of boron ions into a solid substrate, which broke the lattice sites. In some boron implant cases, the ions could not activate and diffuse inside the semiconductor leading to the unsuccessful replacement of boron atoms in lattice points. This resulted in the less of electrical carriers in this semiconductor and the ineffective boron doping process.

The amorphization process regularly creates a bunch of extended defects. It also causes end-of-range (EOR) defects beyond the amorphous/crystalline interface. EOR defects produced during amorphization are more abundant in self-interstitials compared with extended defects created under non-amorphization conditions. The formation of the amorphous layer is investigated related to using high-dose boron implants. Suppose the implanted boron doses are more enormous than required. In that case, it produces a high density of defects (silicon interstitial and vacancies) that can accumulate into defect clusters and trigger the amorphization process. The interstitials are highly mobile; the amorphization process that occurs at low energy implantation can cause an out-diffusion of boron atoms and interstitials to the surface, which limits the fabrication of shallow junction. The loss of interstitials results in the failure of recrystallization during annealing [23, 24]. Boron implantation with a high dose causes the enhancement of the boron diffusion, which leads to inactive boron in silicon. Boron atoms are found that gather into clusters and substituted silicon atoms at a supersaturation condition of a boron implantation dose that was greater than 1.1×10^{19} ions/cm^2. The boron diffusivity appeared in silicon that lowered the activation of boron, but annealing at high temperatures from 800°C to 1000°C in silicon retarded the boron diffusion in silicon and increased active boron concentration [25].

Moreover, the high implantation doses used to obtain high boron concentrations can cause amorphization of the implanted region [26]. In a study by Aradi et al., the significant increase of defect concentration at higher boron ion fluence of 1×10^{17} ions/cm^2 caused a lattice disorder resulting in amorphization of h-BN material [27, 28]. Similarly, a report on implanting boron in Ge showed that high boron concentration exceeded the solid solubility limit and caused boron atoms to be immobile even after annealing treatment at high temperatures. Some research indicated that implantation induced defects that increase the diffusion of boron rather than retard the diffusion. Furthermore, using a high dose of boron can lead to precipitation of excess boron, which may reduce the boron diffusion. However, it still leaves boron atoms inactive because of the combination of boron with defects [20]. The defect clusters arise from the dissolution and erosion resulting from the recombination and out-diffusion of defects. Temperature conditions in boron implantation are also a factor affecting amorphization. The lattice damage at negative implantation temperatures is recorded that is more severe than implantation at room temperature. It was reported that the level of lattice disorder could be 20–30 times lower in room temperature implants than those implanted under cold conditions, for instance, −150°C. Lui et al. also found that boron implanted at a cold temperature of −100°C caused more implant damage by boron self-amorphization no matter the dose and implant energy [23, 29].

2.1.2 Post-annealing process and boron activation

The post-annealing is an important process to repair the principal damage created by ion implantation, restore the lattice site to a perfect lattice state, and activate dopants into substitutional sites [9, 30]. After ion implantation, the semiconductor is usually so severely damaged; therefore, its electrical behavior is controlled by deep-level electron and hole traps where carriers are captured and increase the resistivity of the semiconductor. The subsequent annealing process is required to heal lattice damage and reside dopant atoms in substitutional positions. A suitable annealing treatment is very important, which resolves problems after implantation, including recrystallization, dopant activation, and diffusion depth. There are mainly two types of post-annealing: furnace thermal annealing and rapid thermal annealing (RTA) for ion implantation. During post-annealing, the repair and diffusion processes coincide, but their speeds vary depending on the annealing's temperatures and time. The furnace thermal annealing is satisfied to supply a high temperature but requires a time furnace annealing of at least 15 min to ensure a practical operation. Therefore, the furnace annealing typically causes unnecessary boron diffusion. Rapid thermal annealing is used to heat implanted materials by different methods (with various heating-based lamps) in a rapid period from a hundred seconds to nanoseconds, which allows for minimizing the boron diffusion. The mechanism of thermal annealing to repair the lattice damage depends on damage levels in materials after implantation, and it relates closely to the boron activation.

At the beginning stages of annealing, the vacancy clusters and interstitial clusters are disbanded to release vacancies and interstitials. Most of the Frenkel pairs are removed in the initial stages of annealing, leaving interstitial-type defects, which freshly released after dopant atoms occupy lattice sites and kick Si interstitials out. These Si interstitials condense quickly into characteristic rod-like defect {311} clusters on annealing at temperatures over 400°C. When annealing at 900°C, the density of these {311} defects can increase rapidly to reach the peak and start to dissolve upon ongoing annealing due to the evaporation of Si interstitials [6].

If the damage is not severe, these rod-like defects dissolve absolutely, and the crystal structure recovers perfectly. Above severe damage level, the larger {311} defects can turn into stable dislocation loops, which are very strenuous to remove. These loops are secondary defects and remain after the primary damage is disappeared utterly. Higher-dose implants create a large number of stable dislocation loops, which trigger the silicon amorphous. The high density of these loops locates at the interface region between amorphous and crystalline silicon (amorphous/crystalline interface) by a solid-phase epitaxy growth process. These defects are referred to as the end-of-range (EOR) defects situated at the amorphous/crystalline (a/c) interface, as depicted in **Figure 3(b)**. This is because a large amount of damage locates below the threshold of amorphization beyond the a/c interface. The amount of damage beyond the a/c interface can be possible depending on the damage limitation that crystal can contain without being amorphous. This damage includes the most significant amount of {311} defects and a range of dislocation loops in a narrow area just below the a/c interface on the crystalline side [6, 31].

The secondary damage is very stable, even annealing by RTA anneal at a temperature of 1000°C. The loops increased the size from 10 nm to around 20 nm of radios during annealing; this happens to conserve the total number of interstitials trapped in the loops and make these loops hard to remove. When the temperature of RTA is high enough, the EOR dislocation loops can be removed; for example, it is revealed that these loops disappeared at 1100°C for 60 s [6, 32].

To activate the electrical activity, implanted boron atoms must reside in substitutional sites in the semiconductor material lattice. Moreover, the broken bonds in the lattice matrix must be cured to return the mobility of the electrical carrier [6, 20]. This is a principle to achieve high levels of dopant activation. The activation of ions, therefore, depends on the level of damage in the lattice after implantation and post-annealing treatment. The levels of damage can be classified into three types: low levels of damage, very high levels of damage that occur in amorphization, and the mid-levels of damage below the amorphization threshold where partial disorder occurs. Depending on the ion implantation conditions, the primary damage is often at a low level that the annealing process can repair completely, and the high dopant activation levels are reached. For instance, post-annealing increases boron implant performance in a diamond. Yuhei Seki et al. carried out the B doping by ion implantation in diamond by 60 keV at room temperature followed by thermal annealing at 1150°C for 2 h. They reported that an excellent doping efficiency reached approximately 80% with the maximum boron concentration of 3.6×10^{19} ions /cm^3 (around 200 ppm) [30]. In addition, boron was doped into graphene film assisted by a stopping layer of polymethyl methacrylate (PMMA) on top to control the B distribution centered on the graphene sheet. The electrical properties of graphene were enhanced by the increase in charge carrier density corresponding to the rise of concentration ranging from 5 to 50 $\times 10^{10}$/cm^2. The roughness of the graphene surface was also increased after the doping process. Moreover, the post-annealing at 1000°C for 10 s improved the boron doping performance by increasing approximately 13 times the boron activation in graphene, which proves the importance of the annealing step after ion implantation [13].

Oppositely, at extremely high levels of damage that take place in amorphization, the annealing cannot treat and deal with the amorphous region, so a nearly practical method to remove damage and recrystallize lattice to achieve high dopant activation is solid-phase epitaxial regrowth (SPER) [6]. SPER can regrow the lattice of the substrate, which is amorphous by layer-by-layer epitaxial restructure starting from the amorphous/crystalline interface. Its mechanism is similar to the crystallization process in which a crystal solid is formed from either a melted liquid phase or gas phase deposited onto a crystalline substrate, except that SPRE occurs from a solid phase rather than a liquid or gas phase. The regrowth eliminates the damage in the amorphous area and limits the dopant diffusion at a low temperature. Most of the dopant atoms' broken bonds are recovered onto lattice sites in the amorphous regions during the SPER process, increasing the activated dopant concentration to create electrical carriers [6, 26, 33–35]. For example, the experiment simulation about the effect of low-temperature SPER with boron activation in pre-amorphized Si was carried out by Aboy et al. They calculated the active B concentration reached up a few times 10^{20} cm^{-3} and the minimal diffusion after effective SPER treatment [26]. However, the boron activation levels can be dropped drastically as the boron dose is increased [33]. At a high concentration of implanted boron, it is challenging to recrystallize amorphous layers and fully active boron ions in pre-amorphized silicon [35]. However, it has been found that a fully amorphized region is much easier in many cases to repair than a partially damaged region.

The third type of damage level, which lies below the amorphization threshold, is much more difficult to be cured by annealing because this region contains secondary defect forms that make the annealing treatment more complex [6, 36]. The activation process behaves correspondingly to a temperature that indicates the complex interactions between the dopants and the defects [37–39]. At very low doses, boron ions are almost activated even after annealing at a very low temperature, and it is quickly fully active after increasing temperature. However, it is very slow to activate boron in higher doses. A publication by Chang et al. described boron activation at low temperatures below 400°C and concluded that boron activated increasingly during annealing, but the active boron percentage was dropped with increasing implant doses [37]. The observation of boron activation was investigated in the research of Seidel TE et al. Boron implantation process creates deep-level traps of damage that increase the resistivity of silicon, and a fraction of these traps is disappeared after annealing at 400°C, decreasing resistivity and enhancing the boron activation, but a "reverse annealing" phenomenon occurred between at 450°C and 500°C, which reduced the carrier concentration in silicon. This phenomenon is explained by the competition between silicon interstitials and boron atoms in institutional lattice sites or by the pairing of boron atoms with interstitials to form inactive complexes. Then annealing at temperatures beyond 550°C, the activation process gained a gradual rise to reach the complete activation level at the highest anneal temperatures [40]. The complexation of the thermal annealing is affected by the diffusion of boron atoms in silicon. The damage can exist longer at low temperatures and increase the boron diffusion, whereas, at high temperatures, the damage is eliminated faster [36]. Huang et al. described the "reverse annealing" phenomenon at low-temperature annealing (525–800°C) in boron-implanted silicon and explained that the occurrence of reverse annealing is due to the formation of boron-silicon interstitial complexes and enhancement of boron diffusion that related to silicon self-interstitials [41]. In addition, the enhancement of boron diffusion is also observed during annealing treatment at a higher temperature, which is caused by the complex damage in the region below the amorphous/crystalline interface [38, 42].

Although the annealing is essential for boron implantation, annealing at higher temperatures to active implanted boron atoms can cause the diffusion process, which makes it difficult to generate the shallow junctions. The restraining of boron diffusion is important to obtain higher boron activation. However, the presence of the excess interstitials causes a transient enhancement in the dopant diffusion called transient-enhanced diffusion (TED) [43, 44]. TED often occurs during annealing at low temperatures, wherein boron atoms diffuse faster than annealing at a higher temperature. Jain et al. found out that annealing boron implanted Si substrate at 800°C made boron diffusing much faster than normal thermal diffusion. This enhanced diffusion is temporary and stops when it reaches saturation. They explained that during low-temperature annealing, Si interstitials kicked the substitutional boron atoms out of lattice sites; boron atoms can diffuse easily. Besides, the combination of interstitials and boron atoms created highly mobile complexes. Therefore, the main reason that caused the diffusion of boron is the excess Si interstitials resulting from implant damage and surface oxidation. Suppose the annealing process is conducted at higher temperatures. In that case, the interstitial and interstitial-boron clusters are unstable, and the pairing of interstitials and boron atoms is decreased, leading to the retardation of TED. Therefore, in post-annealing processes, rapid thermal annealing is likely to prevent the TED phenomenon [19, 38, 45, 46]. Two typical analysis methods are used to determine the quality of semiconductors after boron implantation and post-annealing treatment, which are thermal wave measurement [47] and the sheet resistance measurement [48]. However, both methods are ineffective in measuring the thickness of ultra-shallow junction produced at low-energy implantation due to the beyond resolution limit [49].

Most implant energies range from 30 keV to 200 keV; fabricating a junction shallower than 100 nm usually requires low energy, for example, below 100 eV. At low energy, it is difficult to implant ions into the substrate. It requires an economically feasible approach and the progression of technology generations. The reason most implants cover the range above 30 keV is that this is a low-energy limit that is required for extraction voltage for the ions from the source plasma. Moreover, extracted ions are usually accelerated to higher energies; a deceleration can cause tricky problems in engineering and require optimization for machine design. However, the high doses cannot be implanted at very low energies because of sputtering off surface atoms of the incoming ions and resulting in a self-limiting dopant dose. Besides, the profiles are affected by transient enhanced diffusion (TED), which reduces activated ions in materials, and it can be recovered by annealing and still can obtain junctions with the depths around a few tens of nanometers [50]. Collart et al. reported that the boron atoms are difficult to activate in silicon if implanting at lower energies. During ion implantation at the low ion energy of 100–1 eV, the boron penetrated the silicon creating a profile with a depth of around 100–200 nm. However, most of the profile depth disappeared after the rapid thermal annealing at around 1000°C. This is explained by the fact that boron is trapped and deactivated at the surface during the implanted process, and annealing enhances the diffusion of boron, leading to the removal from the substrate [19]. On the other hand, implant with high energy with MeV range is often achieved simply. This technique is applied to form the deep well in CMOS technology to achieve super-junction power [6, 51].

2.1.3 Advantages and disadvantages

Ion implantation is a doping process conducted at low temperatures, in specific areas, and with an exact dopant dosage. It is easy to turn the depth/ions selection. By changing fluence and accelerating the energy of the ion beam, the dosage and implant energy can be controlled and modified for requirements. Besides precise dose control, the dopant profile (peak depth and spread range) can also be adjusted better than the diffusion method, in which peak concentration is always defined near the surface. Ion implantation has been known as an exceptionally clean surface treatment technique. There is truly little or no contamination during implantation because boron ions were collected from beam analysis, and other contaminant ions were removed before penetrating the target. Moreover, it normally operates in a high vacuum environment, so the atmosphere's impurities cannot affect the surface. Boron ions penetrate and replace the lattice sites of materials to activate the electrical properties. Therefore, the implanted substrates are not sensitive to either surface treatment or surface cleaning procedures.

The ion implantation process requires specialized and relatively expensive equipment, such as a modern ion implanter, which costs about 2–5 million dollars depending on the model and size (the price reported in 2003) [8]. The costs of operation and maintenance for ion implantation are also high because it demands a high vacuum environment during operating and periodic maintenance to avoid contamination and technical issues [52].

However, ion implantation with a larger amount of dose at high energy causes severe damage to semiconductor material lattice, for example, the amorphization in silicon or the graphitization in a diamond that cannot repair by a normal post-implant anneal. For example, using a high-energy boron beam at 8 MeV of B^{3+} and fluence of 570 × 10^{14} ions/cm^2 caused the total disappearance of diamond peak by loss of diamond structure, and the annealing at 1000°C for 1 h is reported to not be able to heal its structure [20]. Moreover, a boron implant at a higher dose is reported to create the secondary defects as stable dislocation loops, which can remain and can trigger the silicon amorphization after annealing at 1000°C [6].

Boron implantation is the most convenient method recently applied to dope boron in semiconductors. However, it is very difficult or sometimes impossible to obtain very shallow. Because the shallow implantation is very complicated and requires an optimal process of ion implantation and appropriate post-annealing to control various phases involving the collision between doped ions and lattice matrix, destruction of the matrix, projection of implanted ions, and the restructure (recrystallization) and dopant atom activation and diffusion [53]. In particular, boron implantation is challenging to create ultra-shallow junction because of two main impediments: transient enhanced diffusion and Si interstitial/boron-interstitial typed clusters because the increase of excess interstitials in silicon lattice leads to enhancement of the boron diffusion rate, which related directly to boron inactivation and the loss of boron out of substrate [3, 19, 54]. Around 20% of the implanted boron resides at substitutional lattice sites, and the rest of the boron ions produce pure boron clusters and silicon-boron clusters [12, 54]. Both are caused by silicon interstitial supersaturation, which is a consequence of implant damage and creates extended defects that tend to agglomerate and form interstitial silicon clusters [3, 12]. Therefore, forming an ultra-shallow junction requires not only the optimization for the implanter to control low energy in the implantation process but also the need to manage the boron diffusion and defect clusters during annealing.

Ion implantation is a standard method that typically introduces ions into the top side of the flattened substrates or films. However, it is very directional. Therefore, it can introduce boron ions into the sidewall of multi-gate devices such as fin field-effect transistors (FinFETs) by tilting the incident ray to implant ions. But there are some limitations of implantation on the sidewall: (1) the boron dose retained after implantation is very sensitive to the incoming angle of the ion beam. The high tilt angle can implant ions sidewall easily and increase sidewall boron storage. (2) It is difficult to implant at a high tilt angle for dense structures in which transistors are located close to each other on the wafer, (3) the severe implantation damage is hard to repair, the silicon structure is not able to recrystallize even after rapid thermal annealing at high temperature [55, 56].

2.2 Boron monolayer doping

2.2.1 Current development of boron monolayer doping

The dimensions of electronic devices have been shrunken to the nanoscale following the semiconductor generation node. The traditional planar structure devices are hard to realize generation nodes (<10 nm). 3D finFET structure device is proved to achieve better performance and minimize the fabrication difficulties. During progressive doping, ion implantation is typically used for FinFET fabrication, but it faces challenges from crystal damage for such fin structure and limitations of dimensional geometry [57]. Monolayer doping (MLD) was suggested first by Javey et al. in 2008 as a substitute doping technique to obtain ultra-shallow junctions [58, 59]. They successfully fabricated sub-5 nm junction depths, which can be down to approximately 2 nm of depth with low sheet resistance (lowest value ~825 Ω/sq) via phosphorous monolayer doping method using diethyl 1-propylphosphonate (DPP) to obtain 70% active phosphorous dopant after RTA with temperatures ≥950°C.

The monolayer doping process consists of two main stages: self-assembly of molecules onto the surface to form monolayers and thermal annealing process to diffuse and active dopants. In the self-assembled monolayer phase, the dopant-containing molecules are grafted onto a semiconductor surface via a covalent bond between the terminated functional groups of molecules and the termination modified surface. Next phase, a capping layer was applied to prevent uncontrolled loss of the dopant molecules upon heating. A thermal annealing process was conducted to drive the dopants into the semiconductor substrate that simultaneously activates dopant atoms. The masking layer was then removed to obtain a thin doped layer or junction [58, 60]. MLD demonstrated that it causes no lattice damage and is capable of doping impurities into dimensional structures due to the conformal nature of the monolayer assembly process that avoids the shadow effects occurring in ion implantation. There are various elements that were doped into semiconductor substrate by MLD to obtain ultra-shallow doping including phosphorus [58, 59, 61–64], boron [58, 60, 65–69], nitrogen [70], sulfur [71–73], arsenic [74], antimony [75].

Monolayer contact doping (MLCD) is an innovative method based on monolayer doping (**Figure 4a**). In this MLCD technique, the dopant-containing monolayer is formed onto a thermal oxide wafer (Si + SiO_2) as a donor substrate by a self-assembly process. The donor substrate is then brought into contact with the target substrate (typically with intrinsic silicon substrate), afterward, annealed using the RTA process. Under the annealing process, the molecular monolayer occurs by the thermal decomposition, and dopant atoms from monolayer fragments diffuse into donor and target substrates. This indicates that both monolayer and contact doping arise simultaneously onto donor and target substrate, respectively. Due to direct contact between two substrates during annealing, MLCD does not need a capping layer of SiO_2 to avoid the out-diffusion of dopant atoms. MLCD can apply to conventional top-down or

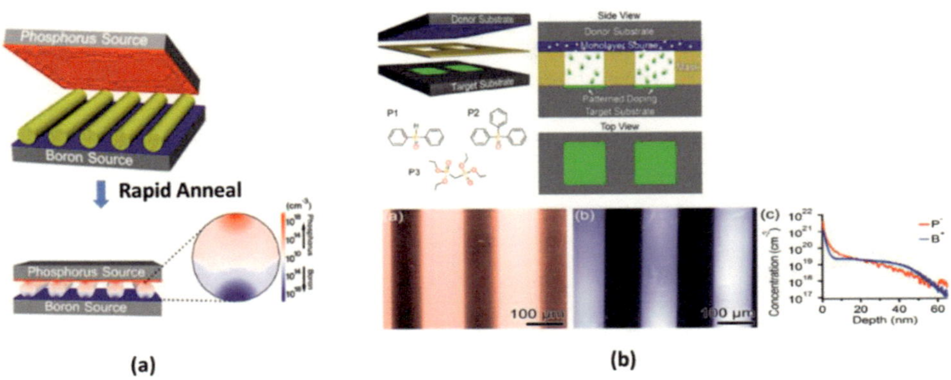

Figure 4.
Schematic of (a) parallel p-n junction configuration formation across oriented NWs by a one-step contact doping process printed with permission from ref. [79]. Copyright 2014 American Chemical Society; (b) remote monolayer contact doping process with phosphine oxides molecule with a photoresist mask printed with permission from ref. [80]. Copyright 2017 American Chemical Society.

bottom-up semiconductor processes and doping impurities in nanoscale structures such as silicon nanowires. This method allowed control of surface doping with nanometer-scale structures. The first report on MLCD was published by Hazut et al. in 2012. They used phosphorus-containing molecules (phosphine oxides) for MLCD onto the target silicon substrate. They obtained a level of dopant concentration higher than 5×10^{20} cm^{-3} with a depth of dopant profile around 30–40 nm and sub-10 nm at short annealing times [76]. Subsequently, MLCD is utilized widely to dope materials such as phosphorus [77], boron [66, 76], sulfur [78] to obtain an ultra-shallow doping layer with nanometer scales for semiconductor applications. However, to achieve a high dopant concentration in the target substrate, the minimization of dopant diffusion in the donor substrate is required to focus dopant atoms on the target substrate. MLCD was applied to fabricate parallel p-n junctions on NWs by one-step doping. Boron and phosphorus were doped simultaneously onto two sides of NWs, achieving high dopant concentrations with P-doped and B-doped poles respectively of 2.6×10^{19} cm^{-3} and 1.0×10^{20} cm^{-3} concentration [79].

To control the doping areas, remote monolayer doping (R-MLD) is developed with the principle of monolayer contact doping, but there is a distinct feature that R-MLD is performed without the contact between donor and target substrate. In R-MLD, the target substrate is covered partially by a thin separator mask with microscale thickness. There are unmasked areas and masked areas on the target substrate. Therefore, the donor substrate with dopant-containing monolayers cannot contact directly with the target substrate due to having a gap between these substrates. During the rapid thermal annealing, the monolayer source is fragmented at elevated temperature to generate volatile fragments, which subsequently evaporate into the gas phase to react with the oxide surface at the substrate surface. Annealing with the RTA process causes dopant diffusion through the native oxide and is activated and incorporated into the semiconductor surface [68, 80]. Hazrat et al. described the R-MLD process using diphenyl phosphine oxide for phosphorus doping with a silicon wafer in which the target substrate was patterned by an AZ4562 photoresist as a separator mask. RTA process was implemented at 1000°C in 6 s and 30 s for additional annealing. Although the diffusion of gas-phase dopant between the mask and target substrate was observed, the phosphorus incorporation efficiency into the target silicon substrate reached 70%. Moreover, boron doping using phenylboronic acid was carried out with the same procedure to compare with phosphorus doping using tetraethylmethylene diphosphonate (40% of incorporation efficiency). The SEM of doping profiles showed a higher contrast for boron onto the target silicon wafer compared with phosphorus. This indicated that tiny boron atoms are diffused into the mask layer during R-MLD. R-MLD process is shown in **Figure 4(b)** [80].

A modification of MLD reported by Ye et al. is monolayer contact doping (MLCD). They modified the MLD technique by forming boron-containing SAM onto a thermal oxide silicon substrate instead of directly onto the target substrate. This source substrate is subsequently brought into contact with the target substrate, upon which the dopant is driven into the target substrate by thermal annealing. Therefore, the thermal oxide substrate was an efficient capping layer for annealing. Carboranyl-alkoxysilane was used as a boron-rich source and easily created SAM without using harsh reaction conditions owning to the active silane headgroups. The higher boron-doped concentration was achieved compared with normal MLD by carborane alkene under the same RTA condition (more than two times) [65]. Moreover, the MLCD method reduced the boron diffusion to only 2%, which is advantageous for reusing the source substrate [66].

An investigation by Park et al. demonstrated that surface states of the target substrate significantly influence the boron doping efficiency using monolayer doping. The good boron doping levels were achieved with a non-damaged clean surface, but the boron incorporated level dropped approximately an order of magnitude on the damaged surface. However, treatment processes to heal the surface state effectively boron doping by MLD. The doping levels on these treated surfaces were much higher than the damaged surface but still lower than the pristine and undamaged surface. The different orientations of silicon substrate also affect the boron doping performance. The 100-oriented silicon was observed as a two times higher doping level than the 110-oriented silicon. That is because of the dependence of the ratio of hydrogen terminations on orientations. The (110) surface has a lesser number of active reaction sites for monolayer formation compared with the (100) surface [81].

2.2.2 Formation of dopant-containing self-assembled monolayers (SAM)

Self-assembled monolayers (SAMs) are monolayers formed by the self-organization of organic molecules in a solution or vapor environment onto the solid substrate through chemical interaction between head groups of molecules and functional groups of solid surfaces [82]. Self-assembly is a process in which molecules graft spontaneously onto a semiconductor substrate by chemical adsorption between head groups of molecules and specific terminations on the substrate surface. During assembling, the tail (back bond) of molecules interacted with each other under a balanced state to create a well-organized and stable monolayer [83]. Therefore, depending on the head groups of dopant-containing molecules, the semiconductor surface requires particular and suitable terminations. For instance, terminal alkene (C=C) or alkyne (C≡C) (unsaturated organic compounds) can attach to the hydrogen-terminated surface, and alkyl silane groups ($Si-(OR)_3$) can bond with the hydroxyl-terminated surface. If the semiconductor substrate is a silicon wafer, these processes with hydrogen and hydroxyl terminations as known as hydrosilylation [57, 84] and silanization [66, 85], respectively. In some cases, the SAM formation can create by the non-covalent interaction of head groups of molecules with terminated groups of a substrate. For example, phosphine oxide groups of the phosphorus-containing molecules can form the phosphorus dopant SAM by a non-covalent bond on to hydroxyl-terminated substrate [76].

In monolayer doping on silicon, the hydrosilylation process primarily conducts the self-assembled monolayers. In this process, silicon must be cleaned and the native oxide removed to create hydrogen termination by an aqueous solution of HF or NH_4F [86]. The silicon wafer was then incubated in the molecular-containing solution. Relying on the molecular type, different conditions, including heating or irradiation with light, were added to promote the reaction. For example, the dopant-alkene molecules bind covalently with hydrogen-terminated silicon to form the C−Si bond onto the silicon surface under a traditional heating condition of 150–200°C or under irradiation with UV light, visible light [87–89]. This hydrosilylation process between saturated compounds and hydrogen-terminated silicon was demonstrated following a radical-chain mechanism [90].

On the other hand, SAM was also produced by the silanization process, a conventional method used to cover the solid substrate with organofunctional alkoxysilane molecules [85]. In this process, the solid substrates are required hydroxyl terminal groups that can react with alkyl silane to form a covalent Si−O−Si bond. The substrate surface must be cleaned to remove organic residues and generate sufficient hydroxyl groups. Numerous methods are used to clean surfaces consisting of a wet etching by combinations of acid, bases, and organic solvents at different temperatures or irritation with UV light and O_2 plasma [91–93]. The most widely used cleaning method is called Piranha cleaning, which is a mixture of sulfuric acid (H_2SO_4) and hydrogen peroxide (H_2O_2). The silane molecules are hydrolyzed into silanol groups, which react with a hydroxyl-terminated surface via the condensation reaction:

$$Si - OH + HO - \text{substrate} \Rightarrow Si - \text{O} - \text{substrate} \qquad (2)$$

The self-assembly using silanization was performed using vapor-phase deposition and solution-phase deposition. The cleaned substrate is dipped in molecular solution in the solution-phase deposition. For the vapor-phase deposition, the hydroxyl-terminated substrate was kept under a vacuum environment where molecular liquid can be evaporated into molecular gases and assembled onto the substrate. The reactivity of molecules with hydroxyl-terminated surfaces depends on the molecule's properties [94, 95].

2.2.3 Thermal annealing in monolayer doping

The thermal annealing is used to decompose the dopant-carrying molecules and drive dopant atoms into the substrate, creating a thin doped surface layer. Ultra-shallow doping by MLD required a higher solid solubility and a lower diffusivity of dopant to prevent the deeper dopant profile. Besides, solubility and diffusivity factors proportionally correlate to the temperature of the annealing process. The enhancement of boron diffusivity happens at elevated temperatures of annealing. Therefore, controlling the annealing process at a suitable temperature and time is essential for MLD.

The thermal annealing techniques include rapid thermal annealing (RTA), furnace thermal annealing (FTA), and microwave annealing (MWA), which RTA is a favorite in MLD. The report of Ho et al. investigated the boron diffusion at different annealing temperatures during the RTA process. The results showed the sharp boron diffusivity at a higher temperature induced a more profound depth of boron profile. For example, the boron profile depth obtained at an annealing temperature of 950°C was around 18 nm, but that was deeper, around 43 nm at 1000°C for 5 s of RTA.

The increase of boron diffusion can cause a decrease in the boron doping level. For instance, the number of the diffused boron atoms into silicon lattice after RTA is estimated at around 33% of the total number of boron atoms onto surface lattice before spike annealing [58]. The high temperature of annealing can promote a more active dopant into the substrate. Besides, the annealing time is also necessary to control the dopant profile depth. The short annealing time can prevent dopant atoms from being driven deep into the substrate. Therefore, RTA at high temperatures with a temperature ramping rate above 50°C/s is favorable in MLD [57, 59]. Furthermore, Ye and coworkers note that the annealing time reported having a smaller effect on the active dopant concentration than the annealing temperature. The doping concentration at 1000°C for 15 s of annealing time was observed to be nearly the same as observed for 6 s. However, increasing the annealing temperature from 1000°C to 1050°C appears to significantly change the highest dopant concentration. This can be explained by the dependency of boron solubility upon temperature: the higher the temperature, the greater the solubility of boron [65].

Hence, an ultra-shallow junction can be obtained by optimization of the RTA process with lower temperatures and shorter times. The report in 2009 by Ho and coworkers exhibited the successful fabrication of shallow junctions using boron-containing molecules to obtain the depts of around 1–2 nm, which is even shallower than phosphorus MLD (sub-5 nm) at the same annealing conditions due to the lower diffusivity of boron compared with phosphorus. The sheet resistance of the boron-doped layer is reported, that is, higher than ~10^4 Ω/sq. [59]. The boron diffusion was reported that is lower than phosphorus atom diffusion, which was investigated by Ye et al. [62, 69]. In the same MLD conditions, the boron can achieve shallower depth around sub-5 nm but phosphorous at nearly sub-10 nm. The surface concentration of boron is higher than the surface concentration of phosphorous. [59].

A furnace thermal annealing at 1000°C for 5 min was used for MLD of the mixture of dopant-containing molecules and blank precursors [62, 69]. The boron profiles were investigated by using dynamic secondary ion mass spectroscopy (D-SIMS). The authors found out that the boron atoms diffuse around 125 nm deeper than RTA at short times of 5 s (43 nm at 1000°C) with boron-containing molecules only. The boron diffusivity is decreased when using the molecules mixture. Several reasons contributed to boron diffusions, such as temperature, annealing time, molecule doses and types, and the contamination in the monolayer. That makes the diffusion of the atoms from the monolayer into silicon a complex process. Despite the lower diffusivity of boron in the SiO_2 capping layer than in silicon substrate, the amount of dopant lost in the capping layer remains unclear and requires a particular investigation.

Hsu et al. did an investigation of the boron dopant profile not only on the silicon substrate but also on the capping oxide. They designed an alternate annealing process using microwave annealing (MWA) for boron monolayer doping to compare with RTA, as shown in **Figure 5**. The boron atoms were found to cannot fully activate after microwave annealing compared with RTA at 900°C. Hence, the insufficient thermal budget of the MWA process limited the replacement of boron atoms in silicon lattice leading to the formation of boron deactivated clusters in silicon. However, the shallower junction is obtained by MLD using microwave annealing with a junction depth of 5.1 nm compared with 7.1 nm of junction depth using RTA annealing. The sheet resistance of the MWA junction is reported that is higher than that of the RTA junction because of the lower boron activation level with MWA. Moreover, they also measured the dopant profile at SiO_2/silicon interface using PCOR-SIMS and calculated that less than 20% of boron atoms diffused in the silicon target substrate [67].

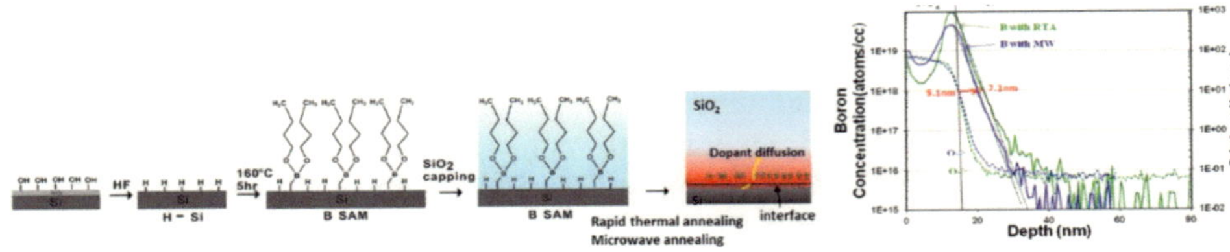

Figure 5.
Schematic of monolayer doping using microwave annealing and rapid thermal annealing. Printed with permission from Ref. [67]. Copyright 2021 Shu-Han Hsu et al. published by American Chemical Society.

This enhanced SIMS technique allows a more comprehensive understanding of the boron dopant distribution at the interface.

2.2.4 The capping layer of boron monolayer doping

After assembling the monolayer, a capping layer of SiO_2 was deposited onto a substrate to block the dopant-containing monolayer from exposing directly during the thermal annealing process that can cause an out-diffusion of dopants. The capping layer is essential in MLD to prevent dopant atoms from escaping into the surrounding environment during thermal annealing [57–59, 69]. Javey et al. investigated MLD without depositing a capping layer and found that boron atoms were lost significantly after annealing [58]. SiO_2 is a typical material used as capping layer in MLD that can be prepared using different deposition techniques including evaporation [58, 59, 68], sputtering [65], and spin coating [69, 96], atomic layer deposition [68]. The capping layer was reported that affects the dopant incorporation in the substrate. The oxygen deficiency in the capping layer, formed during the evaporation and sputtering process, decreases the dopant incorporation. Gao and workers investigated that some oxygen atoms that escaped from the oxide capping layer during annealing can diffuse into the silicon substrate and attach with boron dopants inducing boron deactivation slightly at nearly 1% [60].

The initial reports of boron MLD demonstrated that capping a layer of oxide before the annealing process is required to confine the escapes of dopant atoms from the surface into the surrounding environment [57–59, 69]. However, a recent study stated that the oxide capping layer affects boron activation in the target substrate. It can damage the boron-containing monolayer due to elevated temperatures during the deposition of SiO_2. Therefore, in some instances, a higher doping level can be achieved without employing the capping layer. A series of experiments were conducted by Tzaguy et al. to compare the boron doping levels and the effects of the SiO_2 capping layer on phenylboronic acid (PBA) monolayer doping using different techniques including MLD, MLCD, and R-MLD. The results showed that the doping techniques without a SiO_2 capping layer enabled the lower sheet resistance values than doping with an oxide cap layer. This is because the oxide capping layer in MLD functioned as a barrier to prevent the out-diffusion of boron atoms during the RTA phase and concurrently entrapped a part of boron atoms in the deposited SiO_2 layer. In addition, the PBA monolayer was formed by non-covalent assembly onto the surface. During thermal evaporation deposition of the oxide layer, the PBA monolayer decomposed and evaporated into fragments encapsulated in the oxide capping layer [68].

To avoid the oxide capping during MLD, simplified trends have recently been reported, such as self-capping monolayer doping or non-capping using monolayer contact doping or remote monolayer doping. Self-capping MLD process was studied by Alphazan and workers using hepta-isobutyl-polyhedral oligomeric silsesquioxane triester of phosphorus that provides phosphorus atoms and the silsesquioxane cage as a self-capping layer for phosphorus monolayer doping [61]. In nanoscale doping, a capping layer can cause adverse impacts during fabrication. For example, capping an oxide layer for boron MLD in highly porous nanowires (NWs) was reported that cause surface damage to NWs during the removal step after annealing. Veerbeek and coworkers utilized the MLCD and MLD with an external capping layer as alternative techniques to escape surface damages and obtain higher doping concentrations [97].

2.2.5 Molecules for boron doping

As mentioned above, the molecule types are important and affect the monolayer doping performance. The self-assembly procedure, the monolayer coverage efficiency, and molecular size are the initially critical factors in determining the dopant density on the surface. The boron-containing molecule used first as well as popularly for boron MLD is allylboronic acid pinacol ester (ABAPE) [58–60, 62, 67, 69, 76, 81]. ABAPE precursor possesses a boron atom and a terminal alkene that can form a covalent bond with a hydrogen-terminated semiconductor surface. Ho and coworkers reported the first research on boron MLD using the ABAPE molecule. The authors successfully achieved a high boron doping level of 5×10^{20} cm^{-3} near the silicon surface. The boron atoms rapidly diffused into silicon lattice during the spike annealing process. The sheet resistance of samples decreased around 100 times after MLD. The resistivity was extremely affected by tuning temperature rather than the time of annealing.

The performance of B-MLD depends on the number of boron atoms carried on molecules. A precursor that contains more content of dopant atoms can obtain a higher doping level compared with molecules that hold lower content of dopant. Therefore, the doping levels can increase significantly by designing a specific precursor containing more than one boron atom. For instance, MLD using carborane derivative CB-(Me, allyl) precursor, which has a carborane cluster with 10 boron atoms and alkene groups as boron-containing alkene molecules, was performed by Huskens et al. on hydrogen-terminated silicon (**Figure 6a**). The result of boron activation using carborane derivatives was around 10 times higher boron doping levels compared with using ABAPE molecules that have only a single boron atom [65]. The annealing time does not affect the active dopant concentration, while annealing temperature plays a role. The doping concentration at 1000°C for 15 s of annealing time was observed that stays unchanged compared with using 6 s annealing time. However, increasing the annealing temperature from 1000°C to 1050°C significantly enhanced the successful doping concentration. This can be explained by the dependence of boron solubility upon temperature; the higher the temperature increases the solubility of boron. The sheet resistance was examined by carboranyl molecular doping is lower than 20 times that of ABAPE doping, which indicated higher conductivity obtained by carboranyl molecule [65].

The dose and concentration of boron-bearing molecules impact boron doping efficiency. The areal dose control of boron doping was designed firstly by Ho and coworkers [58]. The different ratios of dopant molecules were controlled by mixing dopant-carrying molecule (ABAPE) with a blank precursor (1-undecene), an alkene containing only C and H, for hydrosilylation, as illustrated in **Figure 6b**. The authors found that the boron concentration on the surface is proportional to the fraction of dopant-containing molecules in the mixture. The sheet resistance of samples

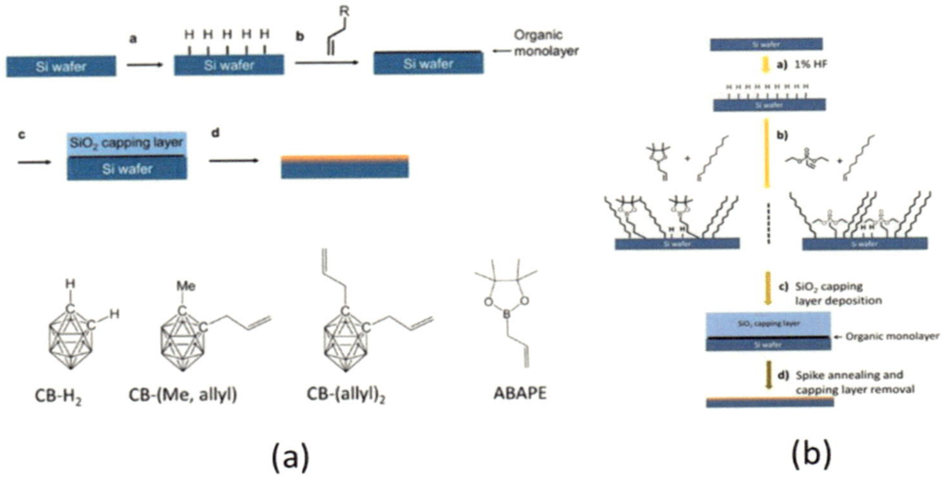

Figure 6.
Schematic depiction of boron monolayer doping process (a) using carborane derivative printed with permission from Ref. [65]. Copyright 2015 American Chemical Society and (b) for the areal dose control by mixing dopant-carrying molecule (ABAPE) with a blank precursor (1-undecene) printed with permission from ref. [62]. Copyright 2015 American Chemical Society.

correlates to the monolayer doping dose and delivers an approach to control the electrical properties of the semiconductor substrate. A more detailed report by Ye et al. about the relation between the precise control of boron dose with the monolayer composition and thermal annealing. It has been found that the monolayer configuration is also proportional to the dose ratio of dopant-carrying molecules. Ye and workers also explored that the higher boron concentration at the surface can prohibit the driving boron atom into the silicon. The boron diffusion from the surface into the substrate increases with the decrease of the concentration of boron-containing monolayer [62].

Similarly, Fu et al. carried out experiments to control the dopant dose and observed the impacts of dopant concentration on the boron activation and photo responses [69]. Reducing the half dose of ABAPE molecules by mixing with 1-undecene decreased the activation rate of boron from 91.4% to 54.2%. Besides, they also reported that the higher ratio of carbon interstitials in silicon contributed by 1-undecene can bound with substitutional boron atoms to form defect clusters. These carbon-boron clusters complex the boron diffusion and prevent the boron occupation in the substitutional sites leading to the reduction of boron activation. Besides, the formation of carbon-boron cluster defects was reported only when the MLD process used the molecular mixture. In a previous study by Gao et al., it was noted that the atmospheric carbon contaminants formed carbon-related defects, including C_sH and C_sOH, which only capture minor electron carriers and have a limited impact on boron activation [60]. However, the effects of carbon contaminants are worse on phosphorus monolayer doping, which can deactivate at least 20% of the phosphorus atoms [98]. They successfully doped boron by MLD, reaching around 95% of electrically active boron atoms with sheet resistance lower than 90 times [60].

Furthermore, monolayer sources have distinct characteristics involving decomposition features, fragmentation details, surface chemistries, and covalent or non-covalent assemblies onto the surface. Therefore, the difference in structure and head groups of dopant-bearing molecules can impact the doping levels at nanometer-scale structures. For example, boron MLCD using phenylboronic acid (PBA) and chlorodicy-clohexylborane (CDB) formed respectively non-covalent monolayer and covalent monolayer, both showed high boron doping levels in silicon nanowires (NWs). The average boron doping level of CDB-MLCD was higher than that of PBA-MLCD. However, the resistivity of the PBA-MLCD-doped NWs was lower compared with CDB-MLCD. The reason was explained because the thermal fragmentation of CDB monolayer was complicated and uncompleted during different periods of thermal annealing that created carbon-boron complexes resulting in the formation of silicon-carbide clusters increasing the boron diffusion [68].

2.3 Advantages and drawbacks of MLD

Monolayer doping allows achieving an ultra-shallow boron dopant profile with a depth of sub-10 manometers applied to fabricate ultra-shallow p-n junction with depths <10 nm that are very difficult to obtain by traditional techniques. The boron atoms are driven into the semiconductor from the surface substrate during the annealing process. Therefore, the boron penetration is shallower and limited depending on the solubility and diffusivity of the boron monolayer, which annealing conditions can control. Due to the self-assembly of boron molecules onto surfaces in a vapor or solution phase by chemical interaction, there is no geometry limitation of boron monolayer doping. It is appliable to fabricate in 3D structures such as FIN with narrow side-well or nanowires with a round shape, which cannot be obtained by ion implantation or CVD techniques. Moreover, monolayer doping is capable of controlling the doping with different doping scales that are highly versatile for various applications from nanoscale fabrication such as FinFETs or nanowire-FETs to larger-scale production of MOSFETs. Ho and coworkers were successful in fabricating p+/n USJs on a 4 inch scale of the silicon wafer. This promises an innovation of boron doping in larger-scale fabrication with uniformity of boron profile [58, 79]. It also can control the specific boron doping areas by novel monolayer contact doping or remote monolayer doping processes to avoid the out-diffusion in the capping layer during annealing. Monolayer contact doping is applied for co-doping to fabricate a parallel p-n junction Si NWs using B and P-containing molecules.

Furthermore, unlike boron ion implantation, boron MLD does not cause any lattice damage during doping and annealing. This damage-free MLD is beneficial for limiting annealing time and avoiding the undesired leakage of boron atoms. The MLD process is a simple method that does not require any specialized and costly equipment. With diverse boron-carrying molecules, the monolayer formation is able to be carried out in different vapor or solution deposition methods that can easily carry out without harsh conditions. The self-assembled monolayers are highly uniform onto silicon substrate due to the self-limiting reaction of molecules, making molecular quantities well-defined with accuracy. The areal dopant dose can be tuned by various methods, such as doping a mixture of two molecules or using different types of molecules. Consequently, the boron monolayer doping can be precise by combining the RTA condition and molecular design of the precursors that are unable to a wide range of doping profiles to satisfy the specific requirements for various applications [58].

However, monolayer doping remains a limitation. Due to the self-limitation source of monolayer onto the surface and escaping of boron atoms into the capping oxide layer, the actual boron concentration is much lower compared with doping by ion implantation and CVD. A low boron doping efficiency of ~33% [58] depending on MLD conditions was measured because of the higher boron diffusivity in oxide capping and self-limiting source of boron monolayer. Due to the dependence of solubility and diffusivity on temperature, the balance of controlling the temperature to achieve higher dopant incorporation and shallower profile depth is challenging. The sheet resistance of junction fabricated by monolayer doping remains quite high, ranging from 10^3 to 10^4 Ω/sq. The carbon or oxygen-related defects form an annealing process that can decrease the boron incorporation into the substrate as well as doping performance. But the boron doping concentration is possible to increase by increasing the number of boron atoms carried in molecules [65, 66]. Moreover, the semiconductor surface's monolayer formation process is highly sensitive and can easily be removed. Therefore, the cleaning and capping removal processes require a careful operation [67].

2.4 Application of monolayer doping

2.4.1 Boron MLD for electronic devices

As explained above, ion implantation is not feasible to dope boron on the sidewalls of finFET structures due to irreparable crystal damage [99]. Homogeneous and conformal doping is required for small dimension devices. CVD doping can be used for 3D structures, but this method needs to control parameters, including the growth temperature, reactor pressure, and precursor dose. Therefore, CVD doping is challenging in mass production to generate uniform thin layers [100]. Due to the ability to doping a thin uniform layer of boron in 3D structures, MLD promises a practical technique applied in the semiconductor industry to fabricate small electronic devices such as CMOS or finFET with affordable expense. Monolayer doping sulfur on CMOS device designed by Barnett et al. using ammonium sulfide, $(NH_4)_2S$ as sulfur monolayer source. A uniformly doped ultra-shallow junction with 9 nm of depth and low sheet resistance of 164 Ω/sq. was achieved without damage to the substrate [99]. Ang and coworkers first applied MLD to fabricate ultra-shallow junction in 20 nm finFET with phosphorous MLD. The authors successfully produced a 5 nm- n+/p junction with a sheet resistance of 8.3×10^3 Ω/sq. [101]. The 3-D finFET devices recently require a channel thickness scaled down to sub-10 nm [101] (**Figure 7a**). Boron and phosphorus co-monolayer doping was used to create a conform thin shell doping on polysilicon junctionless finFET devices [102, 103]. The ultra-shallow doping profiles of n-and p-type were obtained with sub-5 nm of depths. The FinFETs showed excellent gate control with $I_{on}/I_{off} \sim 10^6$, lower off-current, and an exceptional subthreshold slope of 67 mV/dec [102]. The recent publication uses conformal monolayer doping to prepare devices with complex-geometry structures, allowing for the formation of multilayer Ge nanosheet gate-all-around field-effect transistors (**Figure 7b**). This can overcome the limitation of the Wrap-Around Contact method normally used for epi source/drain formation [104].

2.4.2 MLD for solar cells

Electrical energy generation in solar cells depends on splitting holes and electrons efficiently at a p-n junction. Therefore, MLD plays a vital role in the manufacture of silicon solar cells. Boron is introduced in silicon to generate p-type semiconductors that allow the transport of electrons from one atomic layer to another. The boron-doped silicon is used to increase conduction efficiency and lower the production expense of solar panels by focusing on growing surface-to-volume ratios and p-n junction dimensions. Therefore, the solar cells can absorb the larger light converted into energy to separate more electron-hole pairs. Moreover, the non-planar doping

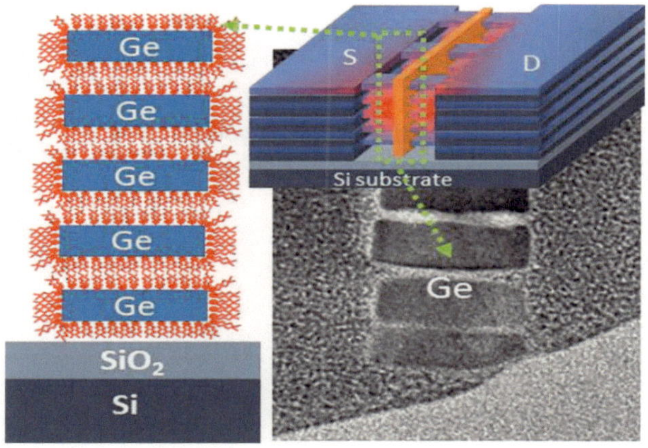

Figure 7.
The graphic diagram of monolayer doping of the five-stacked Ge nanosheets FET printed with permission from Ref. [104]. Copyright 2022 American Chemical Society.

capability of MLD makes it ideal for this application [105]. In the report of Garozzo et al., MLD was utilized to fabricate a doped layer covering the entire nanohole surface of solar cells. The radial junctions were formed inside the nanoholes with a carrier concentration of around 10^{19} cm^{-3} for both n-/p- type doping [106].

3. Boron chemical vapor deposition (CVD) doping

3.1 Introduction to CVD doping technique

Chemical vapor deposition (CVD) or solid-phase doping (SPD) is one technique to grow a thin film layer that involves a chemical reaction of the volatile molecule containing atom precursors. The principle of this method is the interaction between the vapor gas of the precursors with the substrate surface that is heated inside the reaction chamber. Resulting in a condensation layer that grows on the substrate surface and unreacted vapor gas that is later removed. This method is broadly used due to its potential for mass production and flexible controllability of growth parameters (temperature, pressure, precursor concentration, substrates) during the process [107].

The mechanism of this technique is to break the bond between each volatile molecule and leave the targeted atom precursor that is later reassembled as a layer of the atom. A high temperature is needed to break chemical bonding depending on each chemical bonding of the molecules. Then it becomes reasonable why this method needs a quite high temperature.

The formation of the Boron-Si junctions is summarized in **Figure 8**. The H atom on the passive silicon surface (Si−H) is first desorbed to provide a free H-Si dangling bond. H then releases in the form of H_2 after borane deposition due to the B−Si bond that formed. Incoming borane develops bonding with Si-B as a boron cross-link over the silicon surface. Thus, a boron layer formed on the silicon surface.

The junctions of boron-doped silicon can be introduced using two methods: *ex-situ* and *in-situ* methods. The *ex-situ* steps involve removing oxides and contaminants at the Si surface and effectively passivating the surface [108]. First is depositing boron on the Si surface in the form of B_2O_3. The oxide is then reduced on the Si surface by

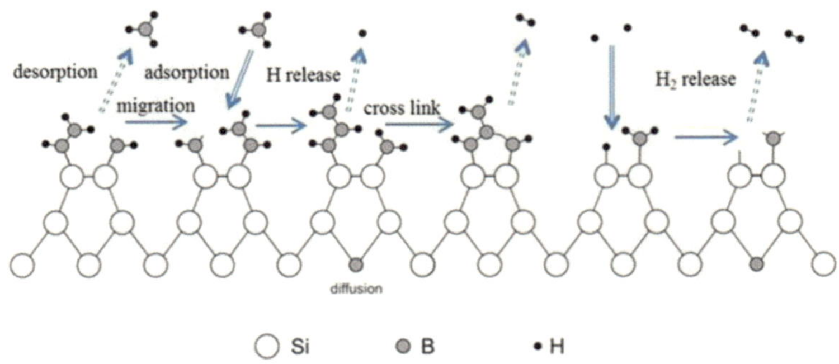

Figure 8.
Chemical interaction scheme of CVD boron deposition printed with permission from ref. [108]. Copyright 2017, Vahid Mohammadi et al.

oxidation, resulting in a boron-rich SiO_x layer to the formation of a shallow p-n junction. The boron is then diffused into the Si and activated (incorporated into a substitutional site) during high-temperature drive-in anneal [109]. The second method uses precursors that containing boron, silicon, and a catalyst in gaseous form to grow both silicon and boron layers simultaneously at high temperatures.

3.2 Current development of boron CVD doping

Boron sources that are usually used are boron hydrides (diborane), boron halides (boron trichlorides), and organoboron (triethyl boron). Boron sources are chosen depending on which precursor and gas environment that used. Even though diborane seems an upcoming boron source, it only contains hydrogen apart from boron. It is known that diborane (B_2H_6) has a toxic, flammable, and explosive nature, so it needs a handful of treatments. Boron halides (BCl_3) are expected to be a safe boron source because it is nonflammable and less toxic. Otherwise, boron trichlorides (BCl_3) will not be suitable as a precursor for BN (boron nitride) since it will produce NH_4Cl as the HCl reacts with NH_3, which can damage the vacuum pump. At the same time, the hydrogen chloride is corrosive to a metallic substrate. The organoboron (such as $B(CH_3)_3$ and $B(C_2H_5)_3$) seems an excellent precursor to obtaining B_4C (boron carbides) because it can act as a boron and carbon source at the same time [110].

Sarubbi et al. demonstrated that diborane has selectively deposited only on Si with ~6 nm thickness at 500°C for 10 min diborane exposure as TEM result does not observe any B deposited on the slope or flat SiO_2 surface. The SIMS profile of the B layer formed by CVD after HNO_3 treatment has a concentration peak of 6×10^{20} cm^{-3} and shows a 5.9×10^4 Ω/sq. sheet resistance. They also mention that it has a 2.44×10^{-2} $A/\mu m^{-2}$ saturation current density and a 13 nm junction depth [111]. Mok et al. in 2013 demonstrated the pure boron deposition using B_2H_6 as a boron source and H_2 as a carrier at 700°C for 9 min of deposit time. It was found that nanometer-thick pure B layers, upon annealing in the presence of oxygen, function as a catalyst for silicon oxide growth. Based on the HRTEM result, the pure B is successfully doped on the surface (100) with 2.9 nm thickness and 2.1 nm on the surface (111) after TMAH texturing. They also reported the effect of oxygen concentration on the oxide forming. The thickness changes to 24.4 nm (100) and 23.4 nm (111) after furnace anneal at 950°C for 30 min in nitrogen ambient. For dry oxidation at 950°C for 30 min in 14% oxygen concentration, the thickness is changed to 37.6 nm (100) and 43.4 nm

(111). They also mention that an ample oxygen supply during annealing results in boron depletion of the boron-doped Si surface due to enhanced oxidization, resulting in a lower surface concentration and higher sheet resistance. The sample that was processed in nitrogen condition and then etched using HNO_3 and HF had 61.2 Ω/sq. sheet resistance compared with the sample that was processed using in the presence of oxygen condition, which had 205 Ω/sq. sheet resistance that measured at 10^5 cm^{-3} of carrier concentration [112]. Higher oxygen concentrations of O_2 lead to increased growth of an intermediate SiO_2 layer, which acts as a diffusion barrier and results in an increase in sheet resistance with increasing O_2 [113].

In 2020, Muroi et al. [114] used BCl_3, H_2, and $SiHCl_2$ as gas precursors on the silicon surface. They observe the deposition and etching behavior at different temperatures. Boron adsorption occurs at a temperature lower than 800°C, the deposition occurs at 900–1000°C, and at a temperature higher than 1000°C, they observe etching behavior due to chlorosilanes that occur in gaseous form. In their further research in 2021 [115], they reported using a similar boron gas source at 800°C. The etching does not occur on the surface based on the HRTEM result that demonstrated the dense film without void. The work that was done by Taniguchi and Inasawa using BCl_3 as a boron source in 2020 showed that the presence of boron-doped silicon nanowires could change sheet resistivity from 10^5 Ωcm to be in the range of 10^{-3}–10^1 Ωcm [116].

B dopants' diffusion can occur under severe conditions, often simultaneous, such as very large concentration gradients, non-equilibrium point defect density, amorphous-crystalline transition, extrinsic doping level, co-doping, B clusters formation and dissolution, ultra-short high-temperature annealing [117]. The vacancies (V) and self-interstitials (I) are intrinsic point defects significant for dopant diffusion. In germanium, both p-type and n-type are mediated by the vacancies. Boron has a slow diffusion rate compared with other p-type dopants, which helps form ultra-shallow doped regions in Ge. The slow diffusion of B is associated with a high diffusion activation enthalpy that exceeds the activation enthalpy of self-diffusion by more than 1 eV. This indicates that B atoms are not likely associated with vacancies, thus meaning that B diffusions are via self-interstitials [118]. Tu et al. [119] successfully introduced a 5 nm thickness of the boron layer in epitaxial Ge on Silicon with a peak surface of 7×10^{21} cm^{-3} boron concentration.

3.3 Typical applications of boron CVD doping

In their report, Liu et al. said a pure boron layer deposited using the CVD method could be used as an a-Si mask to protect from TMAH and KOH etching for long hours of exposure [120]. Other literature also shows the potential ability of boron-doped CVD as anti-corrosion on mild steel [121], used to reduce diamond growth rate to achieve a certain thickness of diamond [122, 123], used to develop boron carbide [124], boron nitride [107, 125], and also to fabricate the uniform p-type doping of silicon nanowires [109, 126], it also found that boron can be used to make a superconductor by heavily doped boron on diamond [127].

3.4 Advantages and shortcomings of the CVD technique

There are many advantages of the boron deposition using the CVD method, namely able to control the growth parameters, it can deposit a single diffusion source only on one side of the wafer, so it can be used to introduce different doping profiles and structures of the diffusion source to achieve dopant concentration profiles next to each other [128]. It also requires fewer steps than other methods and allows better tunings of dopant profiles. It has a lower thermal budget as in-situ B-doped Ge can be grown at low temperatures (400°C), and B is already activated during growth, so it does not need activation annealing [129]. Unlike ion implantation, B doped using CVD does not destroy the structure due to annealing. Other advantages are that it can perform ultra-shallow junction, it can be used to develop boron sheets (2D structure) or boron carbide or boron nitride (3D structure), and the deposit does not depend on the position or flat surface. It is known that it can perform deposits on silicon wires [116, 130]. Furthermore, high-energy boron ion implantation in diamond enhanced the concentration of active boron up to for CVD method 10^{21}–10^{22} ion/cm^3 to reach superconductor, while normally concentration of boron is around 10^{19} for boron ion implantation [30]. Therefore, a nanometer-thin boron amorphous layer can be created on the surface of crystalline silicon through a chemical vapor deposition (CVD) process in the temperature range from 700°C to 400°C [108].

Besides its promising advantages, introducing boron using CVD has a few shortcomings. Such as it is lack of a precursor that is highly volatile and, nontoxic and nonpyrophoric, it needs metal boride compounds that can form on the catalytic substrate and the toxicity of boron gas source that used must be concerned and controlled tightly. The boron will continue to diffuse at higher temperatures, so it must be suppressed [131]. The solid solubility of the dopant at operating temperature also becomes a shortcoming because it will be related to dopant concentration [132]. Unproperly removes oxide and boron-rich layer from the surface leading to poor surface passivation [113].

4. Doping boron in typical semiconductor materials

In semiconductor manufacture, boron doping is a crucial technique to introduce boron atoms into a semiconductor to modify its physical properties. There are intrinsic semiconductor materials, including silicon (Si),germanium (Ge), and compound semiconductors, which is combinations of elements such as group II–VI (ZnSe, ZnTe, CdS, CdTe), group IV-VI (PbS, PbSe, PbTe) of the periodic table, group III–V (AlN, GaAs, InGaN, InP, InGaAlP), or elements in the same group IV–IV (SiC, SiGe), other advanced materials including carbon nanotube, diamond, 2D materials (graphene, hexagonal boron nitride), etc. In intrinsic semiconductors, their atoms connect by sharing electrons to create stable covalent bonds. Generating conduction in a semiconductor requires energy to break the crystal bond and create conduction electrons moving around in a crystal and leaving holes. For example, silicon requires approximately 1.12 eV of energy to free an electron at room temperature. This energy is called bandgap energy or energy gap (Eg), which is necessary energy to excite an electron trapped in the valence band to the electrical conduction band. Silicon doped by boron is introduced a more significant number of conduction electrons and mobile holes that can lift the valence band close to the conduction band, decreasing the bandgap energy of boron-doped silicon to 0.045 eV [133]. The number of holes (positive charge carriers) rises with the increased amount of active boron concentration. In p-type semiconductors, the conduction is attributable to an enormous number of holes; therefore, holes and electrons are referred majority carriers and minority carriers, respectively.

The electrical conductivity of boron-doped silicon depends on the amount of boron and temperature. According to the calculation modeling of hole mobility on

DOI: http://dx.doi.org/10.5772/intechopen.106450

boron concentration of Masetti and coworkers, the hole mobility of boron-doped silicon can be estimated around 424–25 cm^2/Vs, correlating with the range of 10^{14}–10^{21} cm^{-3} for boron concentration. The higher the boron concentration, the lower the carrier mobility and resistivity [134]. Moreover, boron doping improves the hardness property of silicon, the hardness increases with increasing boron-doped concentrations. For example, the hardness at boron concentration of 1.3×10^{20} atoms/cm^3 was 30% higher than that at 2.9×10^{17} atoms/cm^3 [135]. The thermal conductivity of boron-doped silicon (with a B concentration of 5×10^{20} atom/cm^3) was lower than undoped silicon at 300 K. Lee et al. found that the mass disorder effect is the main reason for the thermal transport suppression in boron-doped Si [136]. Like silicon, germanium (Ge) is an intrinsic semiconductor as silicon with a bandgap of 0.67 eV [137]. Introducing boron in Ge causes changes in electrical, mechanical, and thermal properties that are approximate to boron-doped silicon. The carrier mobility in boron-doped Ge monocrystals decreases with the increase of boron concentrations. The elasticity limit of Ge enhances after doping with low boron concentration. The mechanical property of boron-doped Ge at high boron concentration [138]. Si and Ge are primary materials for the semiconductor industry. Boron-doped Si and Ge show highly electrical conductivity that more effective for application in the electronic device fabrication including diodes [139], transistors [104], integrated chips/circuits [140], microcontrollers [141] and other applications for sensors [142, 143], light-emitting diodes (LEDs) [144], energy storage such as solar cells [145–147], photovoltaic devices [139, 148], capacitors [149], etc.

Boron doped in carbon nanotubes using CVD doping method that lowered HOMO-LUMO bandgap, featured for chemical reactivity and kinetic stability, of CNTs from 0.56 eV of original CNTs to Eg ~ 0.44 eV of B-CNTs after doping [150]. Introducing boron into CNTs increases the defects that break inertness and improves the reactivity in CNTs. The changes in electrical properties of CNTs varied depending on the boron concentrations. Yi and coworkers investigated that the acceptor state after doping boron was located at 0.16 eV above the Fermi energy for the ratio of B/C ~ 1/80 [151]. Boron doping improves the metallic property of CNTs. Moreover, the mechanical and thermal properties of CNTs were modified after doping with boron. The rupture stress of the B-CNTs was reduced compared with pristine CNTs, but at higher temperatures, B-CNTs showed drawbacks on maximum stress [152]. The thermal conductivity of B-CNTs depends on the temperature. At low temperatures, the thermal conductivity decreases with a rise in boron concentration in zigzag CNTs. However, the thermal transport enhances with increased boron concentrations at higher temperatures [153]. Boron-doped CNTs were applied in various application from hydrogen energy storage [154–156], catalysis [157], electrocatalysis [158], sensors [157, 159, 160].

Similarly, boron was introduced into graphene to modify its physical, chemical, mechanical, and electrical properties. The nature of graphene structure changes from ductile to brittle after being doped with boron. The thermal conductive property of graphene is reported to weaken after boron doping. Thermal conductivity dropped around 60% after introducing 0.75% boron concentration in graphene. Pristine graphene is a zero-gap semiconductor with semi-metallic property [161]. Boron-doped graphene monolayer shows a p-type semiconductor behavior with a high carrier mobility level of approximately 800 cm^2/Vs at ambient temperature [162]. Wu et al. fabricated B-doped Graphene-based back-gate FETs with mobilities of 450–650 cm^2/Vs [163]. Graphene doped with boron exhibits excellent electrochemical properties for diverse applications, including electrocatalysis [164], energy storage (batteries, supercapacitors) [165], sensors [166], and photovoltaics [167].

B-doped graphene can obtain a small band gap of 0.05 eV combined with n-type silicon to fabricate a p-n junction for solar cell application. The B-graphene/silicon-based solar cell showed a higher short-circuit current density of 18.8 mA/cm [168].

The sp^3-hybridized diamond is an insulation material with a wide bandgap of 5.47 eV and extremely high resistivity of roundly 10^{12} Ω/cm. Doping with boron turns an insulative pure diamond into a conductive p-type semiconductor. After introducing boron into the diamond, the acceptor level is quite deep, around 0.37 eV above the valance band. Boron doped diamond shows high-level conductivity and enhanced electron transport compared to undoped diamond. The average boron doping level in diamond ranges from 10^{18} to 10^{20} atoms/cm^3 [169, 170]. The hole mobility of boron-doped diamond was examined, reaching the maximum of about 2000 cm^2/Vs at ambient temperature [170]. Heavy boron-doped diamond with a higher boron concentration of 10^{21}–10^{23} atoms/cm^3 for superconductivity can obtain at high pressure (10^5 atmospheres) and temperature (2500–2800 K) [171]. The sheet resistance of B-doped diamond was dropped from 10^{14} to about 10^{10} Ω/sq. [30]. Doping boron in diamond also changes its physical and mechanical properties. Similar to B-CNTs and B-doped graphene, B-doped diamond exhibits a comparable tendency in thermal conductivities [172]. The higher the boron-doped concentration, the weaker the thermal transport. The surface area of a diamond is larger after doped with boron. B-doped diamond is electrode material for numerous fields of electroanalysis [173], electrochemical energy storage [174, 175], and sensors [171, 176].

Apart from the above materials, doping boron is applied to improve the mechanical property of semiconductor compounds. Boron doping using ion implantation has been proven to change the roughness, hardness, stress/strain of materials, and other morphological characteristics of materials. The Zinc Selenide (ZnSe) thin films were implanted with boron ions at 75 keV and ranging in doses from 10^{12} to 10^{16} ions/cm^2 in the research of Venkatachalam et al. that revealed the increase of film surface roughness and the decrease in the optical band gap value while increasing the dose of boron ions [177]. The hardness and elastic modulus of the hosts were also increased in some substrates of 60NiTi/NiTi after being doped by boron atoms [9, 178]. This is accounted for by replacing boron atoms in lattice matrix to create new nanocrystals, for example, TiB_2 in B-doped 60NiTi [9]. In addition, a study by Zhu et al. proved boron ion implantation can enhance a hardening effect in the TiAlN. This hardening resulted from of the increase of excess stresses and the formation of new forms (TiB_2 and BN nanocrystals) in the structure after the boron implants [179]. Similarly, boron ion implantation at 150 KeV and a fluence of 1×10^{15} ions/cm^2 in hexagonal boron nitride (h-BN) induced the formation of c-BN nanocrystals due to the collisions of ions with the radical atoms and created the displacement of these atoms out from the lattice positions, which generated an atomic vacancy and temporary accumulation of defects in the interstitial site in h-BN. This increases the stress/strain level in h-BN and increases the electron density in the interatomic and interlayer places in the material [27]. Additionally, boron implants modified the structure of two-dimensional carbon-fiber-reinforced carbon-carbon (C/C) composites to generate the boron carbide composition, improving resistance during exposure to air at high temperature [180].

5. Conclusion and outlook

By doping with different atoms, materials can significantly improve stability or change their properties. Boron doping affects Si, Ge, graphene, boron nitride, etc., especially tuning their electrical properties effectively. With the advancement of the semiconductor device, 3D monolithic integration that employs multiple vertically stacked devices for higher device density appears to have lower power consumption and provides a platform for heterogeneous integration of different active semiconductor layer materials. Therefore, the ability to prepare devices with geometry design is highly desired, such as FinFET, Gate-all-around FET, and nanosheets FET. Proper doping techniques must be decided to dope such dimension channels with shallow junction formation uniformly. Herein, this chapter investigates currently available methods and compares their performance, as shown in **Table 1**. The monolayer layer doping appears to overcome the limitation of the ion implantation for their better conformal doping profile and capable shallow junction formation.

Features	Ion implantation	Monolayer doping	Chemical vapor deposition doping
Profile depth (nm)	<10 nm to 10 μ	<100 nm	<3000 nm
Boron concentration (cm^{-3})	10^{13}–10^{21}	10^{17}–10^{20} (decreasing from the surface)	10^{15}–10^{21}
Boron activation (% in boron activation)	80%	91–54% of boron concentration	N/C
Sheet resistance (Ω/sq)	<10^3	10^3–3×10^4 (depending on B concentration)	<10^5
Advantages	Easy to tune the depth/ions selection	Easy to obtain ultra-shallow junction	Fewer process steps than other methods
	Nonsensitive to surface cleaning procedures	Larger scale fabrication, mass production	In-situ doped Si/Ge
	Higher doping efficiency	Simple and affordable process	
	Good reproductivity on doping profile	No geometry limitations	
		Uniform doping layer	
Disadvantages	Severe damage to substrate (amorphization), high diffusion	Low doping efficiency	Using toxic, pyrophoric gases
	Expensive costs in equipment and operation	Difficult control between diffusion depth and incorporation	Expensive costs in equipment and operation
	Hard to get ultra-shallow or deeper doping	Sensitive to surface treatment	Poor surface passivation after doping
	Geometry limitation (for the sidewall of fins)	Potential contaminations of C/O elements	
	non-uniform doping layer (increased roundness)		

Table 1.
Comparison of three different boron doping techniques.

Acknowledgements

This work was supported by I-Dream Grant (13111F1201) from Taiwan Semiconductor Research Institute (TSRI) and National Applied Research Laboratories (NARLabs).

Author details

Linh Chi T. Cao, Luqman Hakim and Shu-Han Hsu*
Sirindhorn International Institute of Technology, Thammasat University, Pathum Thani, Thailand

*Address all correspondence to: shuhanhsu@siit.tu.ac.th

References

[1] Kim KS, Song YH, Park KT, Kurino H, Matsuura T, Hane K, et al. Novel doping technology for ultra-shallow junction fabrication: Boron diffusion from boron-adsorbed layer by rapid thermal annealing. Thin Solid Films. 2000;**369**(1):207-212

[2] Pérez-Martín AMC, Jiménez-Rodríguez JJ, Jiménez-Sáez JC. Shallow boron dopant on silicon: An MD study. Applied Surface Science. 2004;**234**(1):228-233

[3] Robertson LS, Brindos R, Jones KS, Law ME, Downey DF, Falk S, et al. The effect of impurities on diffusion and activation of ion implanted boron in silicon. Materials Research Society Symposium Proceedings. 2000;**610**:B5.8.1-B5.8.1.6

[4] Michel AE, Kastl RH, Mader SR, Masters BJ, Gardner JA. Channeling in low energy boron ion implantation. Applied Physics Letters. 1984;**44**(4):404-406

[5] Crowder BL, Ziegler JF, Cole GW. Ion implantation in semiconductors and other materials. In: 1st Ion Implantation in Semiconductors and Other Materials. New York: Plenum; 1973. 257 p

[6] Plummer JD, Deal MD, Griffin PB. Silicon VLSI Technology. New Jersey: Prentice Hall; 2000

[7] Ahmad I. Ion Implantation - Research and Application [Internet]. London: IntechOpen; 2017. 152 p

[8] El-Kareh B. Ion implantation. In: Fundamentals of Semiconductor Processing Technology [Internet]. Boston, MA: Springer US; 1995. pp. 353-466. DOI: 10.1007/978-1-4615-2209-6_6

[9] Yan C, Zeng Q, He W, Zhu J. Enhanced surface hardness and tribocorrosion performance of 60NiTi by boron ion implantation and post-annealing. Tribology International. 2021;**155**:106816

[10] Adey J, Goss JP, Jones R, Briddon PR. Identification of boron clusters and boron-interstitial clusters in silicon. Physical Review B: Condensed Matter and Materials Physics. 2003;**67**(24):245325(1-5)

[11] Hwang GS, Goddard WA. Catalytic role of boron atoms in self-interstitial clustering in Si. Applied Physics Letters. 2003;**83**(5):1047-1049

[12] Bhattacharya P, Fornari R, Kamimura H. Comprehensive semiconductor science and technology. In: Comprehensive Semiconductor Science and Technology. Vol. 1-6. Amsterdam, London: Elsevier Science; 2011

[13] Jang CW, Kim JH, Lee DH, Shin DH, Kim S, Choi SH, et al. Effect of stopping-layer-assisted boron-ion implantation on the electrical properties of graphene: Interplay between strain and charge doping. Carbon N Y. 2017;**118**:343-347

[14] Michel AE, Rausch W, Ronsheim PA. Implantation damage and the anomalous transient diffusion of ion-implanted boron. Applied Physics Letters. 1987;**51**(7):487-489

[15] Vos M, Boerma DO, Smulders PJM, Oosterhoff S. Defect and dopant depth profiles in boron-implanted silicon studied with channeling and nuclear reaction analysis. Nuclear Instruments and Methods in Physics Research B. 1986;**17**(3):234-241

[16] Cristiano F. Ion Implantation-Induced extended defects: Structural

investigations and impact on Ultra-Shallow Junction properties. Micro and nanotechnologies/Microelectronics. France: Paul Sabatier University; 2013. tel-00919958

[17] Takeda S. An atomic model of electron-irradiation-induced defects on {113} in Si. Japanese Journal of Applied Physics. 1991;**30**(4):639-642

[18] Ohring M. Materials Science of Thin Films. 2nd ed. San Diego: Academic Press; 2001

[19] Jain SC, Schoenmaker W, Lindsay R, Stolk PA, Decoutere S, Willander M, et al. Transient enhanced diffusion of boron in Si. Journal of Applied Physics. 2002;**91**(11):8919-8941

[20] Ynsa MD, Agulló-Rueda F, Gordillo N, Maira A, Moreno-Cerrada D, Ramos MA. Study of the effects of focused high-energy boron ion implantation in diamond. Nuclear Instruments and Methods in Physics Research Section B: Beam Interactions with Materials and Atoms. 2017;**404**:207-210

[21] Aprà P, Ripoll-Sau J, Manzano-Santamaría J, Munuera C, Forneris J, Ditalia Tchernij S, et al. Structural characterization of 8 MeV 11B implanted diamond. Diamond and Related Materials. 2020;**104**:107770

[22] Uzan-Saguy C, Cytermann C, Brener R, Richter V, Shaanan M, Kalish R. Damage threshold for ion-beam induced graphitization of diamond. Applied Physics Letters. 1995;**67**(9):1194-1196

[23] Mayer JW. Ion implantation. Lattice disorder. Radiation Effects. 1971;**8**(3-4):269-277

[24] Shih Y, Washburn J, Gronsky R, Weber ER. Amorphization of silicon by boron ion implantation. Materials Research Society Symposia Proceedings. 1986;**71**:203-209

[25] Ryssel H, Müller K, Haberger K, Henkelmann R, Jahnel F. High concentration effects of ion implanted boron in silicon. Applied Physics. 1980;**22**(1):35-38

[26] Aboy M, Pelaz L, Barbolla J, Duffy R, Venezia VC. Boron activation and redistribution during thermal treatments after solid phase epitaxial regrowth. Materials Science and Engineering B: Solid-State Materials for Advanced Technology. 2005;124-125:205-209

[27] Aradi E, Naidoo SR, Cummings F, Motochi I, Derry TE. Cross-sectional transmission electron microscopy studies of boron ion implantation in hexagonal boron nitride. Diamond and Related Materials. 2019;**92**:168-173

[28] James WM, Lennart Eriksson JAD. Ion Implantation in Semiconductors. New York: Academic Press; 1971

[29] Liu LJ, Hsu WH, Brumfield K, Padmanabhan R, Morinville W, Qin S, et al. Effects of implant temperature on process characteristics of low energy boron implanted silicon. In: AIP Conference Proceedings. 2012;**1496**:96-94

[30] Seki Y, Hoshino Y, Nakata J. Remarkable p-type activation of heavily doped diamond accomplished by boron ion implantation at room temperature and subsequent annealing at relatively low temperatures of 1150 and 1300°C. Applied Physics Letters. 2019;**115**(7):072103(1-4)

[31] Robertson LS, Jones KS, Rubin LM, Jackson J. Annealing kinetics of {311} defects and dislocation loops in the

end-of-range damage region of ion implanted silicon. Journal of Applied Physics. 2000;**87**(6):2910-2913

[32] Pan GZ, Tu KN, Prussin A. Size-distribution and annealing behavior of end-of-range dislocation loops in silicon-implanted silicon. Journal of Applied Physics. 1997;**81**(1):78-84

[33] Lindfors CD, Jones KS, Rendon MJ. Boron solubility limits following low temperature solid phase epitaxial regrowth. Materials Research Society Symposium Proceedings. 2001;**669**:J8.5.1-6

[34] Aboy M, Pelaz L, López P, Marqús LA, Duffy R, Venezia VC. Physical insight into boron activation and redistribution during annealing after low-temperature solid phase epitaxial regrowth. Applied Physics Letters. 2006;**88**(19):191917.1-6

[35] Lindfors CD, Jones KS, Law ME, Downey DF, Murto RW. Boron activation during solid phase epitaxial regrowth. Materials Research Society Symposium Proceedings. 2000;**610**:B10.2.1-B10.2.6

[36] Liang JH, Wang SC. Boron depth profiles and residual damage following rapid thermal annealing of low-temperature BSi molecular ion implantation in silicon. Nuclear Instruments and Methods in Physics Research Section B: Beam Interactions with Materials and Atoms. 2007;**261**(1-2 SPEC. ISS):651-655

[37] Chang RD, Lin JC, Lee BW. Initial activation behavior of boron at low temperatures with implantation doses below the amorphization threshold. Japanese Journal of Applied Physics. 2020;**59**(9):58-63

[38] Caturla MJ, Johnson MD, Diaz De La Rubia T. The fraction of substitutional boron in silicon during ion implantation and thermal annealing. Applied Physics Letters. 1998;**72**(21):2736-2738

[39] North JC, Gibson WM. Channeling study of boron-implanted silicon. Applied Physics Letters. 1970;**16**(3):126-129

[40] Seidel TE, Mac Rae AU. Isothermal annealing of boron implanted silicon. Radiation Effects. 1971;**7**(1-2):142-144

[41] Huang J, Fan D, Jaccodine RJ. Reverse annealing and low-temperature diffusion of boron in boron-implanted silicon. Journal of Applied Physics. 1988;**63**(11)

[42] Uppal S, Willoughby AFW, Bonar JM, Evans AGR, Cowern NEB, Morris R, et al. Diffusion of ion-implanted boron in germanium. Journal of Applied Physics. 2001;**90**(8)4293-4295

[43] Eaglesham DJ, Stolk PA, Gossmann HJ, Poate JM. Implantation and transient B diffusion in Si: The source of the interstitials. Applied Physics Letters. 1994;**65**(18):2305-2307

[44] Stolk PA, Gossmann HJ, Eaglesham DJ, Poate JM. Implantation and transient boron diffusion: The role of the silicon self-interstitial. Nuclear Instruments and Methods in Physics Research B. 1995;**96**(1-2):187-195

[45] Bracht H, Stolwijk NA, Laube M, Pensl G. Diffusion of boron in silicon carbide: Evidence for the kick-out mechanism. Applied Physics Letters. 2000;**77**(20):3188-3190

[46] Sedgwick TO, Michel AE, Deline VR, Cohen SA, Lasky JB. Transient boron diffusion in ion-implanted crystalline and amorphous silicon. Journal of Applied Physics. 1988;**63**(5):1452-1463

[47] Taylor M, Hurley K, Lee K, LeMere M, Opsal J, O'Brien T.

Thermal-wave measurements of high-dose ion implantation. Nuclear Instruments and Methods in Physics Research B. 1991;**55**(1-4):3188-3190

[48] Smith AK, Perloff DS, Edwards R, Kleppinger R, Rigik MD. The use of four-point probe sheet resistance measurements for characterizing low dose ion implantation. Nuclear Instruments and Methods in Physics Research B. 1985;**6**(1-2):382-388

[49] Brahma S, Liu CW, Lo KY. The evolution of structure and defects in the implanted Si surface: Inspecting by reflective second harmonic generation. Applied Surface Science. 2016;**388**:517-523

[50] Michel AE, Rausch W, Ronsheim PA, Kastl RH. Rapid annealing and the anomalous diffusion of ion implanted boron into silicon. Applied Physics Letters. 1987;**50**(7):851-853

[51] Collart EJH. Characterization of low-energy (100 eV–10 keV) boron ion implantation. Journal of Vacuum Science & Technology B: Microelectronics and Nanometer Structures. 1998;**16**(1):280-285

[52] White NR. Ion sources for use in ion implantation. Nuclear Instruments and Methods in Physics Research B. 1989;**37-38**(C):78-86

[53] Chang FM, Wu ZZ, Lin YF, Kao LC, Wu CT, JangJian SK, et al. Damage and annealing recovery of boron-implanted ultra-shallow junction: The correlation between beam current and surface configuration. Applied Surface Science. 2018;**433**:160-165

[54] Jung MYL, Gunawan R, Braatz RD, Seebauer EG. A simplified picture for transient enhanced diffusion of boron in silicon. Journal of the Electrochemical Society. 2004;**151**(1):G1-G7

[55] Pawlak BJ, Duffy R, De Keersgieter A. Doping strategies for FinFETs. Materials Science Forum. 2008;**573-574**:333-338

[56] Duffy R, Curatola G, Pawlak BJ, Doornbos G, van der Tak K, Breimer P, et al. Doping fin field-effect transistor sidewalls: Impurity dose retention in silicon due to high angle incident ion implants and the impact on device performance. Journal of Vacuum Science & Technology B: Microelectronics and Nanometer Structures. 2008;**26**(1):402-407

[57] Ye L, de Jong MP, Kudernac T, van der Wiel WG, Huskens J. Doping of semiconductors by molecular monolayers: Monolayer formation, dopant diffusion and applications. Materials Science in Semiconductor Processing. 2017;**62**:166-172

[58] Ho JC, Yerushalmi R, Jacobson ZA, Fan Z, Alley RL, Javey A. Controlled nanoscale doping of semiconductors via molecular monolayers. Nature Materials. 2008;**7**(1):62-67

[59] Ho JC, Yerushalmi R, Smith G, Majhi P, Bennett J, Halim J, et al. Wafer-scale, sub-5 nm junction formation by monolayer doping and conventional spike annealing. Nano Letters. 2009;**9**(2):725-730

[60] Gao X, Kolevatov I, Chen K, Guan B, Mesli A, Monakhov E, et al. Full activation of boron in silicon doped by self-assembled molecular monolayers. ACS Applied Electronic Materials. 2020;**2**(1):268-274

[61] Alphazan T, Mathey L, Schwarzwälder M, Lin TH, Rossini AJ, Wischert R, et al. Monolayer doping of silicon through grafting a tailored molecular phosphorus precursor onto oxide-passivated silicon surfaces. Chemistry of Materials. 2016;**28**(11):3634-3640

[62] Ye L, Pujari SP, Zuilhof H, Kudernac T, De Jong MP, Van Der Wiel WG, et al. Controlling the dopant dose in silicon by mixed-monolayer doping. ACS Applied Materials & Interfaces. 2015;**7**(5):3231-3236

[63] Shimizu Y, Takamizawa H, Inoue K, Yano F, Nagai Y, Lamagna L, et al. Behavior of phosphorous and contaminants from molecular doping combined with a conventional spike annealing method. Nanoscale. 2014;**6**(2):706-710

[64] Packard G, Spaulding C, Taylor A, Hirschman K, Williams S, Kurinec S. Selective phosphorus doping of polycrystalline silicon on glass using self-assembled monolayer doping (MLD) and flash anneal. Materials Letters. 2021;**305**:130780(1-4)

[65] Ye L, González-Campo A, Núñez R, De Jong MP, Kudernac T, Van Der Wiel WG, et al. Boosting the boron dopant level in monolayer doping by Carboranes. ACS Applied Materials & Interfaces. 2015;**7**(49):27357-27361

[66] Ye L, González-Campo A, Kudernac T, Núñez R, De Jong M, Van der Wiel WG, et al. Monolayer contact doping from a silicon oxide source substrate. Langmuir. 2017;**33**(15):3635-3638

[67] Hsu SH, Wan CC, Cho TC, Lee YJ. Investigation of boron distribution at the SiO2/Si Interface of monolayer doping. ACS. Omega. 2021;**6**(1):733-738

[68] Tzaguy A, Karadan P, Killi K, Hazut O, Amit I, Rosenwaks Y, et al. Boron monolayer doping: Role of oxide capping layer, molecular fragmentation, and doping uniformity at the nanoscale. Adv mater. Interfaces. 2020;**7**(5):1902198(1-9)

[69] Fu J, Chen K, Chang S, Zhi K, Gao X, Wei H, et al. Dopant activation and photoresponses of boron-doped silicon by self-assembled molecular monolayers. AIP Advances. 2019;**9**(12):125219(1-6)

[70] Guan B, Siampour H, Fan Z, Wang S, Kong XY, Mesli A, et al. Nanoscale nitrogen doping in silicon by self-assembled monolayers. Scientific Reports. 2015;**5**:12641(1-9)

[71] Yum JH, Shin HS, Hill R, Oh J, Lee HD, Mushinski RM, et al. A study of capping layers for sulfur monolayer doping on III-V junctions. Applied Physics Letters. 2012;**101**(25):253514(1-3)

[72] Cho K, Ruebusch DJ, Lee MH, Moon JH, Ford AC, Kapadia R, et al. Molecular monolayers for conformal, nanoscale doping of InP nanopillar photovoltaics. Applied Physics Letters. 2011;**98**(20):203101(1-3)

[73] Ho JC, Ford AC, Chueh YL, Leu PW, Ergen O, Takei K, et al. Nanoscale doping of InAs via sulfur monolayers. Applied Physics Letters. 2009;**95**(7):072108(1-3)

[74] O'Connell J, Verni GA, Gangnaik A, Shayesteh M, Long B, Georgiev YM, et al. Organo-arsenic molecular layers on silicon for high-density doping. ACS Applied Materials & Interfaces. 2015;**7**(28):15514-15521

[75] Popere BC, Russ B, Heitsch AT, Trefonas P, Segalman RA. Large-area, Nanometer-scale discrete doping of semiconductors via block copolymer self-assembly. Advanced Materials Interfaces. 2015;**2**(18):1500421(1-6)

[76] Hazut O, Agarwala A, Amit I, Subramani T, Zaidiner S, Rosenwaks Y, et al. Contact doping of silicon wafers and nanostructures with phosphine oxide monolayers. ACS Nano. 2012;**6**(11):10311-10318

[77] Hazut O, Agarwala A, Subramani T, Waichman S, Yerushalmi R. Monolayer contact doping of silicon surfaces and nanowires using organophosphorus compounds. Journal of Visualized Experiments. 2013;(82):e50770(1-5)

[78] Mor GK, Jones D, Le TP, Shang Z, Weathers PJ, Woltermann MKB, et al. Contact doping with sub-monolayers of strong polyelectrolytes for organic photovoltaics. Advanced Energy Materials. 2014;**4**(13):1400439(1-6)

[79] Hazut O, Huang BC, Pantzer A, Amit I, Rosenwaks Y, Kohn A, et al. Parallel p-n junctions across nanowires by one-step ex situ doping. ACS Nano. 2014;**8**(8):8357-8362

[80] Hazut O, Yerushalmi R. Direct dopant patterning by a remote monolayer doping enabled by a monolayer fragmentation study. Langmuir. 2017;**33**(22):5371-5377

[81] Park CJ, Jung SM, Kim JH, Kim IT, Shin MW. Effect of surface states on monolayer doping: Crystal orientations, crystallinities, and surface defects. Materials Science in Semiconductor Processing. 2018;**82**:67-74

[82] Ter M-SL. Thin films including layers: Terminology in relation to their preparation and characterization (IUPAC recommendations 1994). Pure and Applied Chemistry. 1994;**66**(8):7-78

[83] Schwartz DK. Mechanisms and kinetics of self-assembled monolayer formation. Annual Review of Physical Chemistry. 2001;**52**:107-137

[84] Buriak JM. Organometallic chemistry on silicon surfaces: Formation of functional monolayers bound through Si-C bonds. Chemical Communications. 1999;**12**(12):1051-1060

[85] Brzoska JB, Ben AI, Rondelez F. Silanization of solid substrates: A step toward reproducibility. Langmuir. 1994;**10**(11):4367-4373

[86] Linford MR, Chidsey CED. Alkyl monolayers covalently bonded to silicon surfaces. Journal of the American Chemical Society. 1993;**115**:12631-12632

[87] Effenberger F, Götz G, Bidlingmaier B, Wezstein M. Photoactivated preparation and patterning of self-assembled monolayers with 1-alkenes and aldehydes on silicon hydride surfaces. Angewandte Chemie International Edition. 1998;**37**(18):2462-2464

[88] Sun QY, De Smet LCPM, Van Lagen B, Giesbers M, Thüne PC, Van Engelenburg J, et al. Covalently attached monolayers on crystalline hydrogen-terminated silicon: Extremely mild attachment by visible light. Journal of the American Chemical Society. 2005;**127**(8):2514-2523

[89] Mischki TK, Donkers RL, Eves BJ, Lopinski GP, Wayner DDM. Reaction of alkenes with hydrogen-terminated and photooxidized silicon surfaces. A comparison of thermal and photochemical processes. Langmuir. 2006;**22**(20):8359-8365

[90] Scheres L, Giesbers M, Zuilhof H. Self-assembly of organic monolayers onto hydrogen-terminated silicon: 1-alkynes are better than 1-alkenes. Langmuir. 2010;**26**(13):10924-10929

[91] Cras JJ, Rowe-Taitt CA, Nivens DA, Ligler FS. Comparison of chemical cleaning methods of glass in preparation for silanization. Biosensors & Bioelectronics. 1999;**14**(8-9):683-688

[92] De Vos WM, Cattoz B, Avery MP, Cosgrove T, Prescott SW. Adsorption

and surfactant-mediated desorption of poly(vinylpyrrolidone) on plasma- and piranha-cleaned silica surfaces. Langmuir. 2014;**30**(28):8425-8431

[93] MicroChemicals. Basics of microstructuring: Substrate preparation. Merck. 2021. Available at: www.microchemicals.com/downloads/application_notes.html

[94] Singh M, Kaur N, Comini E. The role of self-assembled monolayers in electronic devices. Journal of Materials Chemistry C. 2020;**8**(12):3938-3955

[95] Matinlinna JP, Lung CYK, Tsoi JKH. Silane adhesion mechanism in dental applications and surface treatments: A review. Dental Materials. 2018;**34**(1):13-28

[96] Puglisi RA, Garozzo C, Bongiorno C, Di Franco S, Italia M, Mannino G, et al. Molecular doping applied to Si nanowires array based solar cells. Solar Energy Materials & Solar Cells. 2015;**132**:118-122

[97] Veerbeek J, Ye L, Vijselaar W, Kudernac T, Van Der Wiel WG, Huskens J. Highly doped silicon nanowires by monolayer doping. Nanoscale. 2017;**9**(8):2836-2844

[98] Gao X, Guan B, Mesli A, Chen K, Dan Y. Deep level transient spectroscopic investigation of phosphorus-doped silicon by self-assembled molecular monolayers. Nature Communications. 2018;**9**(1):118(1-10)

[99] Barnett J, Hill R, Loh WY, Hobbs C, Majhi P, Jammy R. Advanced techniques for achieving ultra-shallow junctions in future CMOS devices. In: IWJT-2010: Extended Abstracts—2010 International Workshop on Junction Technology. New York: IEEE; 2010

[100] Radamson HH, Zhang Y, He X, Cui H, Li J, Xiang J, et al. The challenges of advanced CMOS process from 2D to 3D. Applied Sciences. 2017;**7**(10):1047(1-32)

[101] Ang KW, Barnett J, Loh WY, Huang J, Min BG, Hung PY, et al. 300 mm FinFET results utilizing conformal, damage free, ultra shallow junctions (X j~5nm) formed with molecular monolayer doping technique. In: Technical Digest—International Electron Devices Meeting. IEDM. New York: IEEE; 2011

[102] Lee YJ, Cho TC, Kao KH, Sung PJ, Hsueh FK, Huang PC, et al. A novel junctionless FinFET structure with sub-5nm shell doping profile by molecular monolayer doping and microwave annealing. In: Technical Digest—International Electron Devices Meeting. IEDM; 2015

[103] Lee YJ, Cho TC, Sung PJ, Kao KH, Hsueh FK, Hou FJ, et al. High performance poly Si junctionless transistors with sub-5nm conformally doped layers by molecular monolayer doping and microwave incorporating CO_2 laser annealing for 3D stacked ICs applications. In: Technical Digest—International Electron Devices Meeting. IEDM; New York: IEEE; 2015

[104] Chun-Lin Chu, Guang-Li Luo, Dean Chou, Shu-Han Hsu. Demonstration of monolayer doping of the five-stacked Ge nanosheets field-effect transistor. ACS Applied Electronic Materials. 2022;**4**(7):3592-3597

[105] Elbersen R, Vijselaar W, Tiggelaar RM, Gardeniers H, Huskens J. Fabrication and doping methods for silicon nano- and micropillar arrays for solar-cell applications: A review. Advanced Materials. 2015;**27**:6781-6796

[106] Garozzo C, Bongiorno C, Di FS, Italia M, La MA, Scalese S, et al. Nanofabrication processes for innovative nanohole-based solar cells. Physica Status Solidi (A) Applications and Materials Science. 2013;**210**(8):1564-1570

[107] Liu H, You CY, Li J, Galligan PR, You J, Liu Z, et al. Synthesis of hexagonal boron nitrides by chemical vapor deposition and their use as single photon emitters. Nano Materials Science. 2021;**3**(3):291-312

[108] Mohammadi V, Nihtianov S, Fang C. A doping-less junction-formation mechanism between n-silicon and an atomically thin boron layer. Scientific Reports. 2017;**7**(1):1-10

[109] Pearton S. Ediorial: Doped nanostructures. Nanoscale. 2010;**2**(7):1057

[110] Imam M, Gaul K, Stegmüller A, Höglund C, Jensen J, Hultman L, et al. Gas phase chemical vapor deposition chemistry of triethylboron probed by boron-carbon thin film deposition and quantum chemical calculations. Journal of Materials Chemistry C. 2015;**3**(41):10898-10906

[111] Sarubbi F, Scholtes TLM, Nanver LK. Chemical vapor deposition of α-boron layers on silicon for controlled nanometer-deep p + n junction formation. Journal of Electronic Materials. 2010;**39**(2):162-173

[112] Mok KRC, Vlooswijk AHG, Mohammadi V, Nanver LK. Effects of annealing on chemical-vapor deposited PureB layers. ECS Journal of Solid State Science and Technology. 2013;**2**(9):P413-P417

[113] Meier S, Lohmüller S, Mack S, Wolf A, Glunz SW. Control of boron diffusion from APCVD BSG layers by interface oxidation. AIP Conference Proceedings. 2018;**1999**(August 2018):070003(1-8)

[114] Muroi M, Yamada A, Saito A, Habuka H. Deposition and etching behaviour of boron trichloride gas at silicon surface. Journal of Crystal Growth. 2020;**529**(July 2019):125301

[115] Muroi M, Otani M, Habuka H. Boron-silicon film chemical vapor deposition using boron trichloride, dichlorosilane and monomethylsilane gases. ECS Journal of Solid State Science and Technology. 2021;**10**(6):064006

[116] Taniguchi R, Inasawa S. Formation of boron-doped silicon wires and control of dopant concentration using zinc, SiCl4 and BCl3. Journal of Crystal Growth. 2020;**547**(May):125796

[117] Mirabella S, De Salvador D, Napolitani E, Bruno E, Priolo F. Mechanisms of boron diffusion in silicon and germanium. Journal of Applied Physics. 2013;**113**(3):031101(1-21)

[118] Chroneos A, Bracht H. Diffusion of n-type dopants in germanium. Applied Physics Review. 2014;**1**(1):011301(1-20)

[119] Tu WH, Hsu SH, Liu CW. The pn junctions of epitaxial germanium on silicon by solid phase doping. IEEE Transactions on Electron Devices. 2014;**61**(7):2595-2598

[120] Liu X, Nanver LK, Tom LM. Nanometer-thin pure boron layers as mask for silicon micromachining. Journal of Microelectromechanical Systems. 2017;**26**(6):1428-1434

[121] Nadeem A, Maqsood MF, Raza MA, Ilyas MT, Iqbal MJ, Rehman ZU. Binder free boron nitride-based coatings deposited on mild steel by chemical vapour deposition: Anti-corrosion

performance analysis. Physica B: Condensed Matter. 2021;**602**(March 2020):412600

[122] Zou Y, Larsson K. Effect of boron doping on the CVD growth rate of diamond. Journal of Physical Chemistry C. 2016;**120**(19):10658-10666

[123] Polushin NI, Laptev AI, Spitsyn BV, Alexenko AE, Polyansky AM, Maslov AL, et al. Deposition of boron-doped thin CVD diamond films from methane-triethyl borate-hydrogen gas mixture. PRO. 2020;**8**(6):666(1-13)

[124] Sezer AO, Brand JI. Chemical vapor deposition of boron carbide. Materials Science & Engineering B: Solid-State Materials for Advanced Technology. 2001;**79**(3):191-202

[125] Wang H, Yang H, Peng Y, Zheng Y, Huang D, Hu P, et al. Highly efficient synthesis of boron nitride nanotubes by catalytic chemical vapor deposition of boron/nickel containing precursors. Journal of Materials. 2022:1-6

[126] Kendrick C, Kuo MW, Li J, Shen H, Mayer TS, Redwing JM. Uniform p-type doping of silicon nanowires synthesized via vapor-liquid-solid growth with silicon tetrachloride. Journal of Applied Physics. 2017;**122**(23):235101(1-7)

[127] Creedon DL, Jiang Y, Ganesan K, Stacey A, Kageura T, Kawarada H, et al. Irradiation-induced modification of the superconducting properties of heavily-boron-doped diamond. Physical Review Applied. 2018;**10**(4):1

[128] Terheiden B. CVD boron containing glasses—An attractive alternative diffusion source for high quality emitters and simplified processing—A review. Energy Procedia. 2016;**92**:486-492

[129] Bogumilowicz Y, Hartmann JM. Reduced-pressure chemical vapor deposition of boron-doped Si and Ge layers. Thin Solid Films. 2014;**557**:4-9

[130] Niu JJ, Wang JN, Chen YX. Boron-doped silicon nano-wires. Materials Science & Engineering B: Solid-State Materials for Advanced Technology. 2007;**139**(1):95-98

[131] El Mubarek HAW. Reduction of phosphorus diffusion in germanium by fluorine implantation. Journal of Applied Physics. 2013;**114**(22):535-537

[132] Boureau V, Hartmann JM, Claverie A. Lattice contraction due to boron doping in silicon. Materials Science in Semiconductor Processing. 2018;**87**(July):65-68

[133] Sze SM, Ng KK. Physics of semiconductor devices. Hoboken, New Jersey: John Wiley and Sons; 2006

[134] Masetti G, Severi M, Solmi S. Modeling of carrier mobility against carrier concentration in arsenic-, phosphorus-, and boron-doped silicon. IEEE Transactions on Electron Devices. 1983;**30**(7):764-769

[135] Yu G, Watanabe J, Izumi K, Nakashima K, Jimbo T, Umeno M. Mechanical property characterization of boron-doped silicon by Berkovich-type indenter. Japanese Journal of Applied Physics, Part 2: Letters. 2001;**40**:L183

[136] Lee Y, Hwang GS. Mechanism of thermal conductivity suppression in doped silicon studied with nonequilibrium molecular dynamics. Physical Review B: Condensed Matter and Materials Physics. 2012;**86**(7):075202(1-6)

[137] Martyniuk P, Kopytko M, Rogalski A. Barrier infrared detectors. Opto-Electronics Review. 2014;**22**:2

[138] Chubinidze G, Kurashvili I, Bokuchava G, Chachkhiani Z, Darsavelidze G. Dynamical mechanical properties of boron-doped monocrystalline germanium. Bulletin of the Georgian National Academy of Sciences. 2012;**6**(2):89-93

[139] Kim DK, Hong SB, Jeong K, Lee C, Kim H, Cho MH. P-N junction diode using plasma boron-doped black phosphorus for high-performance photovoltaic devices. ACS Nano. 2019;**13**(2):1683-1693

[140] Zhao C, Xu B, Wang Z, Wang Z. Boron-doped III-V semiconductors for Si-based optoelectronic devices. Journal of Semiconductors. 2020;**41**:011301

[141] Auden EC, Quinn HM, Wender SA, O'Donnell JM, Lisowski PW, George JS, et al. Thermal neutron-induced single-event upsets in microcontrollers containing boron-10. IEEE Transactions on Nuclear Science. 2020;**67**(1):29-37

[142] Xu M, Feng Y, Han X, Ke X, Li G, Zeng Y, et al. Design and fabrication of an absolute pressure MEMS capacitance vacuum sensor based on silicon bonding technology. Vacuum. 2021;**186**:110065

[143] Cui Y, Wei Q, Park H, Lieber CM. Nanowire nanosensors for highly sensitive and selective detection of biological and chemical species. Science (80-). 2001;**293**(5533):1289-1892

[144] Sobolev NA, Shtel'makh KF, Kalyadin AE, Aruev PN, Zabrodskiy VV, Shek EI, et al. Electroluminescence properties of LEDs based on electron-irradiated p-Si. Semiconductors. 2016;**50**(2):252-256

[145] Fathi E, Vygranenko Y, Vieira M, Sazonov A. Boron-doped nanocrystalline silicon thin films for solar cells. Applied Surface Science. 2011;**257**(21):8901-8905

[146] Vicari Stefani B, Kim M, Wright M, Soeriyadi A, Andronikov D, Nyapshaev I, et al. Stability study of silicon heterojunction solar cells fabricated with gallium- and boron-doped silicon wafers. Solar RRL. 2021;**5**(9):2100406

[147] Hou C, Jia R, Tao K, Jiang S, Zhang P, Sun H, et al. Boron-rich layer removal and surface passivation of boron-doped p-n silicon solar cells. Journal of Semiconductors. 2018;**39**(12):122004

[148] Saini V, Li Z, Bourdo S, Kunets VP, Trigwell S, Couraud A, et al. Photovoltaic devices based on high density boron-doped single-walled carbon nanotube/ n-Si heterojunctions. Journal of Applied Physics. 2011;**109**(1):014321

[149] Suliman SA, Venkataraman B, Wu CT, Ridley RS, Dolny GM, Awadelkarim OO, et al. Electrical properties of the gate oxide and its interface with Si in U-shaped trench MOS capacitors: The impact of polycrystalline Si doping and oxide composition. Solid State Electronics. 2003;**47**(5):899-905

[150] Peng S, Cho K. Ab initio study of doped carbon nanotube sensors. Nano Letters. 2003;**3**(4):513-517

[151] Yi JY, Bernholc J. Atomic structure and doping of microtubules. Physical Review B. 1993;**47**(3):1708-1711

[152] Fakhrabadi MMS, Allahverdizadeh A, Norouzifard V, Dadashzadeh B. Effects of boron doping on mechanical properties and thermal conductivities of carbon nanotubes. Solid State Communications. 2012;**152**(21):1973-1979

[153] Rezania H. The effect of boron doping on the thermal conductivity of zigzag carbon nanotubes. International Journal of Modern Physics B. 2015;**29**(5):1550025

[154] Liu P, Liang J, Xue R, Du Q, Jiang M. Ruthenium decorated boron-doped carbon nanotube for hydrogen storage: A first-principle study. International Journal of Hydrogen Energy. 2019;**44**(51):27853-27861

[155] Liu B, Sun X, Liao Z, Lu X, Zhang L, Hao GP. Nitrogen and boron doped carbon layer coated multiwall carbon nanotubes as high performance anode materials for lithium ion batteries. Scientific Reports. 2021;**11**(1):5633

[156] Sawant SV, Yadav MD, Banerjee S, Patwardhan AW, Joshi JB, Dasgupta K. Hydrogen storage in boron-doped carbon nanotubes: Effect of dopant concentration. International Journal of Hydrogen Energy. 2021;**46**(79):39297-39314

[157] Sawant SV, Patwardhan AW, Joshi JB, Dasgupta K. Boron doped carbon nanotubes: Synthesis, characterization and emerging applications—A review. Chemical Engineering Journal. 2022;**427**:131616

[158] Pei Y, Song H, Liu Y, Cheng Y, Li W, Chen Y, et al. Boron–nitrogen-doped carbon dots on multi-walled carbon nanotubes for efficient electrocatalysis of oxygen reduction reactions. Journal of Colloid and Interface Science. 2021;**600**:865-871

[159] Talla JA. First principles modeling of boron-doped carbon nanotube sensors. Physica B: Condensed Matter. 2012;**407**(6):966-970

[160] Adjizian JJ, Leghrib R, Koos AA, Suarez-Martinez I, Crossley A, Wagner P, et al. Boron- and nitrogen-doped multi-wall carbon nanotubes for gas detection. Carbon N Y. 2014;**66**:662-673

[161] Liang Q, Wei Y. Molecular dynamics study on the thermal conductivity and thermal rectification in graphene with geometric variations of doped boron. Physica B: Condensed Matter. 2014;**437**:36-40

[162] Wang H, Zhou Y, Wu D, Liao L, Zhao S, Peng H, et al. Synthesis of boron-doped graphene monolayers using the sole solid feedstock by chemical vapor deposition. Small. 2013;**9**(8):1316-1320

[163] Wu T, Shen H, Sun L, Cheng B, Liu B, Shen J. Nitrogen and boron doped monolayer graphene by chemical vapor deposition using polystyrene, urea and boric acid. New Journal of Chemistry. 2012;**36**(6):1385-1391

[164] Fazio G, Ferrighi L, Di Valentin C. Boron-doped graphene as active electrocatalyst for oxygen reduction reaction at a fuel-cell cathode. Journal of Catalysis. 2014;**318**:203-210

[165] Naresh V, Bhattacharjee U, Martha SK. Boron doped graphene nanosheets as negative electrode additive for high-performance lead-acid batteries and ultracapacitors. Journal of Alloys and Compounds. 2019;**797**:595-605

[166] Srivastava S, Jain SK, Gupta G, Senguttuvan TD, Gupta BK. Boron-doped few-layer graphene nanosheet gas sensor for enhanced ammonia sensing at room temperature. RSC Advances. 2019;**10**(2):1007-1014

[167] Agnoli S, Favaro M. Doping graphene with boron: A review of synthesis methods, physicochemical characterization, and emerging applications. Journal of Materials Chemistry A. 2016;**4**:5002-5025

[168] Li X, Fan L, Li Z, Wang K, Zhong M, Wei J, et al. Boron doping of graphene for graphene-silicon p-n junction solar cells. Advanced Energy Materials. 2012;**2**(4):425-429

[169] McCreery RL. Advanced carbon electrode materials for molecular electrochemistry. Chemical Reviews. 2008;**108**(7):2646-2687

[170] Pernot J, Volpe PN, Omnès F, Muret P, Mortet V, Haenen K, et al. Hall hole mobility in boron-doped homoepitaxial diamond. Physical Review B: Condensed Matter and Materials Physics. 2010;**81**(20):205203(1-7)

[171] Muzyka K, Sun J, Fereja TH, Lan Y, Zhang W, Xu G. Boron-doped diamond: Current progress and challenges in view of electroanalytical applications. Analytical Methods. 2019;**11**:397-414

[172] Williams G, Calvo JA, Faili F, Dodson J, Obeloer T, Twitchen DJ. Thermal conductivity of electrically conductive highly boron doped diamond and its applications at high frequencies. In: Proceedings of the 17th InterSociety Conference on Thermal and Thermomechanical Phenomena in Electronic Systems, ITherm 2018. New York: IEEE; 2018

[173] Kondo T. Recent electroanalytical applications of boron-doped diamond electrodes. Current Opinion in Electrochemistry. 2022;**32**:100891

[174] Wang J, He Z, Tan X, Wang T, He X, Zhang L, et al. Hybrid supercapacitors from porous boron-doped diamond with water-soluble redox electrolyte. Surface and Coatings Technology. 2020;**398**:126103

[175] Kondo T, Kato T, Miyashita K, Aikawa T, Tojo T, Yuasa M. Boron-doped diamond powders for aqueous supercapacitors with high energy and high power density. Journal of the Electrochemical Society. 2019;**166**(8):A1425-A1431

[176] Cobb SJ, Ayres ZJ, Macpherson JV. Boron doped diamond: A designer electrode material for the twenty-first century. Annual Review of Analytical Chemistry. 2018;**11**(1):463-484

[177] Venkatachalam S, Kanno Y, Mangalaraj D, Narayandass SK. Effect of boron ion implantation on the structural, optical and electrical properties of ZnSe thin films. Physica B: Condensed Matter. 2007;**390**(1-2):71-78

[178] Lee DH, Park B, Saxena A, Serene TP. Enhanced surface hardness by boron implantation in nitinol alloy. Journal of Endodontics. 1996;**22**(10):543-546

[179] Zhu YC, Fujita K, Iwamoto N, Nagasaka H, Kataoka T. Influence of boron ion implantation on the wear resistance of TiAlN coatings. Surface and Coatings Technology. 2002;158-159:664-668

[180] Zhu YC, Ohtani S, Sato Y, Iwamoto N. Influence of boron ion implantation on the oxidation behavior of CVD-SiC coated carbon-carbon composites. Carbon N Y. 2000;**38**(4):501-507

Chapter 9

Boron Compounds for Neutron Capture Therapy in the Treatment of Brain Tumors

Shinji Kawabata, Naonori Hu, Ryo Hiramatsu, Kohei Tsujino, Naokado Ikeda, Naosuke Nonoguchi, Motomasa Furuse and Masahiko Wanibuchi

Abstract

Boron neutron capture therapy (BNCT), which uses the capture reaction between neutrons and boron-10, an isotope of boron, is rapidly gaining interest. The reason for this is the successful development of a compact accelerator-type neutron generator that can be installed in a hospital and launched into the clinical setting. BNCT, which provides selective radiotherapeutic effects at the cellular level, is expected to be effective against invasive cancer. We have been investigating BNCT applications in various types of malignant brain tumors, especially malignant gliomas, as medical applications. Recently, we have conducted clinical trials using the developed accelerator neutron source. Research on pharmaceutical applications of compounds that transport boron to cancer cells is expected to be in even greater need. Currently, the only boron agent used in cancer therapy is BPA (Borofaran 10B), which takes advantage of the demand for essential amino acids, but the research and development of boron agents are an absolutely key technology to further improve the precision of this treatment modality. This chapter summarizes and discusses the results of BNCT in the treatment of brain tumors.

Keywords: boron neutron capture therapy, Borofaran (10B), brain tumor, glioma, accelerator-type neutron generator, nuclear reactor

1. Introduction

Boron neutron capture therapy (BNCT) for head and neck cancer was approved in Japan in 2020, using the world's first accelerator-type neutron generator, (BNCT treatment system NeuCure® Sumitomo Heavy Industries, Ltd.) along with the boron drug for BNCT (Borofaran (10B), and Steboronin® Stella Pharma Co., Ltd.) [1]. It has attracted a lot of interest due to its potential for advancement and widespread use in general medical practice.

Recent developments in quantum-based medicine are remarkable worldwide, and representative particle therapy devices are expected to expand their application range

 IntechOpen

due to their excellent beam quality and biological effects, as well as technological improvements in disease adaptability such as beam shaping technology, rotation of gantry, and diagnostic image-guided irradiation and focal tracking irradiation, which precede X-ray therapy devices. Neutron capture therapy is a representative method for applying quantum to medical treatment using neutron capture reactions with atoms, in addition to therapies that control and apply direct cellular damage of these quanta to living organisms. Neutron capture reactions occur between various atoms, but the stable isotope of boron, boron-10 (^{10}B), which is the most suitable condition for medical applications, is used and is called boron neutron capture therapy. Since naturally existing boron consists of two stable isotopes (^{11}B and ^{10}B), where ^{11}B accounts for 80%, special technology and equipment are required to produce concentrated ^{10}B used for BNCT.

BNCT is a particle therapy that biologically targets tumor cells [2]. By selectively introducing boron drugs containing ^{10}B atoms into tumors and irradiating them with thermal neutrons, charged particles are generated by neutron capture ($^{10}B + n \rightarrow \alpha + {}^{7}Li$ or $^{10}B\,(n, \alpha)\,{}^{7}Li$). The resulting alpha particles and recoil lithium (Li) nuclei are high LET (linear energy transfer) particles that emit all their energy over a short range corresponding to the size of a cell. If boron compounds are selectively introduced, the reaction occurs only in cancer cells and is an ideal "cell-selective treatment" in which surrounding normal cells are preserved. The characteristic of this treatment is that the boron compounds to be administered and the neutrons to be irradiated are non- to low-toxic, respectively, and the treatment is completed by a two-step approach, "neutron capture reaction," in which the effects of both compounds are shown for the first time in vivo (**Figure 1**).

Unlike other advances in radiotherapy that spatially add changes to the distribution of doses, it is necessary to note that there are different distributions of biological effects in the same irradiation field in BNCT. In the case of BNCT for glioma, we mainly examine the distribution dose of the tumor and the distribution dose of the normal brain for medical care. In the case of neutron irradiation, it is necessary to add and calculate other doses mixed in the neutron field to be irradiated, as well as the biological effects of the radiation quality and tissue reaction, respectively. There are

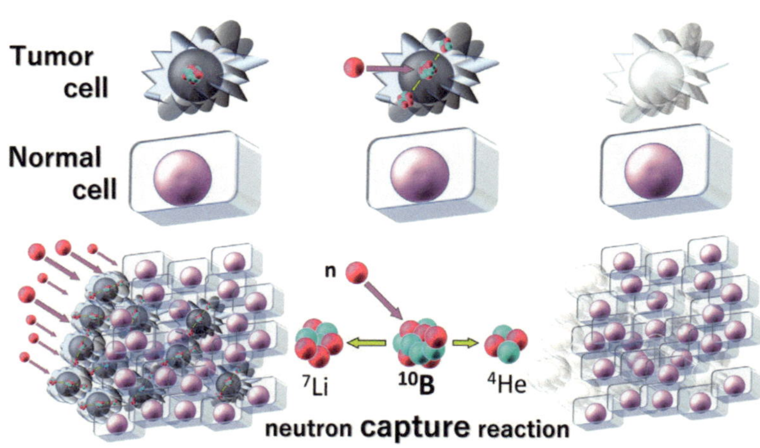

Figure 1.
In boron neutron capture therapy (BNCT), ^{10}B compounds are administered followed by low-energy neutron irradiation, which causes a nuclear reaction between the ^{10}B and the neutrons. The resulting helium nuclei (alpha particles) and lithium recoil nuclei selectively destroy tumor cells from within even in the infiltrated area.

some peculiarities in these calculations, but the results are easy to understand because they are visualized as X-ray equivalent doses (**Figure 2**).

The practice of providing medical care by considering the biological effects of radiotherapy is the same in the current general-purpose radiotherapy devices. The wide range of invasion areas of glioma is sometimes targeted as a tumor or as a risk organ, and different biological effects have been induced by the difference of the number of fractions and the dose at one time in the assumed tissues, which are each subject. For details, refer to the guidelines for radiotherapy and the Guidance on Evaluation of Accelerator Neutron Irradiation Device System for Boron Neutron Capture Therapy (BNCT Review Working Group, National Institute of Health

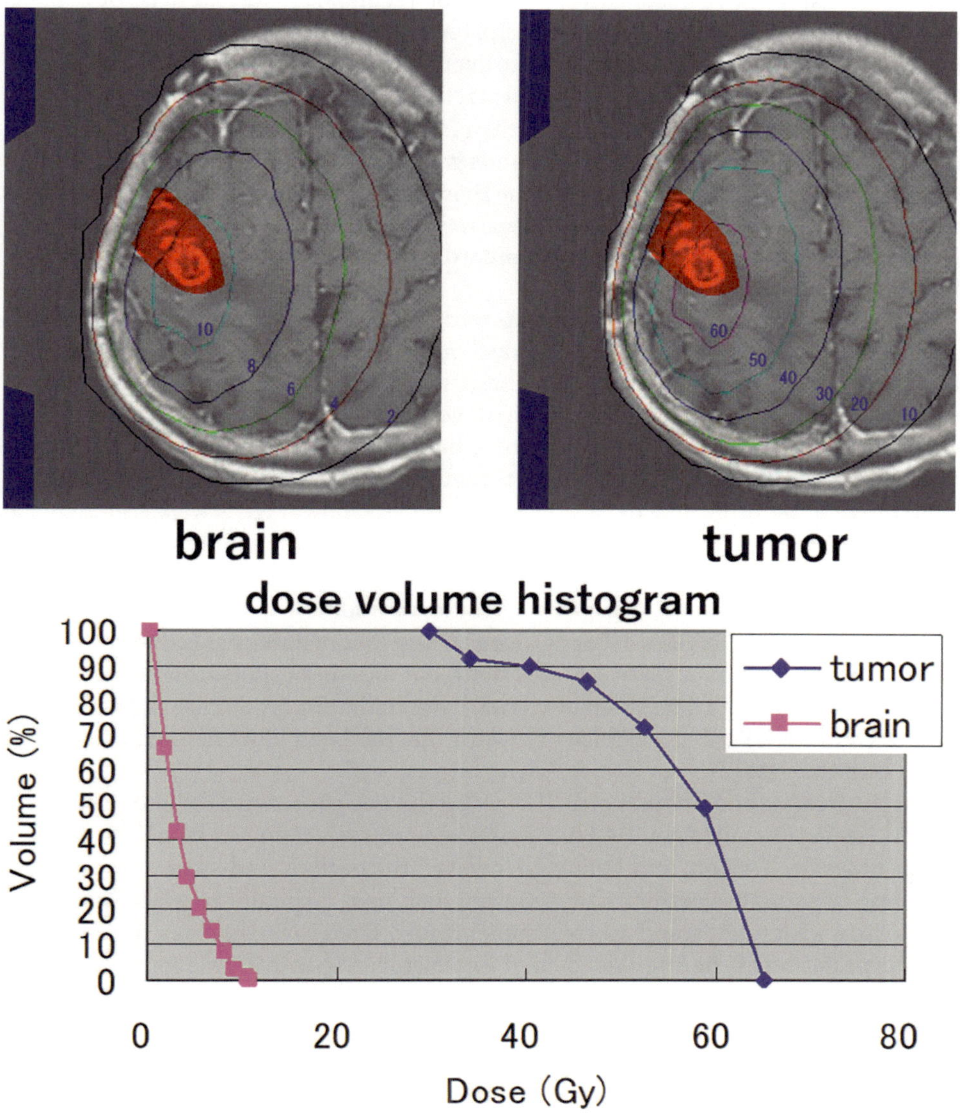

Figure 2.
In BNCT, organ-specific dose distributions are calculated simultaneously (upper: Dose distribution, right: Lower: Dose volume histogram (DVH) for normal and tumor tissue). The SERA calculation engine used in many reactor-based BNCT facilities combines a proprietary Monte Carlo calculation code.

Sciences, Japan) [3]. This knowledge is necessary not only for BNCT, but also for conventional X-ray irradiation (2Gy 30 fractions) (for example, in combination with intensity-modulated irradiation, stereotactic irradiation, reirradiation at the time of recurrence, etc.), and is a sense to be acquired.

2. Background BNCT for brain tumors

In the 1950s, the first clinical trial was initiated at Brookhaven National Laboratory (BNL) in New York, and several low-molecular-weight boron compounds were tested as boron delivery agents [4]. Thereafter, it was progressed on to a full-scale clinical application using the Massachusetts Institute of Technology Nuclear Reactor (MITR), but the results of this study were unanticipated and clinical trials in the United States halted. In 1967, Hatanaka et al., who were deeply involved in research in the United States, launched a clinical trial using thermal neutrons and the boron drug sodium borocaptate (BSH) in Japan, with more than 200 cases treated. Although various tumor tissue types and patient backgrounds were mixed, leading to inconclusive conclusions about the effectiveness of the therapeutic effect on specific diseases and conditions, the results of the standard treatment of refractory malignant glioma at that time were as long as those of the standard treatment and have shown expectations for cure [5].

In the United States, BNCT of patients with brain tumors was resumed in the mid-1990s. Boronophenylalanine (BPA), a novel boron drug, was used in clinical trials for the first time, and BNCT in non-craniotomy was achieved using epithermal neutrons with excellent tissue depth. BPA has been developed to target malignant melanoma with the essential amino acid phenylalanine in the skeleton, but it is a boron drug that exploits the amino acid requirements that are elevated in cancer cells and has been shown to be applicable to various cancer types. Clinical trials were conducted at Harvard University in collaboration with MITR to first treat patients with malignant melanoma of the skin, from which indications were expanded to patients with brain tumors (especially glioblastoma and metastatic melanoma). Twenty-two patients have been treated using BPA, five cutaneous malignant melanomas followed by brain tumor patients. Treatment was well tolerated, but did not outperform the results of conventional X-ray fractionated external beam radiation. A detailed review has conducted of BNCT using the nuclear reactor that has been implemented in Japan and overseas in the past [2].

Recent treatment outcomes of BNCT using the nuclear reactor for glioblastoma (World Health Organization (WHO) Grade 4) have been reported by BNCT research groups in Japan, Sweden, and Finland. However, the background of the targeted cases according to patient selection criteria varied, making it difficult to make a simple comparison with other standard treatment groups. Therefore, there is a limitation of the interpretation of the analysis divided into historical control and recursive partitioning analysis (RPA) subgroup. In our report, we performed a clinical trial using the Kyoto University Nuclear Reactor with a protocol combining BPA (500 mg/kg) and BSH (sodium borocaptate) (100 mg/kg) as "multi-targeted type BNCT," which uses multiple types of target boron drugs at once and irradiates a single neutron, with a median survival time (MST) of 15.6 months (n = 10) with BNCT alone and experiencing long-term survivors (>5 years) [6]. BSH can introduce a large amount of boron atoms, including 12 ^{10}B atoms, but its cell selectivity is low. However, in brain tumors, it remains tissue-selective by exploiting the breakdown of the blood-brain barrier. The MST of BNCT combined with X-ray fractionated external irradiation (20–30Gy) was 23.5 months. No significant toxicity other than hair loss was observed in this protocol, indicating that BNCT with cell selectivity remains highly tolerated in combination with existing radiotherapy (**Figure 3**).

The University of Tsukuba also reported that the median survival time was 27.1 months, the 1-year and 2-year survival rates after BNCT were 87.5% and 62.5%, respectively [7]. In Sweden, the dosage of BPA was increased to 900 mg, and the clinical trial was carried out [8]. BPA was administered over a duration of 6 h, and neutron irradiation was performed from two directions. The mean total brain dose was 3.2–6.1 Gy (X-ray equivalent), and the minimum dose to the tumor ranged from 15.4 to 54.3 Gy (X-ray equivalent). Progression-free survival and median overall survival were reported as 5.8 and 14.2 months, respectively. Adverse events associated with this protocol were only 14%, which was lower than the standard treatment with X-ray fractionated external beam radiation alone or with temozolomide. Our findings also suggested that the combination of BNCT with X-ray fractionated external beam radiation or temozolomide was tolerable and prolonged survival in primary gliomas [9].

Next, the treatment outcome of recurrent malignant glioma (of which the attention is especially high in BNCT of the brain tumor region) was introduced. The prognosis of recurrent malignant glioma is very poor, and especially in the case of radiotherapy, the treatment is difficult. Though surgery followed with radiotherapy has also been carried out, the survival time is approximately 6 months. A prospective trial of BNCT for recurrent malignant glioma has reported 22 treated cases from our institution and 19 cases of Phase I with increased dosage of BPA at the University of Helsinki. The median survival was 10.8 months and 7 months, respectively. Outcomes limited to recurrent glioblastoma included with survival of 9.6 months (n = 19) and 8.7 months (n = 12) after BNCT [10, 11]. To further improve the outcomes of these BNCT-alone treatments, we performed improved efficacy and safety validation when the angiogenesis inhibitor bevacizumab was also used after treatment and has performed very well at the pilot study stage [12].

Clinical trials for brain tumors using the accelerator as a neutron source have been performed for recurrent malignant gliomas, especially for refractory recurrent glioblastoma [13], and are ready for approval in the brain tumor area in addition to the preceding head and neck cancer.

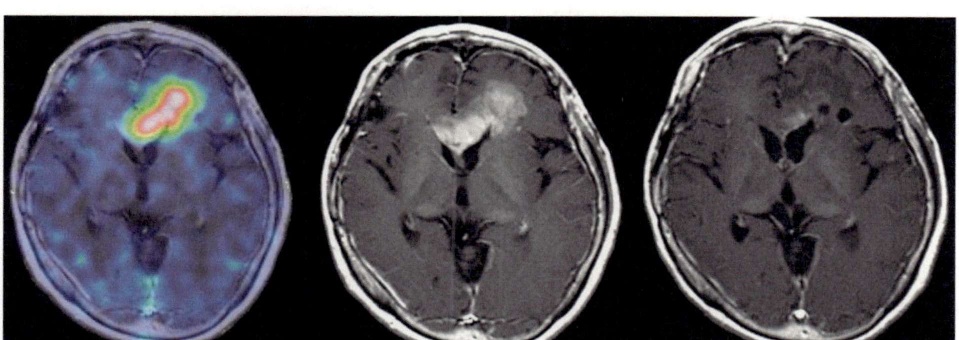

Figure 3.
A case of reactor-based BNCT for malignant glioma with significant response. In this case, favorable boron drug distribution was observed on the pretreatment PET images. (left: FBPA-PET fusion, middle: Before treatment, right: After BNCT of contrast-enhanced MRI T1WI).

3. The practice of neutron capture therapy

3.1 Boron drugs

In many national and international BNCT clinical studies using BPA, neutron beam irradiation has been performed in many cases during the clearance phase after the intravenous infusion of BPA. This, however, required simulation of the in vivo dynamics of boron concentration during irradiation, resulting in an increased importance of subsequent evaluation. In promoting medical applications based on the idea of prescription dose, the accuracy of pre-prediction was pursued, and it was considered that the continuous intravenous infusion during the irradiation used in the clinical research of head and neck cancer using nuclear reactors could solve this problem, and we have taken this approach in our brain tumor treatment. To maintain the concentration of ^{10}B in the tumor tissue that is expected to have a therapeutic effect during thermal neutron beam irradiation, it has been adopted as a dosage and administration method of BPA to maintain the concentration of ^{10}B in whole blood at 20 ppm or higher.

In 15 patients with recurrent head and neck tumors, BPA (fructose solution) 500 mg/kg (400 mg/kg was administered at a constant rate of 200 mg/kg/hr. for 2 h, followed by reducing the infusion rate of the remaining 100 mg/kg to approximately 100 mg/kg/hr. at a constant rate until the end of the irradiation) was administered at the reactor of the Kyoto University Institute for Integrated Radiation and Nuclear Science [14]. In BNCT of malignant glioma carried out by Osaka Medical University adopting the same protocol, an average result of 27 ppm in the whole blood boron concentration was obtained. On the other hand, the whole blood boron level immediately after irradiation in patients treated with BPA (fructose solution) 250 mg/kg could not be maintained at 20 ppm, and even in patients treated with 500 mg/kg, the mean value of whole blood boron level immediately after irradiation decreased to 19.5 ppm, and the fluctuation before and after irradiation was greater compared with 30.4 ppm before irradiation [14]. These experiences suggest that a dose of BPA 500 mg/kg (200 mg/kg/hr. × 2 hr. + 100 mg/kg of BPA at a constant rate of approximately 100 mg/kg/hr. to match the end of irradiation time) is expected to exceed 20 ppm of whole blood boron concentration during irradiation, and that the method of administration that satisfies the conditions for maintaining stable concentrations before and after irradiation.

In a phase II clinical trial (BNCT) in Sweden, 30 patients with glioblastoma were treated with a 6-hour infusion of 900 mg/kg of BPA followed by irradiation 2 hours later. Although transient serious adverse events have been observed, irreversible events have not been observed, and the tolerability of BNCT at 900 mg/kg BPA and 2 hours after 6 hours of intravenous infusion was confirmed [14].

3.2 Dose prescriptions

Factors on the radiation side that generally govern the effects on normal tissue include radiation quality, distribution of dose in the tissue and size of the irradiated volume, and on the tumor side include the presence of a history of radiotherapy and the effect of the tumor on the surrounding normal tissue. It is required that the dose in the irradiated volume is as uniform as possible in order to accurately evaluate the relationship between the reaction of the tissue and the dose. In conventional X-ray or particle therapy, it is possible to irradiate evenly a certain volume of radiation,

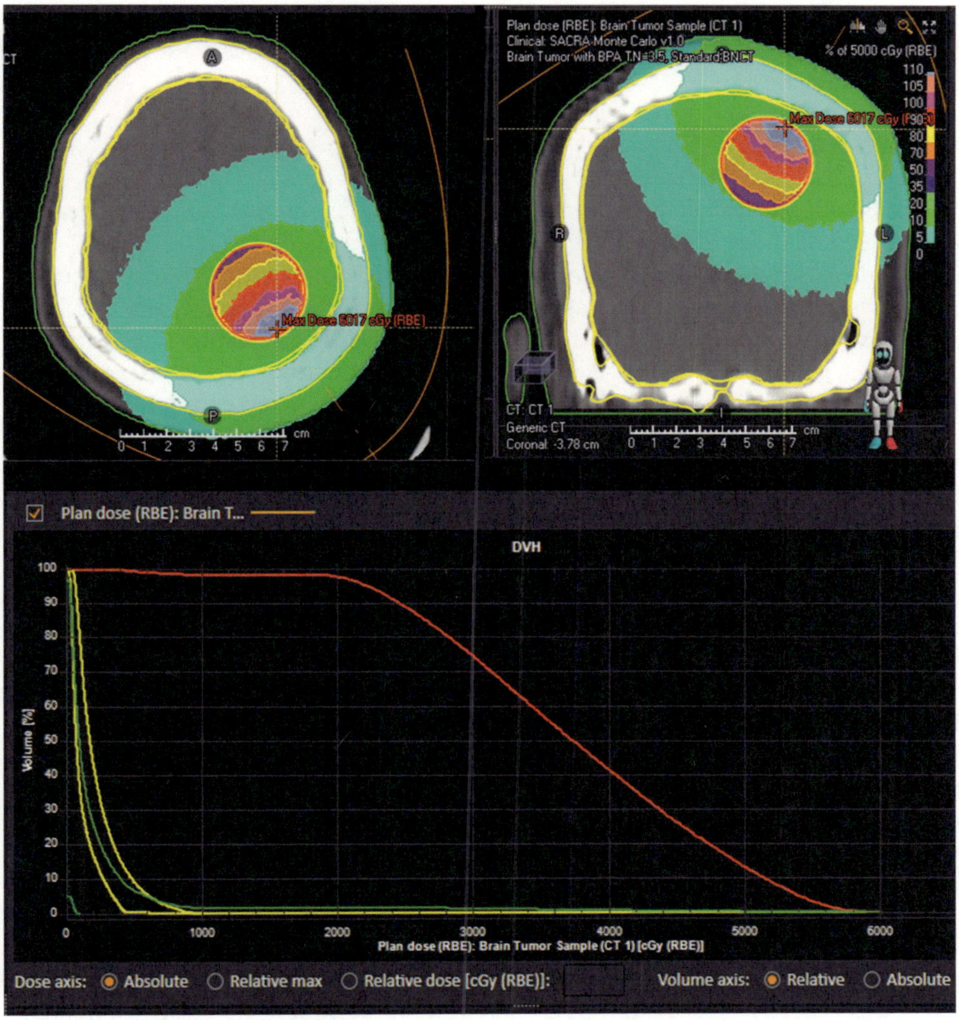

Figure 4.
Example of dose calculation (dose planning for a simulated brain tumor) using the BNCT dose calculation program (NeuCure® dose engine Sumitomo heavy industries, ltd.) the graphical user Interface (GUI) enables detailed visualization of dose distribution, dose volume histogram (DVH) plotting, and reference and modification of various parameters in one GUI.

including the normal brain around the brain tumor. However, for neutron radiation, the attenuation of neutron intensity in the tissue is large and delivering a uniform irradiation becomes difficult.

The skin is a thin layer of tissue that is not affected much by neutron attenuation, so it is possible to define the dose at the skin surface. It's also regarded to be more versatile when it comes to expanding its indication to other organs throughout the body.

Skin dose and dose distribution in the normal brain (maximum dose) can also be associated, and the idea of limiting skin dose so that the maximum dose in the normal brain does not exceed the tolerable dose usually obtained from experience and knowledge of radiotherapy has been adapted. In the case of brain tumors, it has been reported by Mayer et al. that necrosis of normal brain tissue develops when the cumulative dose of initial radiotherapy and re-radiotherapy exceeds 100Gy (2Gy fractionated irradiation) [15]. In clinical practice, the effectiveness and safety of BNCT are considered, and the eligibility is judged, and the plan is made using the skin dose as an index.

3.3 Treatment plan

The BNCT dose calculation program (NeuCure® dose engine Sumitomo Heavy Industries, Ltd.) was developed and approved as a medical device. The Monte Carlo code uses a simulation code called PHITS, which enables BNCT dose calculation, as well as additional functions such as external I/F including security checks, and is developed as a dose engine dedicated to BNCT dose calculation. RayStation (RaySearch Japan Co., Ltd.), a general-purpose radiotherapy program, is used for the user interface, and it has an improved drawing function and excellent operability similar to the state-of-the-art radiotherapy equipment (**Figure 4**).

4. Future development of BNCT

At our institution, we have previously performed a clinical trial of reactor BNCT even for high-grade meningiomas and reported the results of BNCT for recurrent meningiomas [16, 17]. Prior to the treatment, ^{18}F-labeled BPA positron emission tomography (FBPA- PET) was performed using a therapeutic agent, boron compound BPA, labeled with 18F as a tracer, and the tumor-to-normal brain (Tumor/Normal: T/N) ratio averaged 3.8. This value is equal to or better than the value experienced in glioma. The introduction of this PET study proved to be useful for the treatment of recurrent malignant glioma, but it was also indispensable for developing nuclear reactor BNCT for cancers of other organs of the whole body, as well as for the later expansion of indications for head and neck cancer. In high-grade meningiomas, although there was transient enhancement of contrast areas (pseudoprogression) in a few cases, all cases showed a reduction in tumor volume. These were all patients with recurrence after multiple surgeries and radiotherapy, and the treatment outcome after BNCT was generally good, but they are rare tumors, and there are few consolidated reports that can be compared. The major cause of death was metastasis to the whole

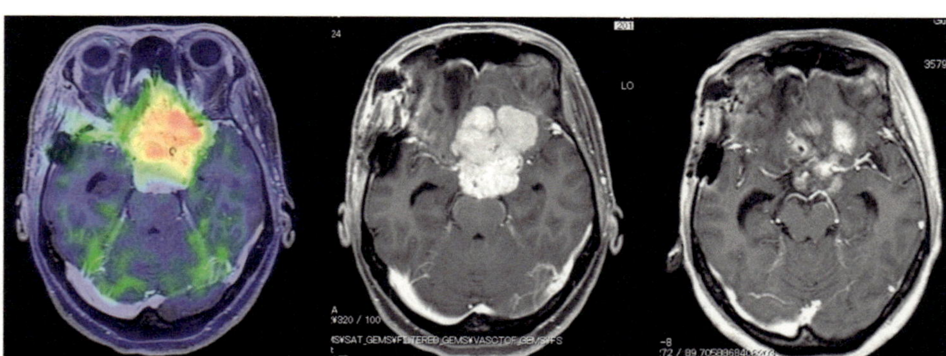

Figure 5.
A case of high-grade meningioma treated with reactor-based BNCT with significant response. The tumor in this case was located in the midline and was a recurrent case of refractory skull base meningioma, although a shrinkage effect was observed. (left: FBPA-PET fusion, middle: Before treatment, right: After BNCT of contrast-enhanced MRI T1WI).

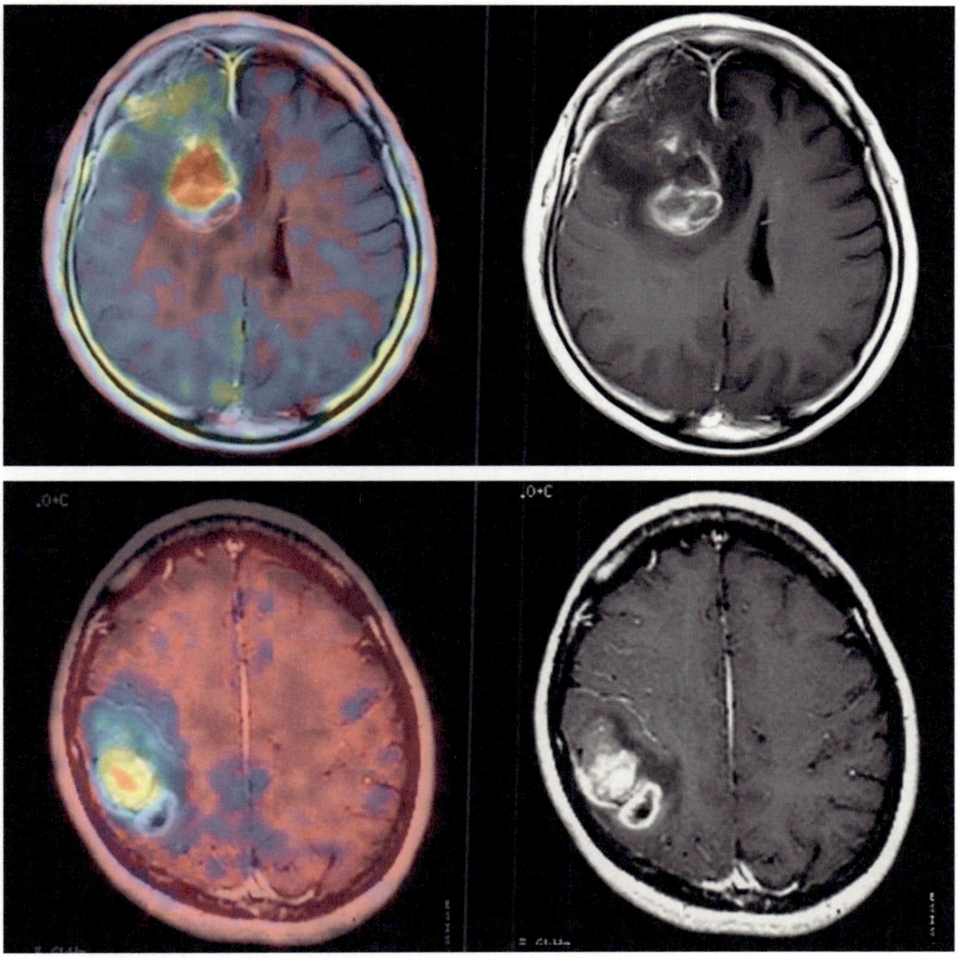

Figure 6.
Contrast-enhanced MRI compared with ^{18}FBPA PET in malignant glioma (left: FBPA-PET fusion, right: Contrast-enhanced MRI T1WI) BNCT uses biological targets and does not cause irradiation of any unnecessary sites because there is no confusion in setting the radiation field. The irradiation field is defined by the boron distribution, and there is no need for the physician to set the borders of the irradiation field.

body, in addition to seeding cancer cells into the cerebrospinal fluid cavity, with good local control of the irradiated area.

Subsequent analysis showed that the method was effective in deep-seated tumors such as the skull base, where the dose is very low (**Figure 5**) [18]. The therapeutic effects of these cases treated using reactors have led to the development of physician-led clinical trials using accelerator-based neutron sources.

Even in situations where nuclear reactors have to be used as neutron sources, proposals for new indications and improved protocols have resulted in better treatment outcomes, and BNCT research has not been interrupted until now since the beginning of clinical research in the 1960s. Under such circumstances, Japan succeeded in BNCT using an accelerator-type neutron generator for the first time in the world [13]. Recently, the attention that exceeds the academic interest from all fields and industries around the world has also increased in keeping with the success of Japan. In the field of brain tumors, global standards are expected to be challenged in the future, such as in combination with standard treatments for new diagnostic cases of glioma. However, improvements in the peripheral environment, such as the worldwide spread of approved devices and the approval of unapproved devices for medicine, are also awaited.

The "treatable" intracranial diseases on BNCT range widely and are absolutely not limited to gliomas. However, diseases such as "BNCT can be expected" are limited. In BNCT, it is one of the most important conditions for boron to maintain contrast with normal tissue and sufficiently accumulate in the tumor, and it is necessary to confirm that the boron drug to be administered shows high accumulation in the tumor using some technique before neutron irradiation. In the nuclear reactor BNCT, as described above, attempts to use PET examination with 18 FBPA have preceded, and clinical research has been conducted as "visible drug" (**Figure 6**). Although this idea is a pioneer in the field of "theranostics" that has been promoted recently, it is expected to expand the application of these combinations to new diseases and to tailor-made BNCT treatment.

5. Summary

Boron neutron capture therapy (BNCT) is a type of radiotherapy, but it has cell selectivity and can be used in combination with other radiotherapy for the treatment of recurrent and irradiated cases. The major adverse events are radiation-induced necrosis and brain edema especially for the patients with history of prior irradiation in the central nervous system.

BNCT using borofalan (10B) has become a general medical treatment by the development of medical equipment using the accelerator as a neutron source.

6. Conclusions

The current status of boron neutron capture therapy (BNCT) for brain tumors was reviewed, including recent advances in BNCT, which is still not mature, although its clinical application has begun following the successful launch of an accelerator-based neutron generation device. The contribution of "boron research" to the further development of BNCT will be important, and many researchers will be involved in the future progress of this field.

Research on the development of boron agents for treatment has also become active, and as introduced here, "multi-targeted BNCT," which simultaneously combines multiple boron agents, is very promising for personalized cancer therapy and the extension of applications.

Acknowledgements

This review was partly supported by the Grants-in-Aid for Scientific Research (B) to Shinji Kawabata (20H03797) from Japan Society for the Promotion of Science (JSPS).

Conflict of interest

The authors declare no conflict of interest.

Notes/thanks/other declarations

The cooperation of many professionals and the participation of patients are necessary for clinical trials and development. We would like to thank all of them.

Author details

Shinji Kawabata[1*], Naonori Hu[2], Ryo Hiramatsu[1], Kohei Tsujino[1], Naokado Ikeda[1], Naosuke Nonoguchi[1], Motomasa Furuse[1] and Masahiko Wanibuchi[1]

1 Department of Neurosurgery, Osaka Medical and Pharmaceutical University, Takatsuki, Osaka, Japan

2 Educational Foundation of Osaka Medical and Pharmaceutical University, Kansai BNCT Medical Center, Takatsuki, Osaka, Japan

*Address all correspondence to: shinji.kawabata@ompu.ac.jp

References

[1] Hirose K, Konno A, Hiratsuka J, Yoshimoto S, Kato T, Ono K, et al. Boron neutron capture therapy using cyclotron-based epithermal neutron source and borofalan ((10)B) for recurrent or locally advanced head and neck cancer (JHN002): An open-label phase II trial. Radiotherapy and Oncology. 2021;**155**:182-187. DOI: 10.1016/j.radonc.2020.11.001

[2] Barth RF, Vicente MG, Harling OK, Kiger WS 3rd, Riley KJ, Binns PJ, et al. Current status of boron neutron capture therapy of high grade gliomas and recurrent head and neck cancer. Radiation Oncology. 2012;**7**:146. DOI: 10.1186/1748-717X-7-146

[3] Boron Neutron Capture Therapy (BNCT) Review Working Group. Guidance on Evaluation of Accelerator Neutron Irradiation Device System for Boron Neutron Capture Therapy [Internet]. 2019. Available from: https://dmd.nihs.go.jp/jisedai/bnct_for_public/2019_BNCT_WG_report.pdf. [Accessed: June 22, 2021]

[4] Farr LE, Sweet WH, Robertson JS, Foster CG, Locksley HB, Sutherland DL, et al. Neutron capture therapy with boron in the treatment of glioblastoma multiforme. The American Journal of Roentgenology, Radium Therapy, and Nuclear Medicine. 1954;**71**(2):279-293

[5] Hatanaka H, Nakagawa Y. Clinical results of long-surviving brain tumor patients who underwent boron neutron capture therapy. International Journal of Radiation Oncology, Biology, Physics. 1994;**28**(5):1061-1066. DOI: 10.1016/0360-3016(94)90479-0

[6] Kawabata S, Miyatake S, Kuroiwa T, Yokoyama K, Doi A, Iida K, et al. Boron neutron capture therapy for newly diagnosed glioblastoma. Journal of Radiation Research. 2009;**50**(1):51-60. DOI: 10.1269/jrr.08043

[7] Yamamoto T, Nakai K, Kageji T, Kumada H, Endo K, Matsuda M, et al. Boron neutron capture therapy for newly diagnosed glioblastoma. Radiotherapy and Oncology. 2009;**91**(1):80-84. DOI: 10.1016/j.radonc.2009.02.009

[8] Henriksson R, Capala J, Michanek A, Lindahl SA, Salford LG, Franzen L, et al. Boron neutron capture therapy (BNCT) for glioblastoma multiforme: A phase II study evaluating a prolonged high-dose of boronophenylalanine (BPA). Radiotherapy and Oncology. 2008;**88**(2):183-191. DOI: 10.1016/j.radonc.2006.04.015

[9] Skold K, Gorlia T, Pellettieri L, Giusti V, HS B, Hopewell JW. Boron neutron capture therapy for newly diagnosed glioblastoma multiforme: An assessment of clinical potential. The British Journal of Radiology. 2010;**83**(991):596-603. DOI: 10.1259/bjr/56953620

[10] Miyatake S, Kawabata S, Yokoyama K, Kuroiwa T, Michiue H, Sakurai Y, et al. Survival benefit of boron neutron capture therapy for recurrent malignant gliomas. Journal of Neuro-Oncology. 2009;**91**(2):199-206. DOI: 10.1007/s11060-008-9699-x

[11] Pellettieri L, HS B, Rezaei A, Giusti V, Skold K. An investigation of boron neutron capture therapy for recurrent glioblastoma multiforme. Acta Neurologica Scandinavica. 2008;**117**(3):191-197. DOI: 10.1111/j.1600-0404.2007.00924.x

[12] Shiba H, Takeuchi K, Hiramatsu R, Furuse M, Nonoguchi N, Kawabata S,

et al. Boron neutron capture therapy combined with early successive bevacizumab treatments for recurrent malignant gliomas - a pilot study. Neurologia Medico-Chirurgica (Tokyo). 2018;**58**(12):487-494. DOI: 10.2176/nmc.oa.2018-0111

[13] Kawabata S, Suzuki M, Hirose K, Tanaka H, Kato T, Goto H, et al. Accelerator-based BNCT for patients with recurrent glioblastoma: A multicenter phase II study. Neuro-Oncology Advances. 2021;**3**(1):vdab067. DOI: 10.1093/noajnl/vdab067

[14] Suzuki M, Kato I, Aihara T, Hiratsuka J, Yoshimura K, Niimi M, et al. Boron neutron capture therapy outcomes for advanced or recurrent head and neck cancer. Journal of Radiation Research. 2014;**55**(1):146-153. DOI: 10.1093/jrr/rrt098

[15] Mayer R, Sminia P. Reirradiation tolerance of the human brain. International Journal of Radiation Oncology, Biology, Physics. 2008;**70**(5):1350-1360. DOI: 10.1016/j.ijrobp.2007.08.015

[16] Kawabata S, Hiramatsu R, Kuroiwa T, Ono K, Miyatake S. Boron neutron capture therapy for recurrent high-grade meningiomas. Journal of Neurosurgery. 2013;**119**(4):837-844. DOI: 10.3171/2013.5.JNS122204

[17] Takai S, Wanibuchi M, Kawabata S, Takeuchi K, Sakurai Y, Suzuki M, et al. Reactor-based boron neutron capture therapy for 44 cases of recurrent and refractory high-grade meningiomas with long-term follow-up. Neuro-Oncology. 2021;**24**(1):90-98. DOI: 10.1093/neuonc/noab108

[18] Takeuchi K, Kawabata S, Hiramatsu R, Matsushita Y, Tanaka H, Sakurai Y, et al. Boron neutron capture therapy for high-grade Skull-Base meningioma. Journal of Neurological Surgery Part B: Skull Base. 2018;**79** (Suppl 4):S322-S3S7. DOI: 10.1055/s-0038-1666837